Environmental and Health Risk Assessment of Heavy Metal Pollution

Environmental and Health Risk Assessment of Heavy Metal Pollution

Editor

Said Muhammad

Basel • Beijing • Wuhan • Barcelona • Belgrade • Novi Sad • Cluj • Manchester

Editor
Said Muhammad
National Centre of Excellence
in Geology
University of Peshawar
Peshawar
Pakistan

Editorial Office
MDPI
St. Alban-Anlage 66
4052 Basel, Switzerland

This is a reprint of articles from the Special Issue published online in the open access journal *Sustainability* (ISSN 2071-1050) (available at: www.mdpi.com/journal/sustainability/special_issues/environmental_risk_heavy_metal).

For citation purposes, cite each article independently as indicated on the article page online and as indicated below:

Lastname, A.A.; Lastname, B.B. Article Title. *Journal Name* **Year**, *Volume Number*, Page Range.

ISBN 978-3-7258-0266-1 (Hbk)
ISBN 978-3-7258-0265-4 (PDF)
doi.org/10.3390/books978-3-7258-0265-4

© 2024 by the authors. Articles in this book are Open Access and distributed under the Creative Commons Attribution (CC BY) license. The book as a whole is distributed by MDPI under the terms and conditions of the Creative Commons Attribution-NonCommercial-NoDerivs (CC BY-NC-ND) license.

Contents

About the Editor vii

Preface ix

Wania Imran and Justin B. Richardson
Trace Element (As, Cd, Cr, Cu, Pb, Se, U) Concentrations and Health Hazards from Drinking Water and Market Rice across Lahore City, Pakistan
Reprinted from: *Sustainability* **2023**, *15*, 13463, doi:10.3390/su151813463 1

Delia B. Senoro, Kevin Lawrence M. De Jesus and Cris Edward F. Monjardin
Pollution and Risk Evaluation of Toxic Metals and Metalloid in Water Resources of San Jose, Occidental Mindoro, Philippines
Reprinted from: *Sustainability* **2023**, *15*, 3667, doi:10.3390/su15043667 15

Yuliang Xiao, Gang Zhang, Jiaxu Guo, Zhe Zhang, Hongyi Wang, Yang Wang, et al.
Pollution Characteristics and Risk Assessments of Mercury in Jiutai, a County Region Thriving on Coal Mining in Northeastern China
Reprinted from: *Sustainability* **2022**, *14*, 10366, doi:10.3390/su141610366 50

Arinze Longinus Ezugwu, Hillary Onyeka Abugu, Ifeanyi Adolphus Ucheana,
Samson Ifeanyi Eze, Johnbosco C. Egbueri, Victor Sunday Aigbodion, et al.
Sequestration of Lead Ion in Aqueous Solution onto Chemically Pretreated *Pycnanthus angolensis* Seed Husk: Implications for Wastewater Treatment
Reprinted from: *Sustainability* **2023**, *15*, 15446, doi:10.3390/su152115446 70

Yu Nong, Xinyi Liu, Zi Peng, Liangxiang Li, Xiran Cheng, Xueli Wang, et al.
Effects of Domestic Sewage on the Photosynthesis and Chromium Migration of *Coix lacryma-jobi* L. in Chromium-Contaminated Constructed Wetlands
Reprinted from: *Sustainability* **2023**, *15*, 10250, doi:10.3390/su151310250 94

Shakeel Ahmad, Fazal Hadi, Amin Ullah Jan, Raza Ullah, Bedur Faleh A. Albalawi and Allah Ditta
Appraisal of Heavy Metals Accumulation, Physiological Response, and Human Health Risks of Five Crop Species Grown at Various Distances from Traffic Highway
Reprinted from: *Sustainability* **2022**, *14*, 16263, doi:10.3390/su142316263 112

Erny Yuniarti, Ida F. Dalmacio, Virginia C. Cuevas, Asuncion K. Raymundo,
Erlinda S. Paterno, Nina M. Cadiz, et al.
Effects of Heavy Metal-Tolerant Microorganisms on the Growth of "Narra" Seedlings
Reprinted from: *Sustainability* **2022**, *14*, 9665, doi:10.3390/su14159665 130

Muhammad Tansar Abbas, Mohammad Ahmad Wadaan, Hidayat Ullah,
Muhammad Farooq, Fozia Fozia, Ijaz Ahmad, et al.
Bioaccumulation and Mobility of Heavy Metals in the Soil-Plant System and Health Risk Assessment of Vegetables Irrigated by Wastewater
Reprinted from: *Sustainability* **2023**, *15*, 15321, doi:10.3390/su152115321 144

Junaid Ghani, Javed Nawab, Zahid Ullah, Naseem Rafiq, Shah Zaib Hasan,
Sardar Khan, et al.
Multivariate Statistical Methods and GIS-Based Evaluation of Potable Water in Urban Children's Parks Due to Potentially Toxic Elements Contamination: A Children's Health Risk Assessment Study in a Developing Country
Reprinted from: *Sustainability* **2023**, *15*, 13177, doi:10.3390/su151713177 163

Nukshab Zeeshan, Zia Ur Rahman Farooqi, Iftikhar Ahmad, Ghulam Murtaza, Aftab Jamal, Saifullah, et al.
Trace Metals in Rice Grains and Their Associated Health Risks from Conventional and Non-Conventional Rice Growing Areas in Punjab-Pakistan
Reprinted from: *Sustainability* 2023, 15, 7259, doi:10.3390/su15097259 183

Hussein K. Okoro, Muyiwa M. Orosun, Faith A. Oriade, Tawakalit M. Momoh-Salami, Clement O. Ogunkunle, Adewale G. Adeniyi, et al.
Potentially Toxic Elements in Pharmaceutical Industrial Effluents: A Review on Risk Assessment, Treatment, and Management for Human Health
Reprinted from: *Sustainability* 2023, 15, 6974, doi:10.3390/ su15086974 199

Aminur Rahman, Kazuhiro Yoshida, Mohammed Monirul Islam and Genta Kobayashi
Investigation of Efficient Adsorption of Toxic Heavy Metals (Chromium, Lead, Cadmium) from Aquatic Environment Using Orange Peel Cellulose as Adsorbent
Reprinted from: *Sustainability* 2023, 15, 4470, doi:10.3390/su15054470 215

About the Editor

Said Muhammad

Dr. Said Muhammad works as an Associate Professor at the National Centre of Excellence in Geology, University of Peshawar. A leading organization in the field of earth and environmental sciences research in the country. He has more than 15 years of experience working in the field and has published over 100 scholarly papers in peer-reviewed journals. His work has earned him several national and international awards. He has supervised several research projects throughout his career and is currently supervising several graduate students. He has served as the peer reviewer of more than 100 journals. Dr. Said Muhammad is mainly interested in the geochemistry of heavy metal contamination and its dynamics in environmental media

Preface

In a world grappling with the pervasive impact of heavy metals on our environment, the need for comprehensive understanding and effective strategies cannot be overstated. Though present in seemingly innocuous concentrations, these non-degradable, persistent, and bioaccumulative metals cast a shadow of potential threat over various life forms, including humans. The intricate web of their occurrence, pathways, sources, and associated risks in our environment, emphasizes the critical importance of dedicated research.

Water, soil, plants, and food emerge as focal points, demanding heightened scrutiny due to their potential implications for human health. This reprint aims to galvanize a collaborative effort in unraveling the complexities surrounding heavy metal pollution. The contributions encompass a spectrum of research areas, including statistical/mathematical modeling elucidating health and ecological effects, treatment methods, and comprehensive risk assessments for water, soil, plants, and food contaminated by pollutants. This collection will foster a deeper understanding of heavy metals. It aspires to transcend geographical boundaries, featuring contributions from studies worldwide that are dedicated to mitigating the impact of pollutants on our environment.

Besides a survey of methods and heavy metal contamination risks, this reprint carries insights based on the modified *Pycnanthus angolensis* seed husks for the sequestration of Pb(II), providing an environmentally conscious and economically viable approach for wastewater treatment. A study contemplated the alleviation of Cr contamination via phytoremediation in constructed wetlands to ensure their sustainable operations.

It is anticipated that the insights shared within this reprint will serve as a cornerstone for future research and policy initiatives. The collective wisdom of the global scientific community, disseminated through these pages, will contribute to a more resilient and sustainable approach to addressing the multifaceted challenges posed by heavy metal pollutants in our environment. The reprint shall also serve as reference material for students, scholars, and aspiring researchers.

Finally, I extend my heartfelt appreciation to the esteemed authors whose valuable contributions have shaped this volume and made the creation of this reprint possible. I am particularly grateful for the invaluable insights provided by my colleague, Dr. Wajid Ali, whose support and thoughtful suggestions have been of immense importance to me.

Said Muhammad
Editor

Article

Trace Element (As, Cd, Cr, Cu, Pb, Se, U) Concentrations and Health Hazards from Drinking Water and Market Rice across Lahore City, Pakistan

Wania Imran [1] and Justin B. Richardson [1,2,*]

1 Environmental Science Program, University of Massachusetts Amherst, Amherst, MA 01003, USA; wimran@umass.edu
2 Department of Earth, Climate, and Geographic Sciences, University of Massachusetts Amherst, Amherst, MA 01003, USA
* Correspondence: jbrichardson@umass.edu; Tel.: +1-413-545-3736

Abstract: Exposure to toxic concentrations of trace elements in rice and drinking water is a serious issue for millions of South Asians, due to rice serving as a large portion of their diets and the geochemical enrichment of trace elements in groundwaters. The overall goal of this study was to evaluate and compare the hazards posed from toxic trace elements through the consumption of commercially available basmati rice and public drinking water sources across Lahore, Pakistan. Drinking water samples (n = 36) were collected from publicly accessible drinking taps from eight administrative towns and the cantonment. Rice samples were obtained from 11 markets (n = 33) across Lahore between December and February 2022–2023. Market rice concentrations exceeded the World Health Organization's (WHO) limits and the Total Hazard Quotient (THQ) values exceeded 1.0 for As, Cu, and Pb, thus indicating multielement contamination. Market rice trace element concentrations and price were not correlated. As, Se, and U concentrations in drinking water were above the WHO's drinking water guidelines and had THQ values exceeding 1.0, showing multielement contamination. Cr, Se, and U concentrations in drinking water were greater for impoverished administrative towns compared to middle and wealthy administrative towns, highlighting socioeconomic inequities in exposure to hazardous concentrations. We conclude that the citizens of Lahore are exposed to rice and drinking water that are hazardous to human health, including As and other lesser studied trace elements.

Keywords: arsenic; heavy metals; Punjab; risk assessment; total hazard quotient; toxic elements

1. Introduction

Rice is a staple food for more than half of the global population, particularly in Asia, where it serves as a primary dietary component for billions of people. The cultivation of rice is an integral part of the South Asian diet and is a staple grain crop for food security [1]. Among the nations in this region, Pakistan ranks at 10th place in rice production, dedicating 3.034 million hectares of land with an overall 7.410 million tonnes of the crop produced, thus making it a prominent rice-producing nation. which is beneficial for both the provision of food within the country and the country's economy [2]. However, concerns continue to mount about the safety of consuming domestic rice in light of toxic levels of trace elements present [3]. In the context of Pakistan, the possible effects of rice crops as the second most consumed grain in the country are pressing [4]. As a significant portion of the population relies on rice as a dietary staple, there is an increased likelihood of exposure to these elements, with ingestion being the primary cause of harmful exposure to toxic trace elements [5]. Furthermore, the water used to grow and irrigate rice crops as well as domestically used drinking water to cook the rice pose additional potential risks to human health [6]. The United States Food and Drug Administration (FDA) notes that cooking and

Citation: Imran, W.; Richardson, J.B. Trace Element (As, Cd, Cr, Cu, Pb, Se, U) Concentrations and Health Hazards from Drinking Water and Market Rice across Lahore City, Pakistan. *Sustainability* 2023, *15*, 13463. https://doi.org/10.3390/su151813463

Academic Editor: Said Muhammad

Received: 7 August 2023
Revised: 29 August 2023
Accepted: 4 September 2023
Published: 8 September 2023

Copyright: © 2023 by the authors. Licensee MDPI, Basel, Switzerland. This article is an open access article distributed under the terms and conditions of the Creative Commons Attribution (CC BY) license (https://creativecommons.org/licenses/by/4.0/).

rinsing rice in clean water is an effective way of limiting arsenic exposure. However, when the water itself contains elevated levels of trace elements, the risk of exposure amplifies. For example, prolonged exposure to arsenic can result in a multitude of severe health risks including but not limited to hypertension, skin lesions, neurodegeneration, cancer and cardiovascular disease [7].

Though essential trace elements are required by plants in small quantities for growth and development, they can be considered a double-edged sword when their concentrations exceed the permissible limits [8]. Toxic trace elements in soils and drinking water in Pakistan can originate from various anthropogenic activities such as mining, industrial production, and municipal consumption and refuse [9,10]. Coal mining, synthetic industries, and burning waste are all anthropogenic sources of trace elements that may lead to them seeping into agricultural production [11,12]. However, one of the dominant sources of trace elements to agricultural soils and drinking water resources is natural sources [10,11]. Alluvial plain deposits in eastern Pakistan originated from the erosion and transport from the Himalayas during the Holocene Period contain elevated trace element concentrations [13]. Trace elements of specific concern, particularly As, Cd, Pb, Ni, U, are found in background to elevated concentrations in soils and near surface aquifers across the region of Punjab [10,13,14]. This is particularly true for the Punjab and Sindh regions, which are the top agricultural-commodity-producing regions within Pakistan [7,10]. As a prime example, of the 110 groundwater samples collected across Punjab in 2017, 40% exceeded the WHO's guideline limit of >50 µg/L for As, and 41% exceeded the WHO's guideline limit of >15 µg/L for U [7]. Though some trace elements serve benefit to human health at lower doses such as Cu, Se, Cr, elevated levels of these elements in rice pose potential health risks when they enter the food chain through an essential part of the diet for millions of Pakistani people [10,14].

The overall goal of this study was to evaluate and compare the hazards posed from toxic trace elements through the consumption of commercially available basmati rice and public drinking water sources across Lahore, Pakistan. First, we hypothesized that basmati rice does not contain hazardous concentrations of trace elements. We expected that commercially available rice would meet international health standards and that hazardous concentrations would not be related to the price of the rice. Second, we hypothesized that drinking water would not contain hazardous concentrations of trace elements as it should be treated and originate from aquifers with permissible trace element concentrations, unlike groundwater resources in rural areas. We expected elevated concentrations of As but that not other trace elements would reach or exceed the World Health Organization's (WHO) limits. Moreover, we expected that water quality would be comparable among administrative towns despite varying socioeconomic levels within Lahore. This information is needed to characterize the daily hazard related to trace element exposure by the denizens of the administrative districts of Lahore and highlight the need for personal safety practices.

2. Materials and Methods

2.1. Description of Lahore, Pakistan Study Area

This study was conducted in the capital of Pakistan's Punjab province, Lahore, which is the second largest city in Pakistan with a population of 11.1 million residents. Lahore presents a great socio-economic divide between administrative areas based on many factors such as unemployment rates, asset possession, literacy rate and contribution to the gross domestic product (GDP). The city of Lahore comprises nine administrative towns and a cantonment (Figure 1). These towns can be divided into three broad socioeconomic status (SES) categories, wealthy class, middle-class and impoverished class based on non-standardized indices (NSI). According to the classification through results from a component matrix, Cantonment and Aziz Bhatti town are classified as wealthy, Gulberg, Samanabad, Data Gunj Baksh and Ravi towns are classified as middle-class, and Shalimar, Wagah, Nishtar and Iqbal towns are classified as impoverished. Middle-class towns present the highest population density, with Ravi having an approximate 1.0 million residents. The north of the city presents the lowest population size, with Wagha and Aziz Bhatti town both having an approximate 0.62 million residents [14].

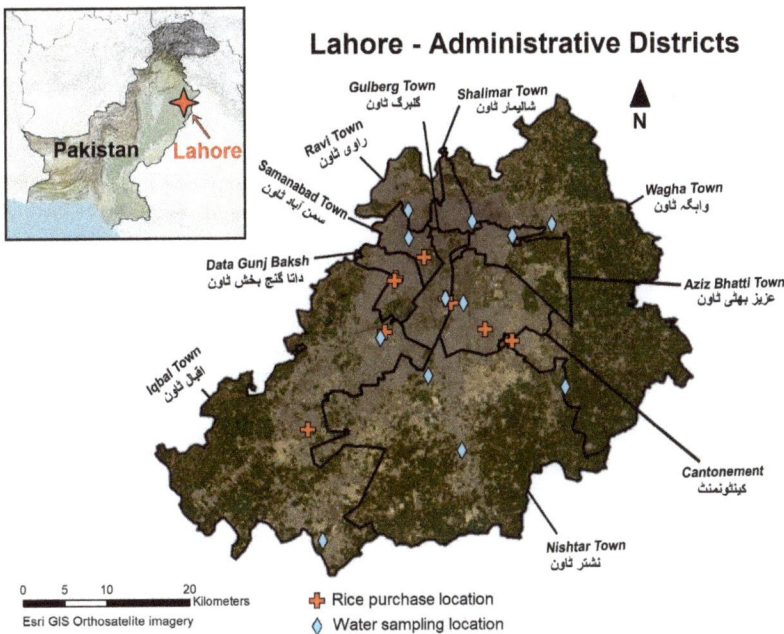

Figure 1. Map of Lahore administrative districts and location of drinking water sampling locations and locations of rice purchases from markets.

2.2. Drinking Water and Market Rice Grain Collection and Processing

Drinking water samples were collected from 12 public locations in triplicate (36 samples total) between December and February 2022–2023. The samples were collected in acid-washed 250 mL polyethylene bottles. During the sample collection of drinking water, each bottle was filled from the tap, dumped out, and re-filled without gaseous headspace. This was repeated within the same day but not in tandem. Water samples were collected from publicly accessible water drinking taps from eight administrative towns and the cantonment. The three towns with an area greater than 100 km^2, Wagha Town, Iqbal Town and Nishtar Town, were divided into north and south regions, with a sample being collected from each.

Market rice was obtained from 11 large commercial markets (giving a total of 33 rice samples) spatially distributed across Lahore between December and February 2022–2023, a product of the kharif season (May–June plantation, harvested in October–December). Only Pusa basmati rice (*Oryza sativa* Linn) varieties were purchased to avoid varietal differences. These rice samples were dehulled and came from one of two rice growing zones in Pakistan: Zone-II in Punjab, located between the Ravi and Chenab rivers, and Zone-III in Sindh, located right at the bank of the Indus River.

2.3. Digestion and Analyses

The drinking water samples were analyzed for pH, electrical conductivity (EC), and oxidation reduction potential (ORP) using Atlas Scientific probes (Ixian industrial kits, Atlast Scientific, Long Island City, NY, USA). The drinking water samples were shipped to the University of Massachusetts Amherst, then acidified with 67% trace metal grade HNO$_3$ to pH 1 and analyzed for trace elements following the EPA method 3005 (Cr, Ni, Cu, Zn, As, Se, Sr, Cd, Sn, Sb, W, Pb, Th, U) with an Agilent 7700x Inductively Coupled

Plasma—Mass Spectrometer (ICP-MS; Agilent Technologies, Santa Clara, California, USA) under no gas mode, pentuplet measurement replicates, check standards at every 10 samples, 12 point calibration curves, and power calibration curves. The limit of detection (LOD) were 0.006 µg/L for As, 0.002 µg/L for Cd, 0.07 µg/L for Cr, 0.001 µg/L for Cu, 0.02 µg/L for Pb, 0.007 µg/L for Se, and 0.001 µg/L for U. Macroelements (Al, Fe, Ca, K, Mg, Na, Mn, P) were measured using an Agilent 5110 Inductively Coupled Plasma Optical Emission Spectrometer (ICP-OES) under axial mode, pentuplet replicates, nine point calibration curves, and power calibration curves. The limit of detection (LOD) for macroelements was 5 µg/L.

Market rice samples were shipped to the University of Massachusetts Amherst and crushed to <0.1 mm diameter using an agate mortar. Rice powder subsamples were digested and analyzed following a modified EPA method 3050B. First, the rice powder was ashed in a muffle furnace at 550 °C for 14 h to remove carbon prepared for digestion using 5 mL of aqua regia (9:1 trace metal grade HNO_3:HCl from Fisher Scientific) in 50 mL centrifuge tubes. The digestate was then heated to 90 °C using a tube rack heater for 1 h. Each sample was then diluted to 50 g and diluted to a 1:3 ratio using 18.2 MΩ de-ionized water. The mass of each sample digest and dilution steps were recorded. Batches of samples included certified reference materials, procedural blanks, and sample duplicates. NIST (National Institute of Standards and Technology) standards were used: peach leaves SRM 1547 and wheat flour SRM 1567b were used as certified reference materials for As, Cd, Cu, Cr, Pb, Se, and U concentrations (National Institute of Standards and Technology Gaithersburg, MD). Like water samples, rice dilutions were analyzed for trace elements using an Agilent 7700x ICP-MS and macroelements Agilent 5110 ICP-OES. Measured macro and trace element concentrations for Peach Leaves SRM 1547 and wheat flour SRM 1567b were 87 to 103% of their certified concentrations. The blanks for macro elements were <LOD and blanks for trace elements were <0.010 µg/L.

2.4. Exposure Risk Assessment

To determine the exposure and risk from trace elements (As, Cd, Cr, Cu, Pb, U, Se) from the market rice and drinking water, the total hazard quotient (THQ) and lifetime cancer risk (LCR) were calculated (for example, see Ali et al. [15]). First, we determined the estimated daily intake (EDI) of rice and water separately, which was calculated using Eq 1. EDI (mg/kg/bw/day) is the amount of trace element consumed by an individual; C (mg/kg) is the concentration of trace element concentration in each rice and water sample; CR (kg/day) is the consumption rate of water and rice in a day; bw is the mean body weight of an adult Pakistani individual (male and female adults). The consumption rate for water was 2 L/capita/day and the consumption rate for rice was calculated to be 46.6 g/day using the annual rice consumption rate provided at 17 kg/capita for Pakistan in the 2022 Annual Country Report—Pakistan by the World Food Program (WFP) [16].

$$EDI = (C \times CR)/bw \qquad (1)$$

The EDI was then used to calculate the target hazard quotient (THQ) using Equation (2). Reference oral doses (RfDs) set for trace elements by the US EPA were used for THQ calculations.

$$THQ = EDI/RfD \qquad (2)$$

THQ < 1 indicates no significant risk of non-carcinogenic effects.

To calculate the risk of consumption over time, especially in terms of Uranium content in water, the lifetime cancer risk (LCR) was also calculated using the EDI and the slope factor (SF = 1.5 mg/kg/day for As, 6.1 mg/kg/day for Cd, 0.5 mg/kg/day for Cr, 0.004 mg/kg/day for Cu, 0.0085 mg/kg/day for Pb, 0.04 mg/kg/day for Se, and

1.5 mg/kg/day for U) based upon values set by the US EPA [17]. The acceptable upper limit set for the LCR is 1.0×10^{-4}.

$$LCR = EDI \times SF \qquad (3)$$

2.5. Data and Statistical Analysis

ArcGIS Pro software was used to map field sites. Matlab R2022b (MATLAB 9.13) was used for data analysis and figure production. Comparisons among towns were calculated using the Kruskal–Wallis test and linear regressions were used to assess significant correlations among the elemental data. Statistical tests between results in our study and other studies utilized two sample *t*-tests. Market rice and trace element concentrations are available in Table S1 and S2 in Supplementary Materials.

3. Results and Discussion

3.1. Market Rice Grain Trace Elements and Risk Assessment

Market rice's mean trace element concentrations were calculated to be 0.77 ± 0.09 mg/kg As, 0.20 ± 0.07 mg/kg Cd, 0.22 ± 0.03 mg/kg Cr, 57 ± 16 mg/kg Cu, 2.3 ± 0.9 mg/kg Pb, 0.09 ± 0.01 mg/kg Se, 0.009 ± 0.003 mg/kg U. The market rice concentrations of As, Cr, and Se exhibited the lowest variation within a factor of 2x (Figure 2). Hence, the exposure rates to trace element concentrations of As, Cr, and Se were consistent across the market rice. Conversely, Cd, Cu, Pb, and U had variability of 3x to an order of magnitude (Figure 2). Thus, there was brand dependence on exposure to Cd, Cu, Pb, and U. Our results match observations noted in previous studies conducted in other regions of Pakistan: Bibi et al. [18] conducted research in Gujranwala, Hafizabad, Vehari, Mailsi, and Burewala within the Punjab region of Pakistan, Sarwar et al. [3] conducted studies across Lahore, Faislabad, Gujranwala, Sargodha, Rawalpindi, Multan, Okara, Swat, Batkhela, Peshawar, Mardan, Shergarh, Karachi and Quetta, and Nawab et al. [19] conducted research in Khyber Pakhtunkhwa, Pakistan. Market rice grain concentrations in our study exceeded the WHO's concentration limitations for As (0.2 mg/kg), Cu (10 mg/kg), and Pb (0.20 mg/kg) [20].

Figure 2 shows that these concentrations were significantly lower than concentrations from other regions in developed nations but were in the range of other studies conducted in Pakistan. These concentrations were significantly higher than those found in white rice grown in California, Louisiana, and Texas in the United States, which presented trace element concentrations of As 0.13 mg/kg, Cd 0.011 mg/kg, Cu 2.5 mg/kg, and Pb 0.006 mg/kg [21]. Similarly, our results were significantly higher than those found for rice grown in Korea, as Jung et al. [22] measured Cd 0.021 mg/kg, Cu 1.9 mg/kg, and Pb 0.21 mg/kg. Our market rice grain concentrations were within the ranges of previous studies in Pakistan but there were significant differences. Market rice grain As concentrations were significantly higher than those reported in Bibi et al.'s [18] (0.4 mg/kg) and Nawab et al.'s work [19] (0.41 ± 0.11 mg/kg). Market rice grain Cd concentrations were significantly lower than those measured by Tariq and Rashid [23] (0.86 ± 0.37 mg/kg) but comparable with those from Bibi et al.'s [18] (0.12 mg/kg) and Nawab et al.'s work [19] (0.09 ± 0.03 mg/kg). Market rice grain Cr concentrations were significantly lower than those found in Tariq and Rashid's [23] (6.9 ± 1.7 mg/kg), Bibi et al.'s [18] (8.0 mg/kg) and Nawab et al.'s research [19] (2.44 ± 1.71 mg/kg). Lastly, market rice grain Pb concentrations were significantly greater than those measured by Nawab et al. [19] (0.26 ± 0.07 mg/kg), comparable with those found in Bibi et al.'s work [18] (4.3 mg/kg), but significantly lower than Tariq and Rashid's reports [23] (46 ± 2 mg/kg).

To investigate whether market rice quality affected trace element concentrations, we compared their concentrations with prices in Pakistani rupees at the market at the time of purchase. We expected a negative correlation between the market price and the trace element concentrations owing to higher standards of agricultural practices commanding higher prices in Pakistani rupees. However, we did not find a significant correlation between the rice market price and As, Cd, Cu, Cr, Pb, Se, or U concentrations (Figure 3).

This suggests that rice quality for potentially toxic trace element concentration exposures is not related to the market price. Instead, the market price of rice is typically controlled by costs for transportation from field to market, communication and business expenses, credit for capital and equipment, and storage facilities [24].

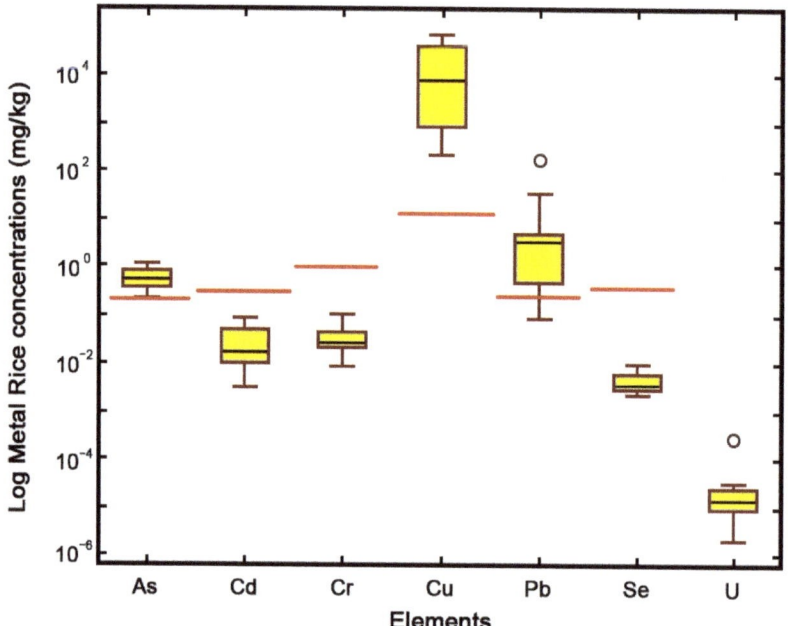

Figure 2. Boxplots of log market rice trace element concentrations collected across Lahore. $N = 33$ and the red lines are the WHO's limits [20]. Boxes represent the interquartile range, bars represent minimum and maximum range and circles represent outliers.

The consumption of rice containing potential toxic trace elements by the millions of denizens is a critical health issue in the developing country of Pakistan. The mean EDI was calculated by taking the mean rice trace element concentrations, annual rice consumption rate provided as 17 kg/capita for Pakistan and 63 kg for the mean body mass for an adult Pakistani. The EDI was further contextualized by using the trace element-specific reference oral dose to determine the THQ and HI, which suggests it is hazardous to consume. Our results show that, as expected, market rice generated As THQ values far exceeding 1.0, indicating that the consumption of rice poses a direct risk to human health. More interestingly, market rice THQ values for Cd, Cu, and Pb also exceeded 1.0 for several market rice samples (Table 1). These results match observations noted by Bibi et al. [18], who posit that the consumption of market rice grains poses a direct hazard to consumer health. These results highlight that market rice at the point of purchase for regular consumers poses a hazard to human health.

Similar to THQ, the consumption of rice containing potential toxic trace elements also poses risk for cancer in Lahore, Pakistan. The trace element-specific cancer slope factor to determine the LCR, which suggests that the rice poses a hazard to cancer development when LCR $> 10^{-4}$ and is thus considered an unacceptable risk for cancer by the US EPA [17]. Our results show that market rice generated As LCR values far exceeding the 1.0×10^{-4} threshold, indicating that the consumption of market rice poses a direct risk to cancer formation (Table 2). Fortunately, the other trace elements, Cd, Cu, Pb, Se, and U, did not exceed the LCR values of 1.0×10^{-4} threshold, implying that only As and Cr may be a cancer risk in the market rice.

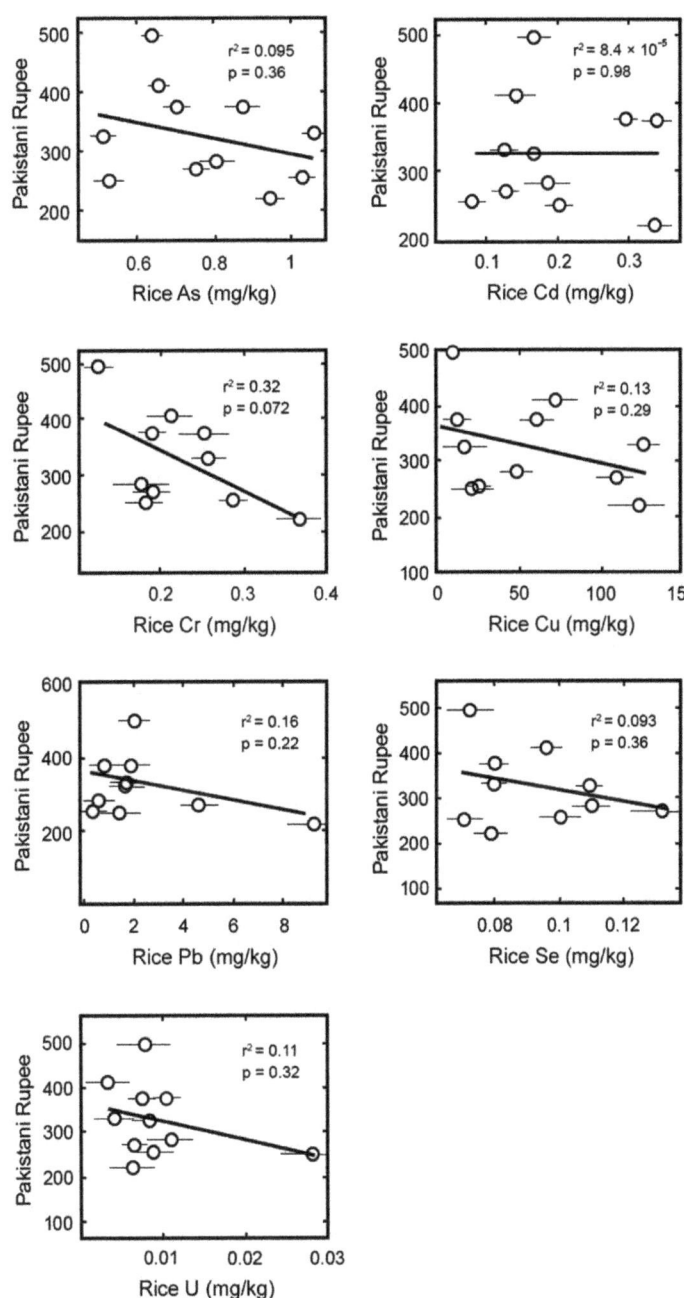

Figure 3. XY plot comparison of mean trace element concentrations and market price (December 2022) for each rice sample with linear regressions. Error bars are standard errors.

Table 1. Rice target hazard quotients (THQ)s, where any value >1.0 is considered a lifetime health hazard from rice consumption. See Section 2.4 for the calculation method.

Market	As	Cd	Cr	Cu	Pb	Se	U	Hazard Index (HI)
1	6.82	0.31	0.19	1.33	0.19	0.39	0.006	9.26
2	7.02	0.49	0.17	6.24	0.98	0.31	0.003	15.2
3	4.22	0.66	0.08	0.49	1.15	0.28	0.005	6.91
4	4.98	0.50	0.12	5.42	2.61	0.52	0.004	14.2
5	5.32	0.74	0.11	2.40	0.31	0.43	0.007	9.35
6	5.79	1.34	0.16	0.65	0.46	0.31	0.007	8.74
7	6.25	1.33	0.24	6.10	5.24	0.31	0.004	19.5
8	4.33	0.56	0.14	3.56	0.39	0.38	0.002	9.38
9	4.65	1.17	0.12	2.99	1.06	0.31	0.005	10.4
10	3.48	0.80	0.12	1.04	0.79	0.28	0.019	6.55
11	3.39	0.65	0.13	0.85	0.95	0.43	0.006	6.43
Mean	5.11	0.78	0.14	2.83	1.29	0.36	0.006	10.53

Table 2. Market rice life cancer risk (LCR) values where any value >1.0×10^{-4} is considered a lifetime hazard for developing cancer. See Section 2.4 for calculation method.

Market	As $\times 10^{-4}$	Cd $\times 10^{-4}$	Cr $\times 10^{-4}$	Cu $\times 10^{-4}$	Pb $\times 10^{-4}$	Se $\times 10^{-4}$	U $\times 10^{-4}$
1	12	0.0	1.1	0.1	0.0	0.0	0.1
2	12	0.0	1.0	0.3	0.1	0.0	0.0
3	7	0.0	0.5	0.0	0.3	0.0	0.1
4	8	0.0	0.7	0.3	0.0	0.0	0.1
5	9	0.1	0.7	0.2	0.1	0.0	0.1
6	10	0.1	0.9	0.0	0.6	0.0	0.1
7	11	0.1	1.4	0.4	0.0	0.0	0.1
8	7	0.0	0.8	0.2	0.0	0.0	0.0
9	8	0.1	0.7	0.2	0.1	0.0	0.1
10	6	0.1	0.7	0.1	0.1	0.0	0.3
11	6	0.0	0.8	0.1	0.1	0.0	0.1

As described in previous soil–plant field studies in this region of Pakistan and beyond, such as Iran [25], Bangladesh [23,26], and India [27], these high trace element concentrations in crops are most likely due to irrigation water (both wastewater/groundwater) containing elevated trace element concentrations [28]. As reported by previous studies, the quality of irrigation water needs to be monitored during rice cultivation in the study area [18].

3.2. Drinking Water Trace Elements and Risk Assessment

Drinking water samples had mean trace element concentrations of 20.5 ± 6.1 µg/L As, 0.02 ± 0.01 µg/L Cd, 0.12± 0.02 µg/L Cr, 2.3 ± 1.0 µg/L Cu, 0.11 ± 0.07 µg/L Pb, 25 ± 14 µg/L Se, 32.1 ± 12.7 µg/L U. Cr and Cu drinking water concentrations exhibited the lowest variation within a factor of 3x (Figure 4). Thus, the exposure to trace element concentrations of Cr and Cu was consistent across the drinking water of Lahore. Conversely, As, Cd, Pb, Se, and U had variability of one or two orders of magnitude (Figure 4). Thus, there was an administrative town- and location-specific dependence on exposure to As, Cd, Pb, Se, and U. Several drinking water concentrations in our study exceeded the WHO's concentration limitations for As (10 µg/L), Se (40 µg/L) and U (30 µg/L) concentrations, but were far below those set for Cd (3 µg/L), Pb (10 µg/L), and Cu (3000 µg/L) [29]. The drinking water in our study should meet the WHO's standards, unlike rural, shallow wells in other studies carried out in the Punjab region of Pakistan [29], but our results highlight that drinking water poses a hazard to human health.

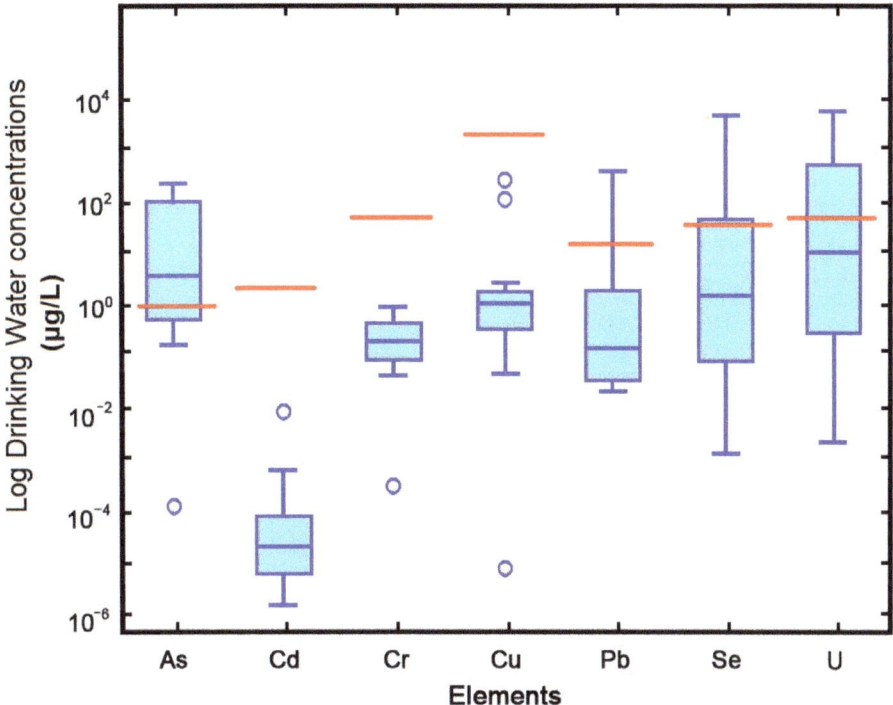

Figure 4. Boxplots of drinking water samples collected across the nine locations of Lahore. $N = 33$ and the red lines are the WHO's limits [20]. Boxes represent the interquartile range, bars represent minimum and maximum range and circles represent outliers.

Our results, shown in Figure 4, were significantly lower than the concentrations reported in other regions in developed nations. The drinking water samples exceeded the US EPA's drinking water standards for As (50 µg/L), Se (50 µg/L), and U (30 µg/L) concentrations, but were far below those set for Cd (5 µg/L), Pb (15 µg/L), and Cu (1300 µg/L). The As concentrations in drinking water in our study were greater than those found in 4547 water samples throughout Punjab as measured by Toor and Tahir [30] (6 to 12 µg/L); however, this study's concentrations were comparable to 48 drinking water samples researched in Lahore, Pakistan, by Akhter et al. [31] for As (36 ± 3 µg/L). The Cd, Cr, and Cu concentrations in the drinking water samples in our study were comparable to ranges observed in shallow and deep drinking water groundwaters across Pakistan, as reviewed by Waseem et al. [32]. The Pb concentrations measured in our drinking water were significantly lower than those found in the 48 drinking water samples in Lahore, Pakistan, by Akhter et al. [31] for Pb (1.8 ± 0.5 µg/L).

To investigate whether socioeconomic factors leading to inequity in drinking water quality affected trace element concentrations, we compared their concentrations across socioeconomic classifications for each town: Middle+Wealthy class and Impoverished class (Aziz et al. [33] defined these as Rich, Middle, Poor). These classes were defined according to literacy, access to education, unemployment, gross domestic production, and household possessions. We expected wealthier areas to have lower trace element concentrations owing to higher standards of water treatment and improved infrastructure for providing water within Lahore, Pakistan. We found significantly lower drinking water concentrations of Cr, Se, and U for drinking water in middle+wealthy administrative towns than those reported for impoverished administrative towns (Figure 5). However, the toxic concentrations of As were not significantly different among the socioeconomic groups, suggesting that As in drinking water is ubiquitous and not associated with lesser treatment of water for consumption. However, our data suggest that Cr, Se, and U toxic trace element concentration exposures are related to factors that vary among socioeconomic areas, such as the quality of groundwater resources, the post-extraction treatment of groundwater, and infrastructure for delivering water to residences and public places (as mentioned by Waseem et al. [32]). As described by Aziz et al. [33], combating inequality in education and income is only part of the building blocks of a successful life, as health and sanitation are also required. While these results show that socioeconomic class does not protect against the consumption of toxic concentrations of trace elements, wealthier individuals can obtain items to further decrease their consumption of this drinking water, while those in lower socioeconomic positions cannot. Even worse, those in more impoverished areas receive the burden of concentrations of Se and U higher by an order of magnitude in their drinking water. In addition to the natural bedrock sources, historical and legacy urban pollution may also be negatively impacting drinking water as emissions and leakages from mine/smelter wastes, phosphate sewage sludge, and municipal waste landfills can contaminate groundwater resources [11,32,34,35].

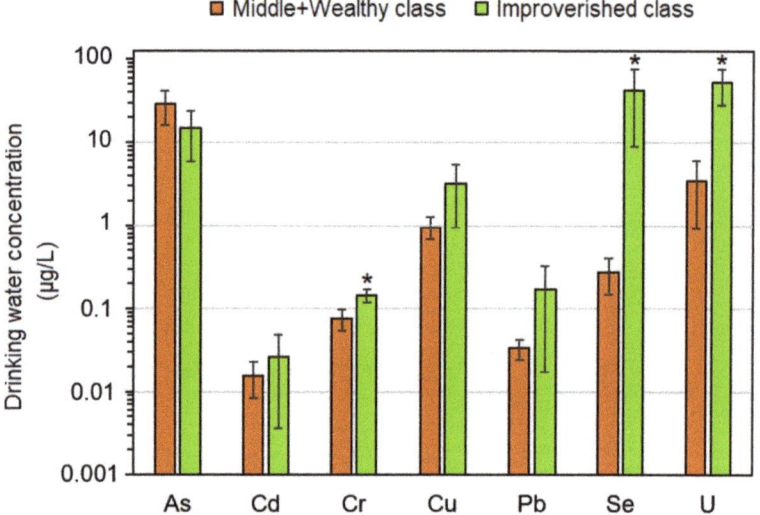

Figure 5. Drinking water samples compared across socioeconomic classes: Middle+Wealthy class administrative towns compared to impoverished class administrative towns [33]. (*) represent a significant difference in drinking water concentrations between socioeconomic groups of administrative towns.

We leveraged the THQ to examine whether trace element concentrations would pose a health hazard for Lahore residents at the noted consumption rates. Our results show that, as expected, the drinking water As THQ values far exceeded 1.0, indicating there is a direct risk to human health (Table 3). More interestingly, the drinking water THQ values for Se, and U also exceeded 1.0 for several drinking water collection sites. These results agree with those of Ali et al. [15], who found As and U THQ in groundwater in the Punjab and Sindh province exceeded the THQ threshold values of 1.0. Our results match the observations noted by Ahmed et al. [36] of Cd and Pb concentrations far below toxic levels for individuals in the Sindh region of Pakistan, who studied school-aged children.

Table 3. Drinking water THQs, where any value >1.0 is considered a lifetime health hazard. See Section 2.4 for the calculation method. Socioeconomic classifications were M+R, indicating middle and wealthy and impoverished towns, based on [33].

Town	SE Class	As	Cd	Cr	Cu	Pb	Se	U	Hazard Index (HI)
Aziz Bhatti	M+W	5.2	0.000	0.32	0.001	0.001	0.01	0.001	5.565
Cantonment	M+W	0.7	0.000	0.53	0.001	0.001	0.04	0.044	1.39
Ravi	M+W	3.3	0.000	0.62	0.001	0.001	0.02	0.130	6.11
Gulberg	M+W	0.02	0.000	0.76	0.000	0.000	0.01	0.000	0.77
Shalimar	IMP	5.5	0.008	0.17	0.001	0.001	0.01	0.001	5.70
Data Gunj Baksh	M+W	5.9	0.001	0.73	0.001	0.001	0.01	0.012	6.71
Nishtar (North)	IMP	0.7	0.000	0.38	0.001	0.000	1.58	0.465	3.09
Nishtar (South)	IMP	1.4	0.001	0.87	0.002	0.009	1.65	0.451	4.37
Wagah (North)	IMP	0.4	0.000	0.33	0.002	0.001	3.81	1.277	5.76
Wagah (South)	IMP	2.4	0.000	0.51	0.001	0.000	0.03	0.053	2.94
Iqbal (North)	IMP	0.3	0.000	0.82	0.002	0.001	0.04	1.254	9.78
Iqbal (South)	IMP	0.5	0.002	0.63	0.002	0.006	12.1	0.421	14.33
Mean		2.2	0.001	0.56	0.001	0.002	1.6	0.342	5.54

Furthermore, LCR was calculated to estimate whether the consumption of drinking water containing elevated trace elements also poses a risk for cancer in Lahore, Pakistan. The trace element-specific cancer slope factor was used to determine the LCR, which suggests it poses a hazard to cancer development when LCR > 10^{-4}, and is thus considered an unacceptable risk for cancer by the US EPA [17]. Our results show that the drinking water contained high enough As, Se, and U LCR values, which exceeded the 1.0×10^{-4} threshold, indicating there is a direct risk to cancer formation from drinking water consumption (Table 4). Fortunately, the other trace elements did not exceed the LCR values of the 1.0×10^{-4} threshold. These results agree with those of Ali et al. [15], who found that the As and U LCR values in groundwater in the Punjab and Sindh province exceeded the LCR cancer threshold of 1.0×10^{-4} as well. Our results match observations noted by Ahmed et al. [36] of Cd and Pb concentrations levels in the Sindh region of Pakistan far below hazard level for cancer development.

Table 4. Water LCR values, where any value >1.0×10^{-4} is considered a lifetime hazard for developing cancer. See Section 2.4 for description.

Town	As $\times 10^{-4}$	Cd $\times 10^{-4}$	Cr $\times 10^{-4}$	Cu $\times 10^{-4}$	Pb $\times 10^{-4}$	Se $\times 10^{-4}$	U $\times 10^{-4}$
Aziz Bhatti	24	0.00	0.00	0.00	0.00	0.00	0.04
Cantonment	3	0.00	0.00	0.00	0.00	0.01	2.2
Ravi	15	0.00	0.00	0.00	0.00	0.00	5.9
Gulberg	0.0	0.00	0.00	0.00	0.00	0.00	0.01
Shalimar	25	0.00	0.00	0.00	0.00	0.00	0.002
Data Gunj Baksh	27	0.00	0.00	0.00	0.00	0.00	0.53
Nishtar (North)	3	0.00	0.00	0.00	0.00	0.02	20.9
Nishtar (South)	6	0.00	0.00	0.00	0.00	0.33	820.3
Wagah (North)	11	0.00	0.00	0.00	0.00	0.01	2.4
Wagah (South)	2	0.00	0.00	0.00	0.00	0.76	57.5
Iqbal (North)	1	0.00	0.00	0.00	0.00	0.01	56.4
Iqbal (South)	2	0.00	0.00	0.00	0.00	2.4	18.9

4. Conclusions

Our study found further evidence that market rice for consumption by the millions of residents across the major city of Lahore, Pakistan, contains trace element concentrations that are above the concentrations considered to be safe for humans. The market rice THQ values for As, Cd, Cu, and Pb also exceeded the hazardous guidelines across several markets and only the LCR values for As suggest a potential hazard for developing cancer over a lifetime of market rice consumption. The widespread contamination of rice in Lahore, Pakistan, is a public health hazard and adds burdens onto communities dealing with potential food insecurity issues. Washing can remove a portion of trace elements, from 3 to 38% [25], but the drinking water also contained elevated trace elements and may be an additional source of trace elements when cooking rice. We assert that the problem begins in the agricultural soils and can be improved with mitigating procedures including safe agrochemical use [22] and safe irrigation water use [32]. Our findings also dispel speculation that more expensive rice is of higher quality and less likely to contain toxic trace element concentrations. There was no correlation between the market price of rice and its concentration of trace elements.

Our study also found that drinking water contained elevated concentrations of As, Se, and U across the administrative towns of Lahore, Pakistan. The THQ and LCR values for As, Se, and U exceeded the thresholds for toxicity to humans and pose a hazard for developing cancer over a lifetime of consumption. Although not addressed explicitly in our study, children who have lower body mass values are likely to be more greatly impacted as, in this study, we only utilized adult body mass. We hypothesized that drinking water quality would be higher in wealthier administrative towns and found some evidence that partially supports this hypothesis. The Cr, Se, and U concentrations in drinking water were higher in impoverished towns, which follows global inequity problems in which individuals living in lower socioeconomic areas hold less economic and sociopolitical power and are subjected to greater environmental health hazards in Pakistan and beyond [37]. As outlined by Waseem et al. [32], ensuring the use of safe aquifers, improvements in post-extraction treatment of groundwater, and a safe infrastructure for transporting the drinking water are needed to address this health hazard.

Supplementary Materials: The following supporting information can be downloaded at: https://www.mdpi.com/article/10.3390/su151813463/s1, Table S1: Market rice element concentrations; Table S2: Drinking water element concentrations.

Author Contributions: Conceptualization, J.B.R. and W.I.; methodology, J.B.R. and W.I.; analysis, J.B.R. and W.I.; writing—original draft preparation, J.B.R. and W.I.; writing—review and editing, J.B.R. and W.I. All authors have read and agreed to the published version of the manuscript.

Funding: This research was funded by University of Massachusetts Amherst, College of Natural Science grant number 11-01 to J.B.R.

Institutional Review Board Statement: Not applicable.

Informed Consent Statement: Not applicable.

Data Availability Statement: All data used in analyses are available in the figures and tables included in the study and in the supplemental materials provided, as well as on the author's website: soilbiogeochemist.com.

Acknowledgments: We would like to thank Mariam Imran, Haider Dawood Khan, and Abbas Khan for aiding in sample collection and support for the project.

Conflicts of Interest: The authors declare no conflict of interest. The funders had no role in the design of the study; in the collection, analyses, or interpretation of data; in the writing of the manuscript, or in the decision to publish the results.

References

1. Chandrasiri, G.U.; Mahanama, K.R.R.; Mahatantila, K.; Arachchige, P.S.P.; Liyanage, R.C.M. An assessment on toxic and essential elements in rice consumed in Colombo, Sri Lanka. *Appl. Biol. Chem.* **2022**, *65*, 24. [CrossRef]
2. Khan, N.A.; Gao, Q.; Abid, M.; Shah, A.A. Mapping farmers' vulnerability to climate change and its induced hazards: Evidence from the rice-growing zones of Punjab, Pakistan. *Environ. Sci. Pollut. Res.* **2021**, *28*, 4229–4244. [CrossRef]
3. Sarwar, T.; Khan, S.; Yu, X.; Amin, S.; Khan, M.A.; Sarwar, A.; Muhammad, J.; Nazneen, S. Analysis of Arsenic concentration and its speciation in rice of different markets of Pakistan and its associated health risk. *Environ. Technol. Innov.* **2020**, *21*, 101252. [CrossRef]

4. Malik, N.; Aboidullah, M.; Chaudhry, M.N. Habits and Practices Regarding Domestic Water Usage in Lahore City. *Pak. Vis.* **2016**, *17*, 244–258.
5. Karalliedde, L.; Brooke, N. Toxicity of heavy metals and trace elements. In *Essentials of Toxicology for Health Protection*; Oxford Academic: Oxford, UK, 2012; pp. 168–186. [CrossRef]
6. Javed, A.; Farooqi, A.; Baig, Z.U.; Ellis, T.; van Geen, A. Soil arsenic but not rice arsenic increasing with arsenic in irrigation water in the Punjab plains of Pakistan. *Plant Soil* **2020**, *450*, 601–611. [CrossRef]
7. Ali, W.; Mao, K.; Zhang, H.; Junaid, M.; Xu, N.; Rasool, A.; Feng, X.; Yang, Z. Comprehensive review of the basic chemical behaviours, sources, processes, and endpoints of trace element contamination in paddy soil-rice systems in rice-growing countries. *J. Hazard. Mater.* **2020**, *397*, 122720. [CrossRef]
8. Cannas, D.; Loi, E.; Serra, M.; Firinu, D.; Valera, P.; Zavattari, P. Relevance of Essential Trace Elements in Nutrition and Drinking Water for Human Health and Autoimmune Disease Risk. *Nutrients* **2020**, *12*, 2074. [CrossRef]
9. Adriano, D.C. Adriano, Trace Elements in Terrestrial Environments. In *Biogeochemistry, Bioavailability, and Risks of Metals*; Springer: New York, NY, USA, 2001.
10. Shakir, S.K.; Azizullah, A.; Murad, W.; Daud, M.K.; Nabeela, F.; Rahman, H.; Rehman, S.U.; Häder, D.-P. Toxic Metal Pollution in Pakistan and Its Possible Risks to Public Health. *Rev. Environ. Contam. Toxicol. Vol.* **2017**, *242*, 1–60. [CrossRef]
11. Alengebawy, A.; Abdelkhalek, S.T.; Qureshi, S.R.; Wang, M.-Q. Heavy Metals and Pesticides Toxicity in Agricultural Soil and Plants: Ecological Risks and Human Health Implications. *Toxics* **2021**, *9*, 42. [CrossRef]
12. Heikens, A. *Arsenic Contamination of Irrigation Water, Soil and Crops in Bangladesh: Risk Implications for Sustainable Agriculture and Food Safety in Asia*; Rap Publication (FAO): Rome, Italy, 2006.
13. Husain, V.I.Q.A.R.; Nizam, H.I.N.A.; Arain, G.M. Arsenic and fluoride mobilization mechanism in groundwater of Indus Delta and Thar Desert, Sindh, Pakistan. *Int. J. Econ. Environ. Geol.* **2012**, *3*, 15–23.
14. Pakistan Bureau of Statistics; Government of Pakistan. Available online: https://www.pbs.gov.pk/ (accessed on 6 June 2023).
15. Ali, W.; Aslam, M.W.; Feng, C.; Junaid, M.; Ali, K.; Li, S.; Chen, Z.; Yu, Z.; Rasool, A.; Zhang, H. Unraveling prevalence and public health risks of arsenic, uranium and co-occurring trace metals in groundwater along riverine ecosystem in Sindh and Punjab, Pakistan. *Environ. Geochem. Health* **2019**, *41*, 2223–2238. [CrossRef] [PubMed]
16. World Food Program (WFP). Annual Country Reports—Pakistan. Available online: https://www.wfp.org/publications/annual-country-reports-pakistan (accessed on 11 July 2023).
17. USEPA. *Exposure Factors Handbook: 2011 Edition*; US Environmental Protection Agency: Washington, DC, USA, 2011.
18. Natasha; Bibi, I.; Niazi, N.K.; Shahid, M.; Ali, F.; Hasan, I.M.U.; Rahman, M.M.; Younas, F.; Hussain, M.M.; Mehmood, T.; et al. Distribution and ecological risk assessment of trace elements in the paddy soil-rice ecosystem of Punjab, Pakistan. *Environ. Pollut.* **2022**, *307*, 119492. [CrossRef] [PubMed]
19. Nawab, J.; Farooqi, S.; Xiaoping, W.; Khan, S.; Khan, A. Levels, dietary intake, and health risk of potentially toxic metals in vegetables, fruits, and cereal crops in Pakistan. *Environ. Sci. Pollut. Res.* **2018**, *25*, 5558–5571. [CrossRef]
20. WHO. *Trace Elements in Human Nutrition and Health*; WHO: Geneva, Switzerland, 1996; p. 361.
21. TatahMentan, M.; Nyachoti, S.; Scott, L.; Phan, N.; Okwori, F.O.; Felemban, N.; Godebo, T.R. Toxic and essential elements in rice and other grains from the United States and other countries. *Int. J. Environ. Res. Public Health* **2020**, *17*, 8128. [CrossRef]
22. Jung, M.C.; Yun, S.T.; Lee, J.S.; Lee, J.U. Baseline study on essential and trace elements in polished rice from South Korea. *Environ. Geochem. Health* **2005**, *27*, 455–464. [CrossRef] [PubMed]
23. Tariq, S.R.; Rashid, N. Multivariate Analysis of Metal Levels in Paddy Soil, Rice Plants, and Rice Grains: A Case Study from Shakargarh, Pakistan. *J. Chem.* **2013**, *2013*, 539251. [CrossRef]
24. Ghafoor, A.B.D.U.L.; Aslam, M.A.N.A.N. *Market Integration and Price Transmission in Rice Markets of Pakistan*; Working Paper; South Asia Network of Economic Research Institute: Dhaka, Bangladesh, 2012.
25. Sharafi, K.; Yunesian, M.; Mahvi, A.H.; Pirsaheb, M.; Nazmara, S.; Nodehi, R.N. Advantages and disadvantages of different pre-cooking and cooking methods in removal of essential and toxic metals from various rice types-human health risk assessment in Tehran households, Iran. *Ecotoxicol. Environ. Saf.* **2019**, *175*, 128–137. [CrossRef]
26. Panaullah, G.M.; Alam, T.; Hossain, M.B.; Loeppert, R.H.; Lauren, J.G.; Meisner, C.A.; Ahmed, Z.U.; Duxbury, J.M. Arsenic toxicity to rice (*Oryza sativa* L.) in Bangladesh. *Plant Soil* **2009**, *317*, 31–39. [CrossRef]
27. Baruah, S.G.; Ahmed, I.; Das, B.; Ingtipi, B.; Boruah, H.; Gupta, S.K.; Nema, A.K.; Chabukdhara, M. Heavy metal(loid)s contamination and health risk assessment of soil-rice system in rural and peri-urban areas of lower brahmaputra valley, northeast India. *Chemosphere* **2021**, *266*, 129150. [CrossRef]
28. Natasha Bibi, I.; Shahid, M.; Niazi, N.K.; Younas, F.; Naqvi, S.R.; Shaheen, S.M.; Imran, M.; Wang, H.; Hussaini, K.M.; Zhang, H.; et al. Hydrogeochemical and health risk evaluation of arsenic in shallow and deep aquifers along the different floodplains of Punjab, Pakistan. *J. Hazard Mater.* **2021**, *402*, 124074. [CrossRef]
29. WHO. *Guidelines for Quality Drinking-Water*, 4th ed.; World Health Organization: Geneva, Switzerland, 2011.
30. Toor, I.A.; Tahir, S.N.A. Study of arsenic concentration levels in Pakistani drinking water. *Pol. J. Environ. Stud.* **2009**, *18*, 907–912.
31. Akhter, G.; Ahmad, Z.; Iqbal, J.; Shaheen, N.; Shah, M.H. Physicochemical characterization of groundwater in urban areas of Lahore, Pakistan, with special reference to arsenic. *J. Chem. Soc. Pak.* **2010**, *32*, 306–312.
32. Waseem, A.; Arshad, J.; Iqbal, F.; Sajjad, A.; Mehmood, Z.; Murtaza, G. Pollution Status of Pakistan: A Retrospective Review on Heavy Metal Contamination of Water, Soil, and Vegetables. *BioMed Res. Int.* **2014**, *2014*, 813206. [CrossRef]

33. Aziz, A.; Mayo, S.M.; Ahmad, I.; Hussain, M.; Nafees, M. Determining town base socioeconomic indices to sensitize development in Lahore, Pakistan. *Tech. J. UET TAxila Pak.* **2014**, *19*, 22–29.
34. Rehman, W.; Zeb, A.; Noor, N.; Nawaz, M. Heavy metal pollution assessment in various industries of Pakistan. *Environ. Geol.* **2008**, *55*, 353–358. [CrossRef]
35. Mary, R.; Nasir, R.; Alam, A.; Tariq, A.; Nawaz, R.; Javied, S.; Zaman, Q.U.; Islam, F.; Khan, S.N. Exploring hazard quotient, cancer risk, and health risks of toxic metals of the Mehmood Booti and Lakhodair landfill groundwaters, Pakistan. *Environ. Nanotechnol. Monit. Manag.* **2023**, *20*, 100838. [CrossRef]
36. Ahmed, J.; Wong, L.P.; Chua, Y.P.; Channa, N.; Memon, U.-U.; Garn, J.V.; Yasmin, A.; VanDerslice, J.A. Heavy metals drinking water contamination and health risk assessment among primary school children of Pakistan. *J. Environ. Sci. Health Part A* **2021**, *56*, 667–679. [CrossRef] [PubMed]
37. Perveen, S.; Haque, A.U. Drinking water quality monitoring, assessment and management in Pakistan: A review. *Heliyon* **2023**, *9*, e13872. [CrossRef] [PubMed]

Disclaimer/Publisher's Note: The statements, opinions and data contained in all publications are solely those of the individual author(s) and contributor(s) and not of MDPI and/or the editor(s). MDPI and/or the editor(s) disclaim responsibility for any injury to people or property resulting from any ideas, methods, instructions or products referred to in the content.

Article

Pollution and Risk Evaluation of Toxic Metals and Metalloid in Water Resources of San Jose, Occidental Mindoro, Philippines

Delia B. Senoro [1,2,3,4,*], Kevin Lawrence M. De Jesus [1,2,4] and Cris Edward F. Monjardin [1,2,3,4,5]

1. Resiliency and Sustainable Development Center, Yuchengco Innovation Center, Mapua University, Intramuros, Manila 1002, Philippines
2. School of Graduate Studies, Mapua University, Intramuros, Manila 1002, Philippines
3. School of Civil, Environmental, and Geological Engineering, Mapua University, Intramuros, Manila 1002, Philippines
4. School of Chemical, Biological, Materials Engineering and Sciences, Mapua University, Intramuros, Manila 1002, Philippines
5. Department of Civil and Environmental Engineering, Spencer Engineering Building, University of Western Ontario, London, ON N6A 3K7, Canada
* Correspondence: dbsenoro@mapua.edu.ph; Tel.: + 63-2-8251-6622

Abstract: Clean and safe drinking water is an integral part of daily living and is considered as a basic human need. Hence, this study investigated the suitability of the domestic water (DW) and groundwater (GW) samples with respect to the presence of metals and metalloid (MMs) in San Jose, Occidental Mindoro, Philippines. The MMs analyzed in the area of study for DW and GW were Arsenic (As), Barium (Ba), Copper (Cu), Chromium (Cr), Iron (Fe), Lead (Pb), Manganese (Mn), Nickel (Ni), and Zinc (Zn). The results revealed that Pb has the mean highest concentration for DW, while Fe is in GW resources in the area. Quality evaluation of DW and GW was performed using Metal Pollution Index (MPI), Nemerow's Pollution Index (NPI), and Ecological Risk Index (ERI). The mean NPI value calculated for DW was 135 times greater than the upper limit of the unpolluted location category. The highest NPI observed was 1080 times higher than the upper limit of the unpolluted site category. That of the ERI observed in the area was 23.8 times higher than the upper limit for a "low" ERI category. Furthermore, the health risk assessment (HRA) of the GW and DW of the study area revealed non-carcinogenic health risks of the MMs analyzed in GW samples, and potential carcinogenic health risks from As, Cr, Pb, and Ni in DW. The use of machine learning geostatistical interpolation (MLGI) mapping to illustrate the PI and health risk (HR) in the area was an efficient and dependable evaluation tool for assessing and identifying probable MMs pollution hotspots. The data, tools, and the process could be utilized in carrying out water assessment, the evaluation leading to a comprehensive water management program in the area and neighboring regions of similar conditions.

Keywords: metals and metalloid; pollution index; ecological risk; health risk; spatial analysis

Citation: Senoro, D.B.; De Jesus, K.L.M.; Monjardin, C.E.F. Pollution and Risk Evaluation of Toxic Metals and Metalloid in Water Resources of San Jose, Occidental Mindoro, Philippines. *Sustainability* **2023**, *15*, 3667. https://doi.org/10.3390/su15043667

Academic Editor: Said Muhammad

Received: 25 January 2023
Revised: 12 February 2023
Accepted: 13 February 2023
Published: 16 February 2023

Copyright: © 2023 by the authors. Licensee MDPI, Basel, Switzerland. This article is an open access article distributed under the terms and conditions of the Creative Commons Attribution (CC BY) license (https://creativecommons.org/licenses/by/4.0/).

1. Introduction

Water is fundamental in sustaining the quality of life in a community, as it is also attributed to the health, food, and economy of the area. Sustainable Development Goal 6 intends to guarantee the access of all people to clean and affordable water sources, especially those in remote areas. However, due to continuous changes in the landscape, land use, and anthropogenic processes, water resources have been compromised, which poses a threat to the people in the nearby communities, especially in remote regions, since these water resources are the only source for domestic and agricultural activities [1]. Due to its possible toxicity and probable health adverse effects, metals and metalloid (MMs) contamination in water resources are of paramount concern. The pollution of water resources is a result of either natural processes, such as weathering of rocks and runoff, or anthropogenic activities, such as mining, industrial, and agricultural activities. A hazardous concentration of MMs could build up and have negative consequences on the environment and human health. As their concentrations increase, these MMs may cause more harm to the environment and water systems. When water supplies and the ecosystem are poisoned with high amounts of MMs, health issues and consequences are imminent [2,3]. The quality of drinking and irrigation water supplies declines as a result of MMs migration to water resources. Furthermore, a high concentration of MMs from both anthropogenic and natural sources has a detrimental effect on domestic water (DW) quality [4]. The DW is defined as water used by the population for drinking, cooking, and bathing. The list of abbreviations and symbols used in this study is presented in Abbreviations Section.

The MMs that are naturally present in the earth's crust are transferred into water resources through weathering and decomposition of metal rock and ores, whereas MMs from anthropogenic activities are released through automobile emissions, improper management of waste, the burning of fossil fuels, the usage of fertilizer and pesticides, untreated wastewater, and atmospheric precipitation from mineral extraction, metal processing, and agricultural operations [5]. Investigations of water quality have been conducted on a regional scale to evaluate the current state of the water resources in reference to the MMs present in the area [6,7]. Table A1 in Appendix A enumerates the MMs pollution and health risk assessment (HRA) studies in various regions of the world.

The HRA is a valuable instrument to assess and appraise the probability of health effects with respect to MMs [8]. The elevated concentrations of arsenic (As) [9], barium (Ba) [10], and manganese (Mn) [11] have adverse effects on human health. Metals like copper (Cu), iron (Fe), nickel (Ni), and zinc (Zn) are necessary for the regular growth and functioning of living organisms [12]; however, excess amounts will lead to adverse health effects, too.

The use of GIS to further reinforce the calculated pollution index and health risks has been implemented to several DW and environmental monitoring studies, including MMs in shallow GW in a lake plain in China [13]; MPI of GW in a city in India [14]; HEI and HRA in a GW plain in Iran [15]; WQI, MPI, and PI in a surface water body in Egypt [16]; WQI, MPI, HEI, and HRA in a river and stream in Turkey [17,18]; and WQI, HPI, HEI, and HRA and identification of pollution hotspots in a river in Ethiopia [19]. A good knowledge of the geochemical origin of the pollutants in water resources was obtained by the mapping of the MMs concentration and associated risk indices [20].

Domestic water quality monitoring with respect to MMs concentration has been a challenge in the Philippines due to several factors such as: cost of portable devices used for on-site detection and analysis, laboratory fees for MMs tests and analysis, proximity of sampling sites to the capable laboratory, limited number of laboratories capable of conducting MMs detection and analysis, and government permits to purchase instruments calibration standard solutions, among related others. In the province of Occidental Mindoro, only a limited number of research works have been carried out to detect and analyze MMs in DW and create spatial concentration maps of MMs and their associated indices that can identify pollution hotspots. Hence, the current work investigated the suitability of the water resources in the municipality of San Jose province of Occidental Mindoro, Philippines for domestic consumption, evaluated the risks of possible pollution, and created spatial distribution maps to determine the pollution hotspots. This is with respect to and degree of concentrations of metalloid arsenic (As) and metals such as Ba, Cu, chromium (Cr), Fe, lead (Pb), Mn, Ni, and Zn. The outcomes of this research offer additional data on the existence of MMs in DW resources and help locals create preventive measures, as well as allowing for environmental health professionals to lessen the adverse impacts of MMs in the water resources. It can also be used as a source of data and benchmark activities to create strategic programs to mitigate the presence of elevated MMs concentrations in the Philippines and other neighboring regions.

The DW is basically the potable freshwater that each household uses for everyday needs. However, due to some malpractices and lack of knowledge, unintentional contamination occurs. The DW is supplied by utility companies that have extracted water underground, as well as some from surface water, but have done traditional treatment before distribution to households and commercial establishments. Some households have their own shallow well for their DW needs.

2. Materials and Methods

2.1. Description of the Study Area

The municipality of San Jose is a coastal municipality in the province of Occidental Mindoro, in which the municipal center is located at 12°21′ N, 121°4′ E with an average natural grade line (ground) elevation of 7.8 m above mean sea level. The municipality has a total land area of 446.70 km², which is 7.63% of the total area of the province of Occidental Mindoro [21]. Geographically, San Jose is 173 km from the municipality of Mamburao, the capital city of the province. The municipalities of Rizal and Calintaan, Mansalay, Bulalacao, Magsaysay, and the Mindoro Strait form the northern, eastern, southern, and western borders of San Jose, respectively [22]. The map of the study area is presented as Figure 1.

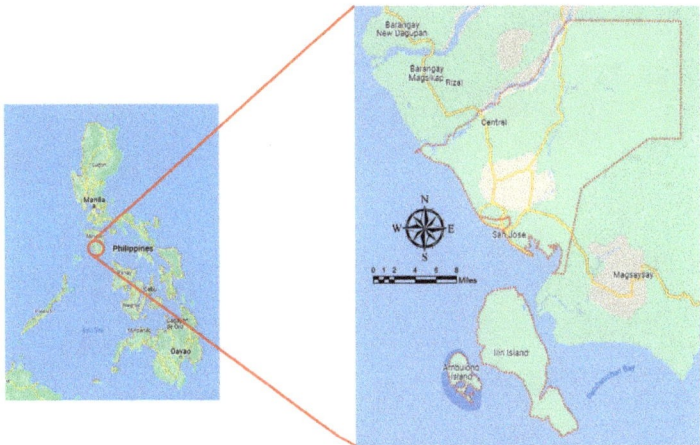

Figure 1. Area of the study.

Based on the 2020 census, the total population of the San Jose municipality was 153,267, which constitutes of 29.17% of the total population of the province. San Jose is comprised of 39 barangays (the smallest administrative unit of the local government), with Barangay San Roque as the most populous barangay with 10.26% of the total population of the municipality [23]. The climate in Occidental Mindoro is Type I, with two distinct seasons. The weather is dry from November to April and rainy the rest of the year [24] with runoff that contributes to the pollution of water resources.

A total of 11 rivers and creeks cut through the town, creating a network that also acts as a natural drainage system. A tributary river system to the Busuanga River exists in the region and is the primary source of irrigation for agricultural land. The barangays of San Jose are divided among four watersheds on the mainland and one on an island, with a combined size of around 626.2 km². These watersheds are Busuanga, Cabariwan, Caguray, and Labangan. Table 1 lists the watersheds along with their coverage area and the included barangays. Watersheds are important in water resources for domestic supply, as they host surface water and groundwater (GW).

Table 1. List of Watersheds in San Jose, Occidental Mindoro [25].

Name of Watershed	Barangays Covered	Area (km²)
Busuanga	Batasan, Camburay, Central, Monteclaro, Murtha, and San Agustin	199.96
Cabariwan	Bayotbot, Labangan Poblacion, Mabini, Mangarin, Mapaya, Natandol, and Pawican	47.34
Caguray	Batasan, Bayotbot, Mapaya, Monteclaro, and Murtha	52.39
Labangan	All barangays except Mapaya	242.92

2.2. Sample Collection, Preparation, and Analysis

A total of 104 water samples (71 DW and 33 GW samples) were collected randomly in different barangays of the municipality of San Jose, Occidental Mindoro. The 104 grab samples were gathered using stainless steel samplers and stored in prepared acid-rinsed one liter polyethylene (PE) bottles. This is to remove the possible contaminants in the PE bottles. The PE bottles were all properly labeled, sealed, and placed temporarily in coolers. The collection, preparation, and storage of the DW and GW samples were in accordance with the EPA No. SESDPROC-301-R3 [26]. The coordinates of every sampling site were recorded utilizing a Garmin Montana 680 GPS. The sampling locations for the DW and GW are presented in Figure 2. The coordinates and corresponding elevations of the sampling locations are enumerated in Appendix B.

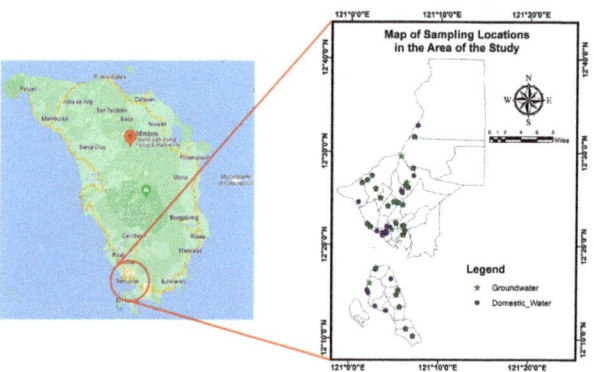

Figure 2. Sampling locations for domestic water and groundwater.

The water samples were transferred from PE bottles into zipper plastic #2 prior to detection and analysis using Olympus Vanta XRF Spectrometer and Accusensing MAS G1. High performance in situ elemental analyzers, such as the Olympus Vanta portable XRF (pXRF) and Accusensing MAS, are suitable for a variety of environmental media, including water samples. Prior to analysis, the pXRF was calibrated utilizing Olympus Vanta blank in zipper plastic #2. The XRF was set to Geochem mode. The reliability and viability on the use of Olympus Vanta XRF and Accusensing MAS for the detection of MMs in water have been discussed in various works [27–32]. The Accusensing MAS G1 was used to analyze MMs that had returned LOD readings in the pXRF [30]. Several metals and metalloid including As, Ba, Cu, Cr, Fe, Pb, Mn, Ni, and Zn were detected and used in the computation of PI and HR indices.

2.3. Pollution Evaluation Indices

2.3.1. Metals Pollution Index (MPI)

The *MPI* can be described as the aggregate extent of the effects of MMs water resources considering human consumption and metal contamination [33,34]. The recommended standard values of the MMs was sourced out from the PNSDW 2017. The *MPI* is based on the weighted arithmetic quality mean method expressed in Equation (1).

$$MPI = \frac{\sum_{i=1}^{n} Q_i W_i}{\sum_{i=1}^{n} W_i} \quad (1)$$

W_i is the weight unit and was calculated as $1/S_i$, wherein S_i is the recommended standard of the MMs, n is the number of estimated metals, and Q_i is the individual quality rating of the MMs, and can be calculated as expressed in Equation (2) [35].

$$Q_i = \frac{C_i}{S_i} \times 100 \quad (2)$$

wherein C_i is the measured concentration value of the MMs. The interpretation of the calculated *MPI* values were divided into three categories, such as low (*MPI* < 90), medium (90 ≤ *MPI* ≤ 180), and high (*MPI* > 180) [30].

2.3.2. Ecological Risk Index (ERI)

The potential *ERI* of the examined MMs was calculated by taking into account the pollution index and T_i. The *ERI* is the summation of the product of the T_i of the MMs, and the pollution index is the ratio between the concentration of the MMs in the sample and the subsequent background values. The *ERI* for each water sample was calculated using Equations (3) and (4).

$$ERI = \sum RI = \sum T_i(P_i) \quad (3)$$

$$PI = \frac{C_s}{C_b} \quad (4)$$

where *ERI* is the potential ecological risk factor of each metal, T_i is the toxic-response factor of the metal, *PI* is the pollution index, C_s is the concentration of the metals in the sample, and C_b is the corresponding background value. The toxic-response factor of the metals was as follows: As = 10, Cu = 5, Cr = 1, Fe = 1, Mn = 1, Ni = 1, Pb = 5, and Zn = 1. The interpretations of the *ERI* values were categorized as follows: low (*ERI* < 95), moderate (95 ≤ *ERI* < 190), considerable (190 ≤ *ERI* < 380), and very high (*ERI* ≥ 380) [36,37].

2.3.3. Nemerow's Pollution Index (NPI)

Based on the single factor pollution index, the *NPI* is an extensive pollution index assessment. It is applied to evaluate the WQ at various sampling locations while also emphasizing the significance of different metals concentration in the water resources. The *NPI* was calculated as the square root of the half of the sum of the squares of average and maximum single factor pollution index (SFPI), shown as Equation (5).

$$NPI = \sqrt{\frac{(SFPI_{max})^2 + (SFPI_{ave})^2}{2}} \quad (5)$$

The SFPI is the ratio between the observed concentration and the evaluation standard of the MMs, shown as Equation (6).

$$SFPI = \frac{C_i}{S_i} \quad (6)$$

The *NPI* is divided into five classes, which includes Class 1–Unpolluted (*NPI* < 1.0), Class 2–Slightly Polluted Water (1.0 ≤ *NPI* < 2.5), Class 3–Moderately Polluted Water (2.5 ≤ *NPI* < 7.0), and Class 4–Heavily Polluted Water (*NPI* ≥ 7.0) [37–39].

2.3.4. Probabilistic Health Risk Assessment

The non–carcinogenic risk linked with oral pathway exposure to MMs, which is the hazard index (HI) shown as Equation (7), was determined by employing the summation of the HQ shown as Equation (8). This is the ratio between the *CDI* and *RfD*. The *CDI* was calculated using the MMs concentration, and *EF*, *ED*, *IR*, *AT*, and *BW* as shown in Equation (9) [40].

$$HI = \sum HQ \tag{7}$$

$$HQ = \frac{CDI}{RfD} \tag{8}$$

$$CDI = \frac{MC \times EF \times ED \times IR}{AT \times BW} \tag{9}$$

where *HQ* is the hazard quotient, *CDI* is the chronic daily intake, *RfD* is the reference dose, *MC* is the metal concentration, *EF* is the exposure frequency, *ED* is the exposure duration, *IR* is the ingestion rate, *AT* is the average time, and *BW* is the body weight.

The following values were utilized in the calculation: EF = 365 days [41], ED = 70 years [42], IR = 2 L/day [43], AT = 25,550 days [44], and BW = 70 kg [45]. The RfD values for each MMs (in mg/kg/day) were: As = 3×10^{-4} [46], Ba = 2×10^{-1} [47], Cu = 0.04 [5], Cr = 0.003 [48], Fe = 7×10^{-1} [49], Pb = 3.5×10^{-3} [48], Mn = 1.4×10^{-1} [49], Ni = 0.02 [5], and Zn = 0.3 [5].

The excess lifetime carcinogenic risk (*ELCR*) approach was utilized to calculate the risk assessment of the carcinogenicity of MMs, which utilized the product of the *SF* and the *CDI* [43]. The *SF* values employed were As = 1.5 [46], Cr = 0.5 [48], Pb = 8.5×10^{-3} [5], and Ni = 8.4×10^{-1} [5]. The *ELCR* was calculated using Equation (10) [43].

$$ELCR = CDI \times SF \tag{10}$$

2.4. Spatial Mapping Using Machine Learning Geostatistical Interpolation (MLGI) Approach

Utilizing the MMs concentrations detected from the water samples and their corresponding PI and MPI, spatial maps were generated by employing the MLGI method that utilized the NN-PSO algorithm and integrating it to an EBK interpolation technique. The MLGI method is the integration of a hybrid artificial neural network (ANN)–particle swarm optimization (PSO) model to the empirical Bayesian kriging method to map the MMs' concentration and pollution indices. The connection weights of the neural network models for the concentrations and pollution indices were optimized using the PSO approach, utilizing MATLAB2021a. The Levenberg-Marquardt algorithm was utilized in the algorithm training for ANN-PSO models, since it is the fastest method for moderately sized networks [38], such as used in this study. The transfer function employed was the hyperbolic tangent sigmoid function [50]. The number of iterations for the model development was 2000 iterations [51]. Figure 3 presents the chart of the process for the MLGI approach employed in this study.

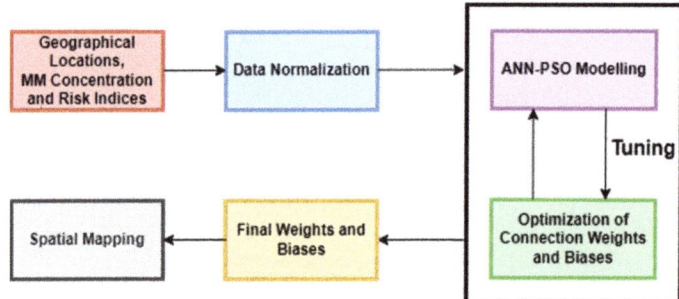

Figure 3. The MLGI approach employed in the study.

The spatial maps produced provided a cleared view of the state of MMs pollution in the research area, together with the associated risk it posed to the area [30].

2.5. Contamination Area Calculation

Using the generated spatial concentration maps by the MLGI approach, the contamination area for each MM and associated risk indices were calculated. Using the reclassification and raster to polygon conversion tool in GIS, the contamination area was calculated in reference to the reference guidelines and index reference values.

2.6. Statistical Analysis

Descriptive statistics were accomplished employing the Microsoft Excel Data Analysis Tool, MATLAB R2021a, and R Studio software. The evaluation for the normality of data were evaluated employing the Shapiro-Wilk test for GW (less than 50 samples) and the Kolmogorov-Smirnov test for DW samples (greater than 50 samples) [49]. Moreover, the variability of the GW and DW samples was analyzed using the coefficient of variability (CV), wherein CV less than or equal to 15% is low; CV greater than 15% but less than or equal to 35% is intermediate; and CV greater than 35% is high [26]. Furthermore, the mean concentration was compared to the PNSDW 2017 standard values.

The MMs concentrations correlations of water samples were analyzed using a multivariate Pearson correlation matrix. In order to demonstrate the correlation concerning the investigated parameters, particularly with regard to their sources, Pearson's correlation analysis was performed on the data using MATLAB R2021a and R Studio software [52]. Parameters with a correlation coefficient of $0.90 \leq R \leq 1.00$ indicate a very strong correlation; $0.70 \leq R \leq 0.89$ suggests a strong correlation; $0.40 \leq R \leq 0.69$ denotes a moderate correlation; $0.10 \leq R \leq 0.39$ signifies a weak correlation; and $0.00 \leq R \leq 0.09$ indicates a negligible correlation [53]. Additionally, pairings with strong and moderate coefficients suggest considerable risk factors, whereas those with weak coefficients suggest low risk factors [54].

3. Results and Discussion

3.1. Metal Concentration

The basic statistical parameters of the MMs in DW samples from various locations in San Jose, Occidental Mindoro are presented in Table 2. The trend of MMs' mean concentration detected in DW was Pb > As > Fe > Cr > Cu > Ni > Ba > Mn. The Kolmogorov-Smirnov test showed that the MMs concentration in DW in the research area was not uniformly distributed, since all the metals have $p < 0.05$ [55]. Moreover, the CV values for all the detected MMs exhibited values higher than 35%. This suggests that the acquired datasets for the DW samples have high variability.

Table 2. Metals and metalloid concentrations (in mg/L) in DW.

	As	Ba	Cu	Cr	Fe	Pb	Mn	Ni
N	71.000	71.000	71.000	71.000	71.000	71.000	71.000	71.000
Max	6.680	0.050	0.800	0.890	1.820	15.160	0.020	0.720
Min	0.000	0.000	0.000	0.000	0.000	0.000	0.000	0.000
Mean	0.792	0.024	0.093	0.110	0.138	1.152	0.003	0.045
SD	1.762	0.016	0.157	0.225	0.279	3.302	0.004	0.152
Skewness	2.228	−0.629	2.493	2.172	3.895	3.538	1.592	3.314
Kurtosis	3.716	−1.107	7.019	3.878	19.340	12.349	2.552	9.867
CV%	222.640	66.560	168.250	205.150	202.610	286.580	137.570	339.760
PNSDW	0.010	0.700	1.000	0.050	1.000	0.010	0.400	0.070

The mean concentrations of the MMs were compared to the PNSDW 2017 as shown in Figure 4. The DW dataset obtained revealed that As, Cr, and Pb were above the threshold standards set by the PNSDW 2017 by 79.2, 2.2, and 115.2 times, respectively. About 33.8%

of the sampling site was above the PNSDW 2017 threshold limit for As, while that for Cr, Fe, Pb, and Ni were 26.8%, 2.8%, 31%, and 8.5%, respectively. Additionally, all sampling sites for Ba, Cu, and Mn were under the PNSDW 2017 standard limits.

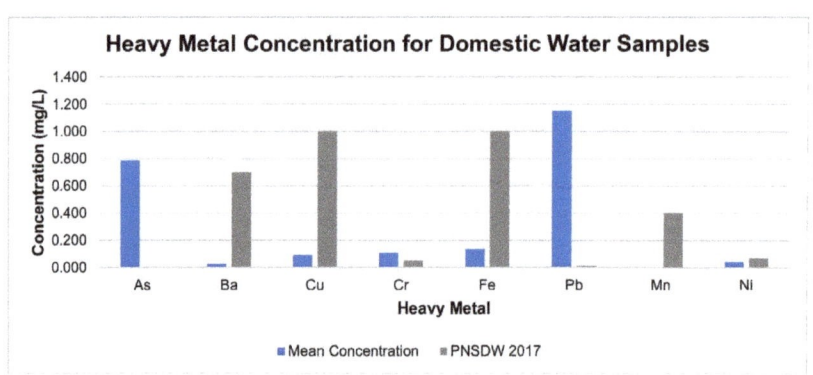

Figure 4. Comparison of MMs' concentrations in DW with PNSDW 2017 standard values.

The highest Fe concentration in DW was 1.8 times greater than the standard value and was detected in a deep well water source in Brgy. Iling Proper. Elevated Fe concentration may lead to Fe poisoning. Localized iron poisoning is characterized by gastrointestinal bleeding, nausea, vomiting, and abdominal discomfort. Injuries to the liver and cardiovascular system lead to systemic toxicity [56]. For the detected Ni concentration in DW samples, the highest detected concentrations were recorded in Barangay II Poblacion. The water samples were collected from a water refilling station (WFS) with a concentrations 10.3 times greater than the standard limit set by PNSDW. Exposure to elevated Ni concentration can have neurological side effects such as giddiness, fatigue, and headaches. The effects of drinking nickel-tainted water on the gastrointestinal system includes nausea, cramping in the abdomen, diarrhea, and vomiting. Additionally, impacts on the musculoskeletal system, including muscular discomfort, have been linked to drinking nickel-contaminated water [57].

Moreover, the highest concentrations of Pb was detected in Brgy. Camburay, collected from a WRS with concentrations of 1516 times greater than the PNSDW limit. The Pb exposure has many hazardous consequences, but might be reversed if discovered promptly. However, chronic high-level Pb exposure has the potential to permanently harm the kidneys, central nervous system, and peripheral nervous system [58]. Acute lead poisoning can cause a number of different signs and symptoms, such as abdominal discomfort, constipation, joint pain, muscular aches, headaches, anorexia, reduced libido, difficulties focusing and short-term memory deficiencies, irritability, excessive tiredness, sleep disturbances, and anemia [59,60].

Highest Cu concentrations were detected in 3 different locations. These are Brgy. Iling Proper, Brgy. Ipil, and Brgy. Murtha. The highest concentrations of Mn and Zn were recorded in Brgy. Central and Brgy. San Agustin, respectively. However, the concentration levels of Cu, Mn, and Zn detected were inside the PNSDW 2017 limit. The highest Ba levels in GW were observed in Brgy. Batasan, and the detected concentration was 2.15 times greater than the PNSDW 2017 limit.

Table 3 exhibits the descriptive statistics of the GW samples in the research area. The trend of mean concentrations for the GW samples was Fe > Zn > Ba > Mn > Cu. The Shapiro-Wilk test revealed that the data array for the GW samples were not normally distributed with $p < 0.05$ [61]. Additionally, all the CV values were greater than 35%, which indicates that the data variability is high.

Table 3. Metal and metalloid concentrations (in mg/L) in GW.

	Ba	Cu	Fe	Mn	Zn
N	33	33	33	33	33
Max	0.020	0.010	5.700	0.020	0.050
Min	0.000	0.000	0.000	0.000	0.000
Mean	0.0051	0.0007	0.3716	0.0047	0.0056
SD	0.006	0.002	1.213	0.005	0.012
Skewness	1.209	2.868	3.897	0.995	2.207
Kurtosis	1.149	6.654	14.832	0.455	4.511
CV%	114.31	310.56	326.63	102.27	220.23
PNSDW	0.700	1.000	1.000	0.400	5.000

The mean concentrations for GW samples were compared to the PNSDW 2017, as shown in Figure 5. All mean concentrations of the GW samples were below the threshold values set by the PNSDW 2017. Among the MMs detected in the GW samples, only the Fe concentration was recorded above the PNSDW 2017 standard values in various sampling locations.

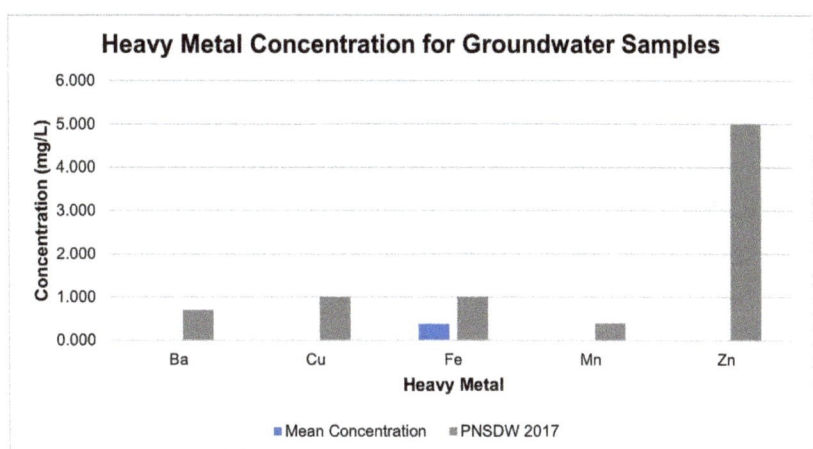

Figure 5. Comparison of MM concentrations and PNSDW 2017 threshold values for groundwater samples.

The extent of the relationship between the MMs in water samples was determined using the Pearson correlation. The correlation plots for the (a) DW and (b) GW samples are presented in Figure 6. Findings demonstrated that a positive correlation was shown between Cr and Cu in DW samples. Similar findings were observed in water quality studies in Saudi Arabia [62], India [63], Sri Lanka [64], Egypt [65], Burkina Faso [66], Greece [67], Iran [68], and Turkey [69]. Likewise, a positive correlation between Fe and Zn was observed and comparable to the result of the studies by Atangana and Oberholster [70], Esmaeili et al. [71], Duggal et al. [72], Varghese and Jaya [73], and Karthikeyan et al. [74]. The correlation analysis that was conducted revealed a positive association between Cr and Cu and Fe and Zn. This is attributed to a potential common origin of these MMs and could be the controlling factor of the MMs' concentration in the water resources [75]. The findings of the study are an evidence of the potential contamination of water resources and/or water supplies lines and present a direction of necessary treatment strategy. These findings should be utilized to create and develop remediation programs to mitigate the effects of these MMs in the community.

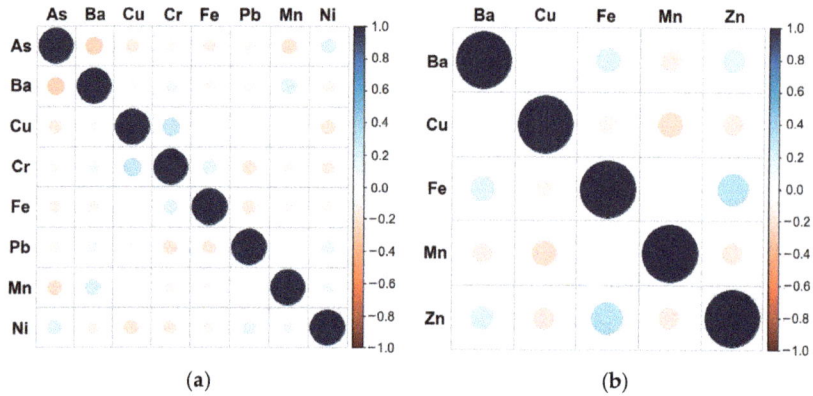

Figure 6. Correlation plots for metals and metalloid in (**a**) DW; and (**b**) GW samples.

3.2. Pollution Evaluation Indices

Using the MMs' concentrations detected in the water samples, the *MPI*, *NPI*, and *ERI* were calculated. The mean *MPI* value calculated in DW samples was 90 times greater than the upper limit of the "low" pollution *MPI* category, while the sampling point observed with the highest *MPI* value was 701 times larger. Considering the GW, the mean *MPI* value was below the upper boundary of the "low" pollution category. The mean *NPI* values for DW revealed that the average calculated *NPI* values were 135 times greater than the upper threshold for the unpolluted location criterion. Moreover, the highest *NPI* calculated was 1080 times the unpolluted threshold limit. On the other hand, the GW mean *NPI* values were within the unpolluted category. The mean *ERI* value calculated for DW was 3.5 times the upper limit of the "low" *ERI* category, while that of the highest *ERI* observed was 23.8 times larger. For the GW samples, the mean *ERI* value calculated was within the "low" *ERI* category.

3.3. Probabilistic HRA

The mean *HQ* with regards to the MMs was calculated from DW samples. It was found out that the *HQ* for As, Cr, and Pb have mean *HQ* values greater than 1. This suggests that potential non-carcinogenic consequences may arise [76]. The mean *HI* for the DW samples was 86 times larger than the limit, which indicates that there is a greater possibility of harmful consequences [77]. The trend of the mean *HQ* values in DW samples was As > Pb > Cr > Cu > Ni > Fe > Ba > Mn, wherein As can be attributed to 88% of the mean *HQ* value.

The mean *HQ* values in GW samples for Ba, Cu, Fe, Mn, and Zn were below 1. Moreover, the mean *HI* value was also less than 1. This means that there was non-carcinogenic risk for Ba, Cu, Fe, Mn, and Zn in GW. The highest mean *HQ* observed was for Fe, but the calculated value was still below 1. The trend of the mean *HQ* values in GW samples was Fe > Mn > Ba > Zn > Cu. The concentration of Fe contributes to the 85% of the mean *HQ* value. The contribution of each MM to the *HQ* is shown in Figure 7a,b. It was observed that *HQ* for As was the significant contributor to the total *HI* value.

The *ELCR* was only calculated for DW samples, since the metal concentrations of As, Cr, Pb, and Ni were not detected in GW samples. The trend of the *CR* values observed in DW was As > Cr > Ni > Pb. The As concentration contributed to the 92.06% of the *ELCR* value.

The summary of the number of sampling locations exceeding the *HQ* threshold limit is presented in Figure 8. About 33.8% of DW sampling sites had the *HQ* value of As above the threshold, while that of Cr, Pb, and Ni were 23.9%, 28.2%, and 1.4%, respectively. No sampling point recorded an *HQ* value above the threshold value for GW sampling sites.

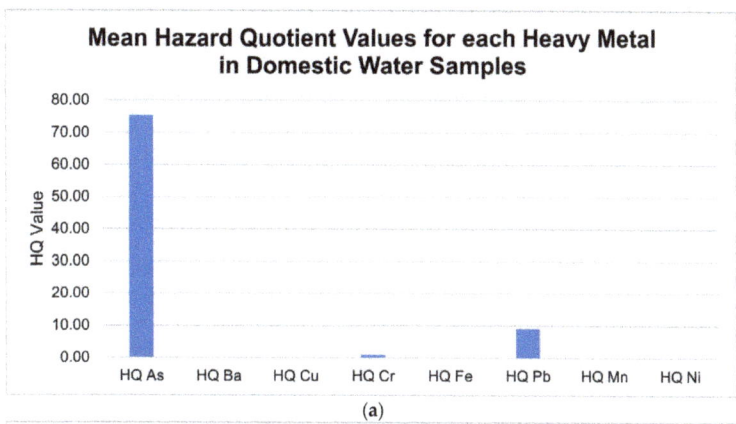

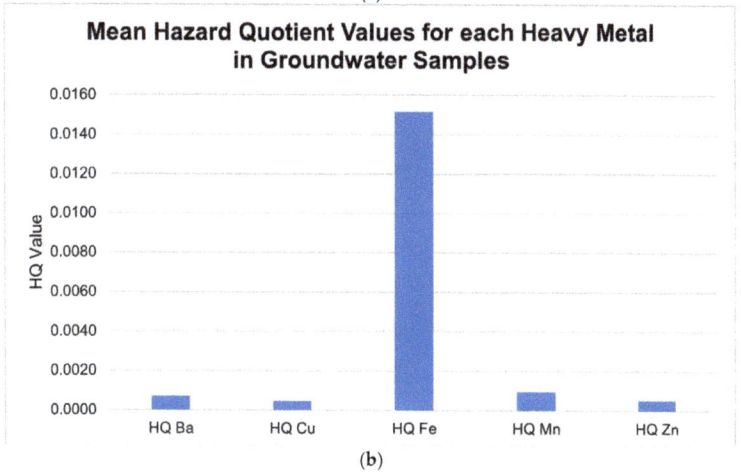

Figure 7. Mean HQ values for each MM in (**a**) DW; and (**b**) GW samples.

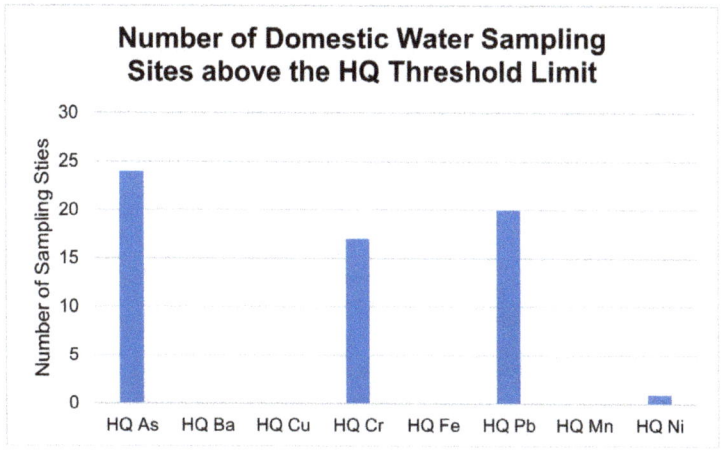

Figure 8. Number of DW sampling sites above the HQ threshold limit.

Considering the *ELCR* values, the summary of the number of sampling sites categorized to have medium to very high cancer risk is shown in Figure 9. About 35.2% of the sampling sites for DW had *CR* categorized as "medium to very high cancer risk". Additionally, Cr, Pb, and Ni were 26.8%, 29.6%, and 8.5%, respectively, and categorized as "medium to very high cancer risk".

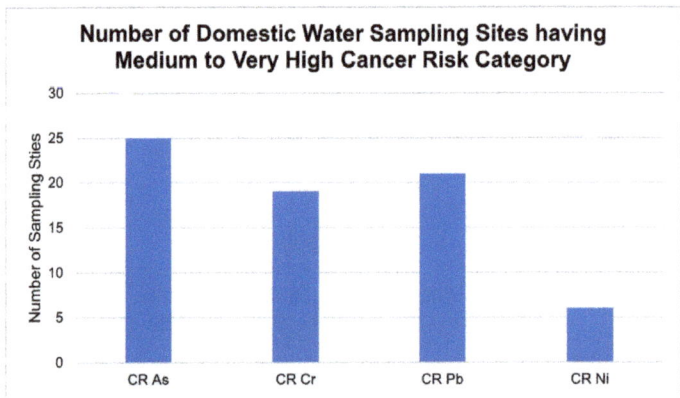

Figure 9. Number of DW sampling sites categorized to have medium to very high cancer risk.

3.4. Spatial Mapping Using MLGI Approach

Spatial maps were produced using the MLGI technique, which combines NN-PSO using the MM concentrations observed in the water samples and their link with PI and HR indices. Table 4 summarizes the results of the simulation for the MLGI mapping of the MMs in DW samples using the hybrid NN-PSO+EBK approach. Figure A1 in Appendix C presents the produced spatial maps for the MMs in DW samples.

Table 4. NN-PSO simulation results for metals and metalloid in DW samples.

	HN	NP	NI	ET (s)	Validation	Testing
As	25	9	2000	124.18818	0.98617	0.99656
Ba	29	6	2000	124.47877	0.95023	0.98621
Cr	29	3	2000	160.77287	0.96793	0.98507
Cu	30	7	2000	121.85391	0.98758	0.94860
Fe	22	7	2000	128.33642	0.95715	0.97733
Mn	25	2	2000	125.32753	0.94777	0.93157
Ni	28	7	2000	153.38283	0.99861	0.99971
Pb	29	9	2000	122.88486	0.99728	0.98994

The results of the simulation for the hybrid NN-PSO+EBK method to the MLGI mapping of the pollution indices in the DW samples are displayed in Table 5. Figure A3 in Appendix C displays the spatial maps for the pollution indices in the DW samples.

Table 5. NN-PSO simulation results for pollution indices in DW samples.

Index	HN	NP	NI	ET (s)	Validation	Testing
HPI	27	8	2000	128.95811	0.98384	0.99973
NPI	29	5	2000	130.91300	0.99203	0.99400
ERI	30	1	2000	140.66034	0.98862	0.98091

The highest concentrations for Ba in DW samples were detected in Brgy. San Agustin and Brgy. Mangarin, while the highest concentrations for Cu and Mn were detected in Brgy.

Monteclaro and Brgy. San Roque. However, the highest Ba, Cu, and Mn concentrations detected were within the limit of PNSDW 2017.

The pollution hotspot for As in DW was observed in Brgy. Bubog, and the water samples were collected from WRS with an As concentration 668 times greater than the PNSDW 2017 standard value. Exposure to As has been associated with several health consequences such as skin and neurological diseases, cancers, and non-communicable diseases, including hypertension and diabetes mellitus [78]. The highest concentration of Cr was detected in a spring from Brgy. Labangan Iling with a concentration 17.9 times greater than the PNSDW 2017. Exposure to elevated concentrations of Cr may decrease the glycemic tolerance factor and raise the risk of cardiovascular disease [79]. Oral intake of Cr usually causes various symptoms such as stomach ulcers, nausea, vomiting, fever, vertigo, diarrhea, liver damage, and even death at doses from 1 to 3 g [80,81].

Table 6 shows the outcomes of the simulation for the MLGI mapping of the MMs in the GW samples utilizing the hybrid NN-PSO+EBK approach. The spatial maps for the MM in the GW samples are presented as Figure A2 in Appendix C.

Table 6. NN-PSO simulation results for metals and metalloid in GW samples.

	HN	NP	NI	ET (s)	Validation	Testing
Ba	25	6	2000	125.76638	0.92786	0.97901
Cu	27	7	2000	132.61975	0.99957	0.99987
Fe	30	2	2000	141.61058	0.99868	0.99706
Mn	29	3	2000	143.39959	0.99152	0.99922
Zn	26	1	2000	150.94246	0.99997	0.99994

Ingestion is one the mechanisms that allow Ba to enter the human body [82]. Acute toxicity of Ba consists of gastrointestinal, metabolic, cardiovascular, musculoskeletal, and neurological effects. Gastrointestinal effects include gastric pain, nausea, vomiting, and diarrhea [83]; metabolic effects include hypokalemia, ventricular tachycardia, and hypertension or hypotension [84]; cardiovascular effects include changes in the heart rhythm and increased or decreased blood pressure [85]; skeletomuscular effects include numbness, muscle weakness, and paralysis [86]; and neurological effects including tremors, seizures, and mydriasis. Another, the hotspot area for Fe was detected in a deep well in Brgy. Bubog, which is 5.7 times greater than the PNSDW threshold. It should be noted that Fe poisoning is frequently fatal due to shock or liver failure [87].

Considering the measure of the cumulative impact of MMs on DW resources, with respect to the human consumption and metal pollution, the *MPI* hotspot was observed in Brgy. Camburay. The DW quality at various sampling locations, with the emphasis on the importance of different MMs in the water resources, was measured using the *NPI*. The highest *NPI* was likewise observed in Brgy. Camburay. Moreover, the highest *ERI* values were observed in a residential DW source in Brgy. Bubog.

Table 7 displays the outcomes of the simulation for the hybrid NN-PSO+EBK approach to the MLGI mapping of the pollution indices in the samples of GW. The spatial maps for the pollution indices in the GW samples are displayed in Figure A4 in Appendix C.

Table 7. NN-PSO simulation results for PI in GW samples.

Index	HN	NP	NI	ET (s)	Validation	Testing
HPI	26	2	2000	149.06667	0.99962	0.99984
NPI	29	3	2000	141.51668	0.99869	0.99781
ERI	28	4	2000	142.93539	0.98216	0.99958

The *HPI*, *NPI*, and *ERI* hotspots were found in a deep well at Brgy. Bubog. The simulation results for the MLGI mapping of the *HQ/HI* values in the DW samples are demonstrated in Table 8. Figure A5 in Appendix C displays the spatial maps for the pollution indices in the GW samples.

Table 8. NN-PSO simulation results for hazard quotient/hazard index in DW samples.

Index	HN	NP	NI	ET (s)	Validation	Testing
HQ (As)	29	5	2000	131.72678	0.98983	0.99024
HQ (Ba)	27	2	2000	119.00130	0.96164	0.93469
HQ (Cr)	29	5	2000	114.47954	0.98674	0.98510
HQ (Cu)	28	7	2000	116.09329	0.99268	0.99534
HQ (Fe)	26	7	2000	114.40573	0.99709	0.99748
HQ (Mn)	29	9	2000	115.17782	0.98861	0.99820
HQ (Ni)	26	7	2000	121.11026	0.97369	0.94806
HQ (Pb)	27	8	2000	126.85262	0.98794	0.96551
HI	28	2	2000	118.55610	0.98692	0.98708

The calculated HQ for Ba, Cu, Fe, and Mn was observed to be less than 1 for DW samples. The highest HQ for As was observed in Brgy. Bubog, which is 637 times greater than the recommended HQ of less than 1. The HQ of Cr was observed in Brgy. Labangan Iling from a spring. It was observed to be 8.5 times higher than the recommended HQ value. The HQ for Ni was 1.03 times higher than suggested HQ value and was detected in Barangay II Poblacion, and was collected in a WRS. The HQ for Pb was 120.3 times higher than the recommended HQ value and was observed in Brgy. Camburay, also collected from WRS. Moreover, the HI for DW was calculated to be 647 times above the standard value. This was observed in Brgy. Bubog, collected from WRS.

Table 9 displays the simulation results for the MLGI mapping of the GW samples' HQ/HI values. The spatial maps for the pollution indices in the GW samples are presented as Figure A6 in Appendix C.

Table 9. NN-PSO simulation results for HW/HI in GW samples.

Index	HN	NP	NI	ET (s)	Validation	Testing
HQ (Ba)	26	7	2000	137.11112	0.99029	0.97838
HQ (Cu)	27	10	2000	136.08892	0.99998	0.99997
HQ (Fe)	28	10	2000	123.77389	0.99662	0.99922
HQ (Mn)	27	3	2000	134.97920	0.99599	0.99213
HQ (Zn)	29	5	2000	132.29242	0.99781	0.99719
HI	26	9	2000	135.89116	0.99918	0.99589

The HQ observed in GW samples for Ba, Cu, Fe, Mn, and Zn were all lower than the recommended HQ value of 1. Consequently, the overall HI in all sampling locations was below 1, which suggests that there is non-carcinogenic risk of exposure to Ba, Cu, Fe, Mn, and Zn in GW.

The MLGI mapping of the simulation results for the carcinogenic risk (CR) index values of DW samples is shown in Table 10. Figure A7 in Appendix C shows the spatial maps for the CR values by As, Cr, Pb, Ni, and the total carcinogenic risk evaluation in the GW samples.

Table 10. NN-PSO simulation results for CR index in DW samples.

Index	HN	NP	NI	ET (s)	Validation	Testing
CR (As)	27	5	2000	139.80123	0.98699	0.99963
CR (Cr)	23	1	2000	139.84580	0.99420	0.99058
CR (Pb)	30	6	2000	131.91650	0.99112	0.99518
CR (Ni)	28	9	2000	179.67048	0.97320	0.97238
TCR	26	4	2000	138.49763	0.99742	0.99289

The hotspot for the CR with respect to As and Ni concentrations was observed in Barangay Población II. The highest *CR* for Cr concentration was observed in Brgy. Labangan Iling. The *CR* for Pb was recorded at Brgy. Camburay. All these highest *CRs* observed were considered to have a very high *CR*. The mean *CR* index levels for As, Cr, and Ni in the study area were classified to have a very high *CR*, while that for the *CR* of Pb was considered to have a high carcinogenic risk.

3.5. Contamination Area Calculation

Pollution indexes such as *MPI*, *NPI*, and *ERI* were calculated to establish the effect of the group of MMs in the DW and GW resources of San Jose, Occidental Mindoro. Considering the DW in the research area, the hotspots for *MPI* and *NPI* were observed in Brgy. Camburay, which inhabited by about 1900 people. On the other hand, Brgy. Bubog was observed to be the *ERI* hotspot, considering the DW and GW in the research area. It is also the hotspot for *MPI* and *NPI* with reference to the groundwater resources in the research area. Brgy. Bubog has a population of about 10,800 people, comprising about 7% of the total population of the municipality of San Jose.

Using the generated spatial maps of MMs and their associated risk index values, the contamination area was calculated using the reclassification and raster to polygon tools of ArcGIS 10.8.1. The contamination area percentages, considering the MMs in DW and GW samples, were shown in Figure 10. It was observed that 100% of DW samples in the entire study area were below the PNSDW 2017 limit for Ba, Cu, and Mn. The % area affected by Cr, Fe, and Ni was 19.9%, 4.2%, and 16.0%, respectively. Moreover, 51.1% and 72.7% of the DW samples in the study area were contaminated with As and Pb, respectively. Considering the MMs' contamination % area in GW, 100% of the total area was below the PNSDW 2017 limit for Ba, Cu, and Mn. However, Fe and Zn affected only 4.1% of the total study area.

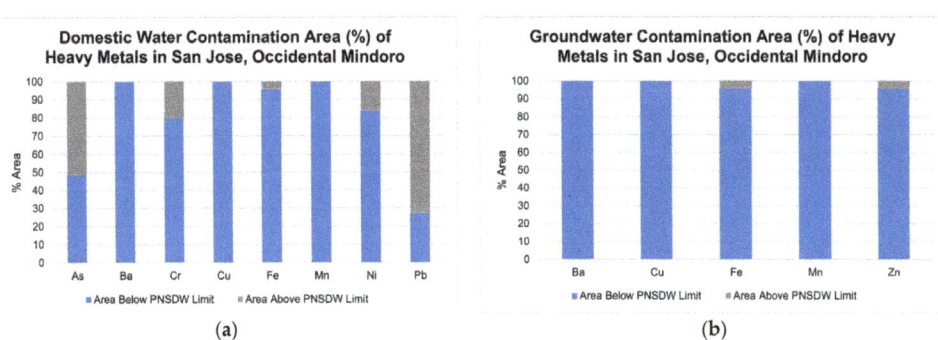

Figure 10. Contamination area (%) of MMs calculated for (**a**) domestic water; and (**b**) groundwater.

The contamination area percentages, considering the pollution risk indices, including *MPI*, *NPI*, and *ERI* in DW and GW samples, are shown in Figure 11a–c. Considering the *MPI* in DW, 38.7% of the study area was classified to have a high MPI, while the GW is recorded as 0.6%. This "high" *MPI* classification suggests that DW quality in San Jose, Occidental Mindoro is unsuitable as DW. About 96.0% and 3.4% of the DW and GW, respectively, recorded an NPI of "moderately to heavily polluted". Moreover, 100% of the GW samples were classified to have "low to moderate" ERI, while 30.0% of the area of DW was classified to have "considerable to very high" ERI.

The contamination area percentage, considering the HW in DW and GW, is shown in Figure 12a,b. It was observed that 100% of DW in the study area has an *HQ* value less than 1 for Ba, Cu, Fe, and Mn. More than 60% of the study area has an *HQ* value greater than 1 with respect to Cr and Pb. Moreover, there was no area with *HQ* greater than 1 in the GW samples as shown in Figure 12b with no color gray.

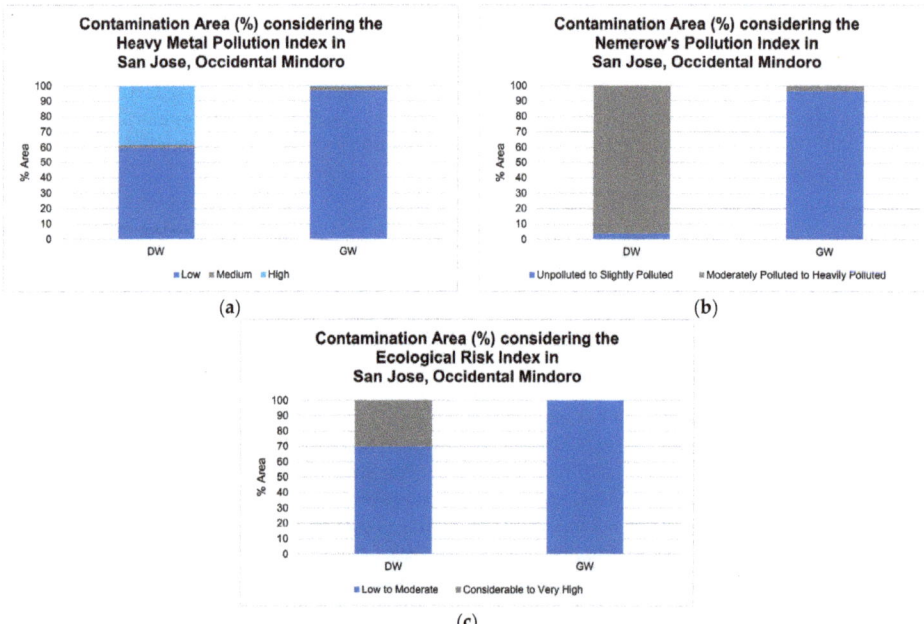

Figure 11. Contamination area (%) calculated for (**a**) *HPI*; (**b**) *NPI*; and (**c**) *ERI* in domestic water and groundwater.

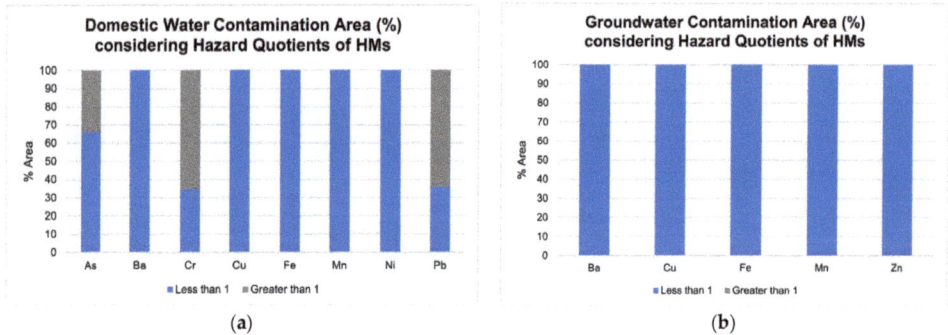

Figure 12. Contamination area (%) of *HQ/HI* in MMs calculated for (**a**) domestic water; and (**b**) groundwater.

Figure 13 presents the area percentage of the GW and DW that was contaminated when the total carcinogenic risk (*TCR*) was taken into account. The calculation showed that 93.3% of the area was classified as "high to very high" TCR. This implies that there is potential carcinogenic risk in the population due to oral exposure of carcinogens from DW supply.

The trend of MMs were noted as Pb > As > Fe > Cr > Cu > Ni > Ba > Mn, while the trend for GW was Fe > Zn > Ba > Mn > Cu.

Results of the correlation analysis showed that Cr and Cu in DW samples had a positive correlation. This was attributed to the eroding of igneous rocks in the region [88,89]. Moreover, Fe and Zn were positively correlated, which implies that the MMs appear to have a geogenic origin [90].

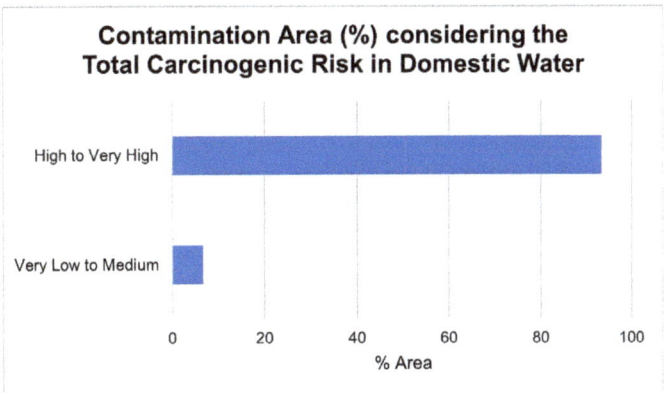

Figure 13. Contamination area (%) of *TCR* in domestic water.

The carcinogenicity of MMs concentration was caused primarily by ingestion [88] and the dose. The *HQ* levels greater than 1 indicate high potential health risks [87]. The *HQ* for As contributed to 87.9% of the hazard index in DW, while Pb and Cr contributed to 10.7% and 1.2%, respectively. The rest of the *HQs* for Ba, Cu, Fe, Mn, and Ni were less than 1%. The *HQ* for GW resources in reference to Ba, Cu, Fe, Mn, and Zn were all less than 1. The carcinogenic risk was calculated for domestic water and was observed to have a trend of As > Cr > Ni > Pb. This implies that the *CR* for As was significantly more elevated than Cr, Ni, and Pb, and was the main carcinogenic MMs area. Considering the calculated *TCR* in the study area, 40 of 71 sites were classified to have a "high" and "very high" cancer risk category.

Based on these results, several sites, including Brgy. Poblacion II, Brgy. Labangan Iling, and Brgy. Camburay, were considered pollution hotspots and have become the priority sites for remediation. Also, it is necessary to conduct regular DW quality monitoring with respect to MMs in the area. Updating of relevant local and national laws, policies, and guidelines becomes necessary. The knowledge of the community of the hazards posed by the MMs through water resources shall be continuously enhanced to reduce the risks and adverse health effects. Therefore, installing and using appropriate water treatment will be beneficial to lessen the HR caused by the MMs content in DW [91]. A thorough assessment of the environmental quality and health cases of the population must be carried out to create more effective health risks reduction strategies and remediation approaches in the area.

It is a requirement of the Philippine government that water utility companies meet the minimum national water quality standards before distributing water to the community; however, monitoring of such should be improved to ensure that the WQ parameters, especially MMs, are within the PNSDW standards. Also, many areas in the country are not receiving water from a utility company, especially those in the far-flung areas. The people in the community typically get the water from deep wells, shallow wells, and springs and is consumed directly without treatment. According to the WHO and UNICEF, as of 2020, only 47% of the Philippines' total population has access to safely managed drinking water services [92].

The use of machine learning techniques and algorithms as predictive models for spatial mapping has become necessary to areas, such as San Jose, Occidental Mindoro, that are remote, and regular water quality assessment and evaluation is a challenge. The application of the machine learning approach and algorithms generated a hybrid method for mapping parameters in various studies, including the use of kernel logistic regression [93], differential flow pollination [94], logistic regression, random forest [95], reduced error pruning trees [96], least squares support vector classification–bat algorithm [97], gradient-boosted trees, and support vector machine [98]. These characteristics of the MLGI become suitable to the mapping needs in this kind of condition. The current study has proven that the use of a hybrid method, the MLGI approach, to map the pollution index and health risks in the region is an efficient and reliable method for evaluating potential sources and hotspots of pollution. The study's findings could serve as the foundation for thorough water management in the area to guarantee safe and viable water resources in the municipality and regions of similar conditions. Research and further studies that would lead to identifying the appropriate remediation technology, policies, and guidelines for regular water quality monitoring of MMs in DW, mitigation measures, and other relevant interlinked laws are necessary. Prompt action from the municipal health and sanitation offices is required.

4. Conclusions

A total of 104 DW and GW samples collected from the municipality of San Juan, province of Occidental Mindoro, Philippines were analyzed for the presence of target MMs such As, Ba, Cu, Cr, Fe, Pb, Mn, Ni, and Zn. Reliable and high-performance portable Olympus Vanta XRF and Accusensing MAS devices were used to detect and analyze the water samples for the target MMs. The detected concentrations of these MMs were compared to PNSDW 2017 limits and used in the calculation of the *MPI, ERI, NPI*, and *ELCR*. This is to determine the suitability of the water resources for domestic use and the risks posed by the elevated concentration of these MMs. Further, this study employed the use of GIS and the MLGI technique to further reinforce the calculated pollution indices to identify the pollution hotspots. Based on the recorded results, the DW samples collected had critical concentrations of MMs compared to the GW samples. The Pb concentration recorded the highest mean among other MMs in DW, while the Fe concentration was the highest mean concentration recorded for GW. The mean *NPI* for DW was 135 times greater than the upper limit of the unpolluted location category. The highest *NPI* recorded was 1080 times higher than the upper limit of the unpolluted site category. The calculated *ERI* in the area was 23.8 times higher than the upper limit for 'low' *ERI* category. The trend of the mean MM concentrations in the DW was Pb (1.152 mg/L) > As (0.792 mg/L) > Fe (0.138 mg/L) > Cr (0.110 mg/L) > Cu (0.093 mg/L) > Ni (0.045 mg/L) > Ba (0.024 mg/L) > Mn (0.003 mg/L). Based on the calculated *MPI, NPI*, and *ERI*, it was observed that the pollution hotspot was noted at Brgy. Camburay and Brgy Bubog. It was recorded that 33.8% and 28.2% of San Jose Municipality had DW containing elevated concentrations of As and Pb, respectively. Also, it was observed that As was contributing significantly to the total carcinogenic health risks in the area. Using the MLGI approach, the contamination area projected by DW samples was calculated and showed that 96.0% of the study area was classified to be moderately to heavily polluted. About 35.2% of the sampling sites for DW have carcinogenic risk category of medium to very high risk specific to elevated concentrations of Cr, Pb, and Ni. The carcinogenic risk hotspots for As, Cr, Pb, and Ni were Barangays Poblacion II for As and Ni; Barangay Labangan II for Cr; and Brgy Camburay for Pb. The conduct of pollution, health risk evaluation, and the use GIS with MLGI mapping to describe the pollution indices and ecological and health risks in specific areas are effective and reliable approaches in evaluating potential pollution sources and hotspots. The extracted data are useful in creating strategic programs for health risk reduction, as well as the mitigation and remediation of polluted areas. The result of this study is useful in water resource management in the area and neighboring regions.

Author Contributions: Conceptualization, D.B.S. and K.L.M.D.J.; methodology, D.B.S., K.L.M.D.J. and C.E.F.M.; software, K.L.M.D.J. and C.E.F.M.; validation, D.B.S., K.L.M.D.J. and C.E.F.M.; formal analysis, K.L.M.D.J. and C.E.F.M.; investigation, D.B.S. and K.L.M.D.J.; resources, D.B.S.; data curation, D.B.S. and K.L.M.D.J.; writing–original draft preparation, K.L.M.D.J. and C.E.F.M.; writing–review and editing, D.B.S.; visualization, K.L.M.D.J. and C.E.F.M.; supervision, D.B.S.; project administration, D.B.S.; funding acquisition, D.B.S. All authors have read and agreed to the published version of the manuscript.

Funding: This research was funded by the Department of Science and Technology Philippine Council for Health Research and Development (DOST-PCHRD). The funded project titled D-HIVE 4B Capital.

Institutional Review Board Statement: Not applicable.

Informed Consent Statement: Not applicable.

Data Availability Statement: All data are contained in the manuscript.

Acknowledgments: This is to recognize the in-kind support of Mapua University, the Occidental Mindoro State College, and the Local Government Unit of San Jose, Occidental Mindoro.

Conflicts of Interest: The authors declare no conflict of interest.

Abbreviations

Abbreviation/Symbol	Description
AT	Average Time
BW	Body Weight
C_d	Degree of Contamination
CDI	Chronic Daily Intake
CI	Contamination Index
CV	Coefficient of Variability
EBK	Empirical Bayesian Kriging
ED	Exposure Duration
EF	Exposure Frequency
ELCR	Excessive Lifetime Cancer Risk
ERI	Ecological Risk Index
GIS	Geographic Information System
GPS	Global Positioning System
HEI	Heavy Metal Evaluation Index
HI	Hazard Index
HM	Heavy Metal
HN	Hidden Neurons
HQ	Hazard Quotient
HRA	Health Risk Assessment
IR	Ingestion Rate
LOD	Limits of Detection
MAS	Metals Analysis System
MLGI	Machine Learning Geostatistical Interpolation
MMs	Metals and Metalloid
MPI	Metal Pollution Index
N	Number of Samples
NCR	Non-Carcinogenic Risk
NI	Number of Iterations
NN-PSO	Neural Network-Particle Swarm Optimization
NP	Number of Particles
NPI	Nemerow's Pollution Index
PE	Polyethylene
PI	Pollution Index
PNSDW	Philippine National Standards for Drinking Water
RfD	Reference Dose
SD	Standard Deviation
SF	Slope Factor
T_i	Toxic-Response Factor
WRS	Water Refilling Station
WQ	Water Quality
XRF	X-Ray Fluorescence

Appendix A

Table A1. Metals and metalloids pollution and HRA studies in various regions of the world.

Continent	Country	Parameters Evaluated	Index Calculated	Results
Asia	Bangladesh	Na, K, Ca, Mg, Mg, As, Fe, Mn, Cl, F, I, CO_3, HCO_3, NO_3, PO_4	WQI, HQ/HI, CR	8.69% (both pre- and post-monsoon) of the sampling sites have very poor WQI category; All sites have very high chronic risk during the pre-monsoon season, while that for post-monsoon is 20 of 23 sites; More than 70% of the sites have very high cancer risk during the pre-monsoon and the post-monsoon season [99].
	Cambodia	Ag, Al, As, Ba, Cd, Co, Cr, Cu, Fe, Ga, Mn, Ni, Pb, Se, U, Zn	HQ/HI	HQ for As is the most significant element observed; Maximum value observed was HI = 3.57, which is greater than the threshold of 1.0 [100].
	China	Cr^{6+}	HQ/HI, CR	NCR shows induced risk for 7.47% and 12.07% of adults and children respectively, while that for CR shows 50.57% for adults and 16.67% for children [101].
	India	Al, As, Cd, Co, Cr, Cu, Fe, Mn, Mo, Ni, Pb, V, Zn	HPI, HQ/HI, CR	Two sites have been identified as not suitable for drinking purposes without prior treatment; As is the most significant element to the NCR and CR in the area [102].
	Iran	As, Cd, Cr	HI/HQ, CR	All sites have HQ > 1 considering As; Only one site has HQ < 1 considering Cr [103].
	Iraq	As, Cr, Cu, Fe, Mn, Mo, Ni, Pb, Zn	HPI, HQ/HI, CR	28% of the sites exhibit HPI values within the category of Medium to Highly Polluted [104].
	Jordan	Cd, Co, Cr, Li, Mn, Mo, Ni, Pb, U, V, Zn	CI, HQ/HI, CR	CR was within the acceptable risk limits [105].
	Kazakhstan	Cd, Co, Cu, Fe, Mn, Ni, Pb, Zn	HPI, HQ/HI, CR	92.86% of the samples were considered to have a high level of pollution [106].
	Korea	As, Cd, Cu, Pb, Zn	HQ/HI, CR	Water consumption on a frequent basis might potentially be detrimental due to long-term As exposure [107].
	Kyrgyzstan	As, F	HQ/HI, CR	Children were observed to have a greater health risk than adults. Specifically, the CR of As via the oral intake pathway was above the permissible limits [108].
	Lao PDR	As, B, Ba, Cr, Cu, Mn, Ni, Pb	HQ/HI, CR	Four sites observed HI > 1 and have CR above the acceptable risk level of 1×10^{-4} [109].

Table A1. Cont.

Continent	Country	Parameters Evaluated	Index Calculated	Results
	Malaysia	SO_4, Cl, Ca, Mg, Na, K, Al, Fe, Mn, Zn, Sr, As, Cr, Cd, Ni, Cu, Co, Pb	WQI	Some sampling sites were categorized as slightly polluted with HM concentrations exceeding the permissible limits [110].
	Nepal	Ba, Cd, Co, Cr, Cu, Pb, Li, Mn, Mo, Ni, Sb, Sr, V, Zn	WQI, HQ/HI	WQI observed in the sampling sites was within the excellent water category; HQ values observed in the sampling sites were all less than 1, which suggested that it poses no adverse health effects to the residents [111].
	Pakistan	F, Cl, Pb, Cd, Ni, Zn, Fe, As	HQ/HI, CR	HQ considering As has a value observed greater than 1, which suggests that it poses potential adverse health effects in the residents; 3.3% of the sites have CR values above the USEPA threshold limit [112].
	Philippines	As, Ba, Cu, Fe, Pb, Mn, Ni, Zn	HPI, NPI, HQ/HI, CR	All sampling locations were categorized to have a high level of pollution; CR via ingestion pathway exceeded the USEPA threshold limit [30].
	Saudi Arabia	Ag, Al, B, Ba, Cr, Cu, Fe, Mo, Ni, Pb, V, Zn	HPI, C_d, HQ/HI	Potential NCR showed that HQ for Al, Mo, Cu, Cr, and Pb exceeded the recommended value of 1, implying that it can cause a potential adverse health effect [113].
	Thailand	Ca^{2+}, Mg^{2+}, Na, K, Fe, SO_4^{2-}, NO_3^-, Cl^-, HCO_3^-	HQ/HI, CR	Study showed that As is the most significant element with respect to the NCR and chronic effects, and was predominant with adults [114].
	United Arab Emirates	Cl^-, SO_4^{2-}, Ca^{2+}, Na^+, K^+, Mg^{2+}, Ba^{2+}, As, Cd, Pb, Cr, Cu, Zn	HQ/HI	30% of the sites were classified to have a very high risk [115].
	Vietnam	As	CR	Skin CR is 11.5 times greater in unfiltered water and 14.8 times higher when consuming water for a lifetime that has been exposed to As [116].
Africa	Algeria	Cd, Cr, Cu, Mn, Ni, Pb, Zn	HPI, HEI	HPI values observed for all samples were below the critical value; No chronic HR was posed by the GW [117].
	Cameroon	As, Cd, Co, Cr, Cu, Ni, Pb, Zn	WQI, HPI, HEI, HQ/HI, CR	WQI suggests that the GW is categorized to be poor to unsuitable; HM concentrations presented a risk for non-carcinogenic health consequences; Ingestion of groundwater also posed a moderate to high risk of developing cancer in both adults and children [118].

Table A1. Cont.

Continent	Country	Parameters Evaluated	Index Calculated	Results
	Egypt	Fe, Mn, Pb, Cd, Ni, Cu, Zn, Cr, As	HQ/HI, CR	Ingestion is observed to be the most significant pathway of exposure; The prevalence of arsenic was stated to be accountable for the highest CR [119].
	Ethiopia	Pb	HQ/HI, CR	Mean HQ levels both for children and adults were below 1; CR values for adults and children were within the permissible levels of USEPA [120].
	Ghana	Na^+, K^+, Ca^{2+}, Mg^{2+}, HCO_3^-, F^-, Cl^-, NO_3^-, PO_4^{3-}, SO_4^{2-}, Pb^{2+}, Co, Cr, Fe, Mn, Zn, Cu, Cd	HQ/HI, CR	HQ due to Mn and Fe was greater than 1, implying that it can cause a potential adverse health effect [121].
	Kenya	Cd, Ni, Pb	HPI, HEI, HQ/HI, CR	All samples had hazard indices that exceeded 1, which indicated a significant risk due to the metal exposure; CR values were greater than the USEPA permissible limit for both children and adults [122].
	Morocco	Cd, Cr, Cu, Fe, Mn, Ni, Pb, Zn	HQ/HI	HQ was significantly influenced by Zn concentration [123].
	Mozambique	Sr, Li, B, Al, Ba, V, Mn, Fe, Hg, Co, Ni, Cu, Pb, Zn, As, Rb, U	HQ/HI	HQ values exceed the maximum permissible limit for children due to ingestion pathway [124].
	Nigeria	Cl, Fe, Zn, Pb, Cu, Ni, Cr, Mn, Cd	WQI, ERI, HEI, HPI, HQ/HI, CR	About 24% of the samples have deteriorated water quality; About 15% of the samples posed moderate ecological risks; Cu is the most significant element to the HI and more than 40% of the samples presented high chronic HR; High Cr risk was observed in 19% of the samples, while that for Cd and Ni is 14% [3].
	Senegal	Pb, Cd, Cu, Fe, Mn	HQ/HI	High risk for infants and children with respect to Pb and Cd [125].
	South Africa	Ag, Al, B, Cd, Co, Cu, Fe, Li, Mn, Li, Mn, Ni, Pb, Si, Zn	HPI, HQ/HI	Two zones presented high contamination classification based on their HPI values; All age groups were exposed to potential health risks due to HQ via ingestion pathway [126].
	Sudan	CO_3^{2-}, HCO_3^-, NO_3^-, NO_2^-, SO_4^{2-}, NH_3, F^-, Cl^-, Na^+, K^+, Mg^{2+}, Ca^{2+}	HQ/HI	About 60% of the sampling sites presented potential health risks for both children and adults [127].
	Argentina	Ca^{2+}, Mg^+, K^+, HCO_3^-, Cl^-, SO_4^{2-}, NO_3^-, Cr(IV)	HQ/HI, CR	HQ value via ingestion pathway is greater than 1 for both adults and children; CR due to Cr(VI) exceeded the permissible limit set by the USEPA [128].

Table A1. Cont.

Continent	Country	Parameters Evaluated	Index Calculated	Results
North and South America	Brazil	Ca, Mg, Na, K, Cl^-, HCO_3^-, SO_4^{2-}, NH_4^+, NO_3^-, F^-, Fe, Mn	WQI, HPI	WQI and HPI values presented samples that were of good quality [129].
	Ecuador	Al, Ba, Ca, Fe, K, Mg, Mn, Na, Sr, Cu, Cr, Pb, Zn	HPI	6.6% of the samples in a sampling region have Pb levels above the threshold value [130].
	Mexico	As, F	HQ/HI	Observed HI > 1 for both adults and children suggests a potential health risk and was attributed to exposure to As and F [131].
	Paraguay/ Uruguay/ Argentina/Bolivia	Li, B, Al, Si, V, Cr(III), Mn, Fe, Co, Ni, Cu, As, Rb, Sr, Cd, Cs, Ba, Pb, U	HPI, HQ/HI	34 sites were categorized as highly polluted sites; HQ value calculated indicated no risk [132].
	Peru	Au, Cd, Co, Cr, Cu, Fe, Mn, Mo, Ni, Pd, Pt, Sc, Ti, V, Zn, Al, Pb, Sn, As, B, Ge, Si, Er, Nd, Yb, P, Se, Ba, Ca, K, Li, Mg, Sr	HQ/HI, CR	As is the most significant CR. HQ via ingestion for As, B, Zn, Cu, Ba, and Sr were major contributors to the NCR [133].
Europe	Czech Republic	Cr_{total}, Zn, Pb, Cd, Hg	HEI, HPI	All samples were below the threshold value [134].
	Greece	Cr(VI), NO_3-N	HQ/HI	HQ was acceptable for all sites considering adults; one site has HQ > 1 for children, suggesting potential adverse health effects; multiple sites have CR greater than the permissible limit by the USEPA [135].
	Italy	Al, As, Ba, Cd, Cr, Cu, Fe, Mn, Ni, Pb, Se, Zn	HPI	HPI values were within the critical pollution level category [136].
	Poland	Pb, Cd, Ni, Cu, Fe, Zn, NH_4	NPI, HQ/HI	High risk of contamination due to Pb and Cu [137].
	Romania	NH_4^+, HCO_3^-, Cl^-, NO_3^-, NO_2^-, SO_4^{2-}, Li, Na, Mg, Al, K, Ca, Sr, Ba, Mn, Fe, Cu, Zn, Ga	WQI, HEI, HPI, C_d, HQ/HI	WQI values suggested water classification as marginal, poor, and very poor quality; Cl^-, Al, and NO_3^- were the most significant parameters in the exceedance of HQ to the permissible limit; Fe and Li were the dominant contributors to HPI value [138].
	Serbia	Al, As, Cu, Zn, Fe, Cr, Cd, Mn, Ni, Pb, Hg	HQ/HI	HQ due to As and Hg in multiple sites suggested potential non-carcinogenic health effects [139].
	Ukraine	Al, As, Cr, Cd, Cu, Mn, Ni, Pb, Zn	HPI, HEI, HQ/HI	17% of the samples have HI > 1, suggesting potential non-carcinogenic health effects [140]
Oceania	Fiji	Cd, Cr, Mn, Ni, Pb, Zn	HQ/HI	HI > 1 for all sites both for children and adults, posing potential health risk; CR of Cd is higher compared to other metals in all sites, exceeding the USEPA threshold limit [141].

Table A1. Cont.

Continent	Country	Parameters Evaluated	Index Calculated	Results
	Papua New Guinea	Ca, Mg, Mn	WQI	14% of the samples have poor quality [142].

Appendix B

Table A2. Coordinates of the Sampling Points for DW.

Sampling Point Code	Barangay	Latitude (° N)	Longitude (° E)	Elevation
SJ-B01-DW1	Bagong Sikat	12.367	121.069	4.0 m
SJ-B04-DW2	Barangay II Pob.	12.353	121.066	7.0 m
SJ-B05-DW3	Barangay III Pob.	12.355	121.061	7.0 m
SJ-B05-DW3	Barangay III Pob.	12.358	121.068	1.0 m
SJ-B05-DW3	Barangay III Pob.	12.356	121.066	8.0 m
SJ-B06-DW4	Barangay IV Pob.	12.351	121.061	5.0 m
SJ-B08-DW5	Barangay VI Pob.	12.353	121.066	7.0 m
SJ-B11-DW6	Batasan	12.527	121.119	85.0 m
SJ-B12-DW7	Bayotbot	12.405	121.098	19.0 m
SJ-B13-DW8	Bubog	12.366	121.038	8.0 m
SJ-B13-DW8	Bubog	12.371	121.033	9.0 m
SJ-B14-DW9	Buri	12.223	121.094	163.0 m
SJ-C01-DW10	Camburay	12.427	121.097	26.0 m
SJ-C01-DW10	Camburay	12.429	121.097	28.0 m
SJ-C01-DW10	Camburay	12.430	121.095	28.0 m
SJ-C04-DW11	Central	12.434	121.048	20.0 m
SJ-C04-DW11	Central	12.457	121.041	27.0 m
SJ-C04-DW11	Central	12.449	121.032	18.0 m
SJ-I01-DW12	Iling Proper	12.245	121.088	144.0 m
SJ-I03-DW13	Ipil	12.253	121.091	17.0 m
SJ-L01-DW14	La Curva	12.410	121.016	6.0 m
SJ-L01-DW14	La Curva	12.403	121.071	13.0 m
SJ-L02-DW15	Labangan Iling	12.292	121.050	15.0 m
SJ-L03-DW16	Labangan Poblacion	12.355	121.069	8.0 m
SJ-L03-DW16	Labangan Poblacion	12.365	121.075	5.0 m
SJ-M01-DW17	Mabini	12.372	121.086	7.0 m
SJ-M01-DW17	Mabini	12.368	121.098	8.0 m
SJ-M02-DW18	Magbay	12.409	121.089	20.0 m
SJ-M03-DW19	Mangarin	12.360	121.102	7.0 m
SJ-M03-DW19	Mangarin	12.352	121.100	0.0 m
SJ-M03-DW19	Mangarin	12.352	121.099	7.0 m
SJ-M06-DW21	Murtha	12.471	121.117	47.0 m
SJ-M06-DW21	Murtha	12.436	121.108	25.0 m
SJ-M06-DW21	Murtha	12.440	121.103	28.0 m
SJ-M06-DW22	Murtha	12.457	121.118	36.0 m
SJ-N02-DW23	Natandol	12.184	121.106	55.0 m
SJ-P02-DW24	Pawican	12.171	121.119	99.0 m
SJ-S01-DW25	San Agustin	12.444	121.022	16.0 m
SJ-S02-DW26	San Isidro	12.417	121.057	13.0 m
SJ-S03-DW27	San Roque	12.358	121.051	4.0 m
SJ-A02-DW01	Ansiray	12.273	121.078	13.0 m
SJ-B01-DW02	Bagong Sikat	12.367	121.069	4.0 m
SJ-B02-DW03	Bangkal	12.224	121.049	5.0 m
SJ-B04-DW04	Barangay II	12.353	121.066	7.0 m
SJ-B05-DW05	Barangay III	12.356	121.066	8.0 m
SJ-B06-DW06	Barangay IV Pob.	12.351	121.061	5.0 m
SJ-B06-DW07	Barangay VI Pob.	12.353	121.066	7.0 m
SJ-B11-DW08	Batasan	12.527	121.118	71.0 m

Table A2. *Cont.*

Sampling Point Code	Barangay	Latitude (° N)	Longitude (° E)	Elevation
SJ-B12-DW09	Bayotbot	12.405	121.098	19.0 m
SJ-B13-DW10	Bubog	12.366	121.039	9.0 m
SJ-C01-DW11	Camburay	12.429	121.097	28.0 m
SJ-C04-DW12	Central	12.433	121.048	16.0 m
SJ-C04-DW13	Central	12.457	121.041	27.0 m
SJ-I01-DW14	Iling Proper	12.253	121.034	20.0 m
SJ-I01-DW15	Iling Proper	12.248	121.036	26.0 m
SJ-I02-DW16	Inasakan	12.216	121.069	16.0 m
SJ-L01-DW17	La Curva	12.410	121.082	18.0 m
SJ-L01-DW18	La Curva	12.403	121.071	13.0 m
SJ-L02-DW19	Labangan Iling	12.290	121.049	52.0 m
SJ-L03-DW20	Labangan Poblacion	12.365	121.075	5.0 m
SJ-M01-DW21	Mabini	12.372	121.086	7.0 m
SJ-M01-DW22	Mabini	12.368	121.100	8.0 m
SJ-M02-DW23	Magbay	12.409	121.089	20.0 m
SJ-M03-DW24	Mangarin	12.360	121.102	7.0 m
SJ-M03-DW25	Mangarin	12.352	121.099	7.0 m
SJ-M05-DW26	Monteclaro	12.547	121.125	75.0 m
SJ-M06-DW27	Murtha	12.471	121.117	47.0 m
SJ-M06-DW28	Murtha	12.440	121.103	28.0 m
SJ-S01-DW29	San Agustin	12.444	121.023	16.0 m
SJ-S02-DW30	San Isidro	12.417	121.057	13.0 m
SJ-S03-DW31	San Roque	12.358	121.051	4.0 m

Table A3. Coordinates of the Sampling Points for GW.

Sampling Point Code	Barangay	Latitude (° N)	Longitude (° E)	Elevation
SJ-B11-GW1	Batasan	12.527	121.119	85.0 m
SJ-B12-GW2	Bayotbot	12.405	121.098	19.0 m
SJ-B13-GW3	Bubog	12.365	121.038	8.0 m
SJ-B13-GW4	Bubog	12.266	121.038	48.0 m
SJ-B13-GW5	Bubog	12.371	121.033	9.0 m
SJ-B14-GW6	Buri	12.223	121.094	163.0 m
SJ-C01-GW7	Camburay	12.492	121.095	144.0 m
SJ-C04-GW8	Central	12.434	121.048	20.0 m
SJ-C04-GW9	Central	12.457	121.041	27.0 m
SJ-C04-GW10	Central	12.449	121.032	18.0 m
SJ-I01-GW11	Iling Proper	12.245	121.088	144.0 m
SJ-I03-GW12	Ipil	12.253	121.091	17.0 m
SJ-L01-GW13	La Curva	12.410	121.082	18.0 m
SJ-L01-GW14	La Curva	12.412	121.085	20.0 m
SJ-L01-GW15	La Curva	12.403	121.071	13.0 m
SJ-L02-GW16	Labangan Iling	12.292	121.050	15.0 m
SJ-L03-GW17	Labangan Poblacion	12.361	121.078	7.0 m
SJ-M01-GW18	Mabini	12.372	121.086	7.0 m
SJ-M01-GW19	Mabini	12.368	121.100	8.0 m
SJ-M02-GW20	Magbay	12.409	121.089	20.0 m
SJ-M03-GW21	Mangarin	12.360	121.102	7.0 m
SJ-M03-GW22	Mangarin	12.352	121.100	0.0 m
SJ-M03-GW23	Mangarin	12.352	121.099	7.0 m
SJ-M06-GW25	Murtha	12.471	121.117	47.0 m
SJ-M06-GW26	Murtha	12.436	121.108	25..0 m
SJ-M06-GW27	Murtha	12.440	121.103	28.0 m
SJ-M06-GW28	Murtha	12.445	121.105	31.0 m
SJ-N02-GW29	Natandol	12.184	121.106	55.0 m
SJ-P02-GW30	Pawican	12.171	121.119	99.0 m

Table A3. Cont.

Sampling Point Code	Barangay	Latitude (° N)	Longitude (° E)	Elevation
SJ-S01-GW31	San Agustin	12.444	121.022	16.0 m
SJ-S02-GW32	San Isidro	12.417	121.057	13.0 m

Appendix C

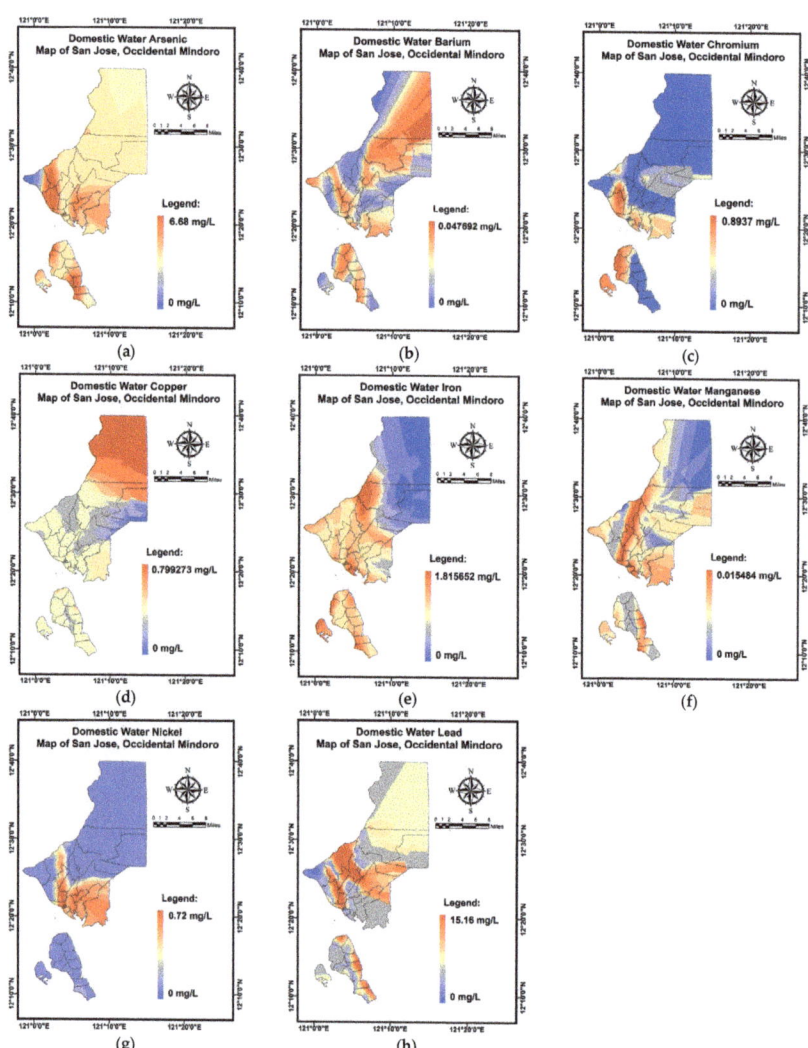

Figure A1. MLGI spatial maps of the (**a**) As; (**b**) Ba; (**c**) Cr; (**d**) Cu; (**e**) Fe; (**f**) Mn; (**g**) Ni; (**h**) Pb in domestic water.

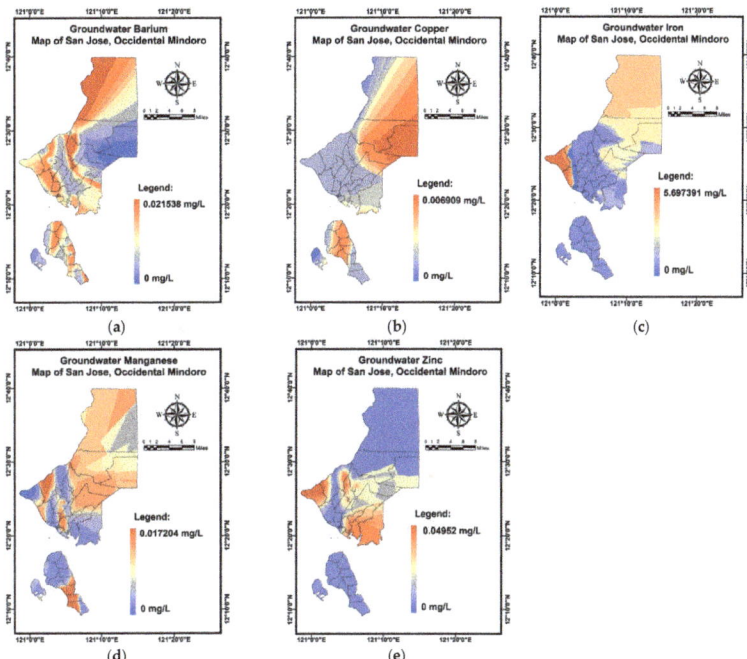

Figure A2. MLGI spatial maps of the (**a**) Ba; (**b**) Cu; (**c**) Fe; (**d**) Mn; (**e**) Zn in GW.

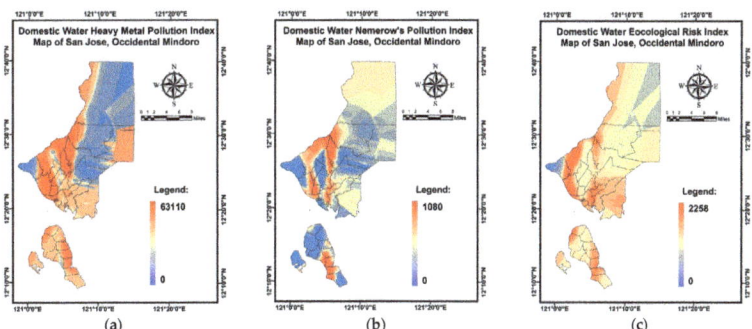

Figure A3. MLGI spatial maps of the (**a**) HPI; (**b**) NPI; (**c**) ERI in DW.

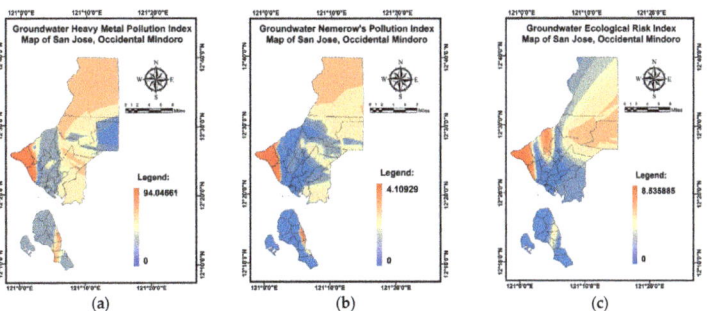

Figure A4. MLGI spatial maps of the (**a**) HPI; (**b**) NPI; (**c**) ERI in groundwater.

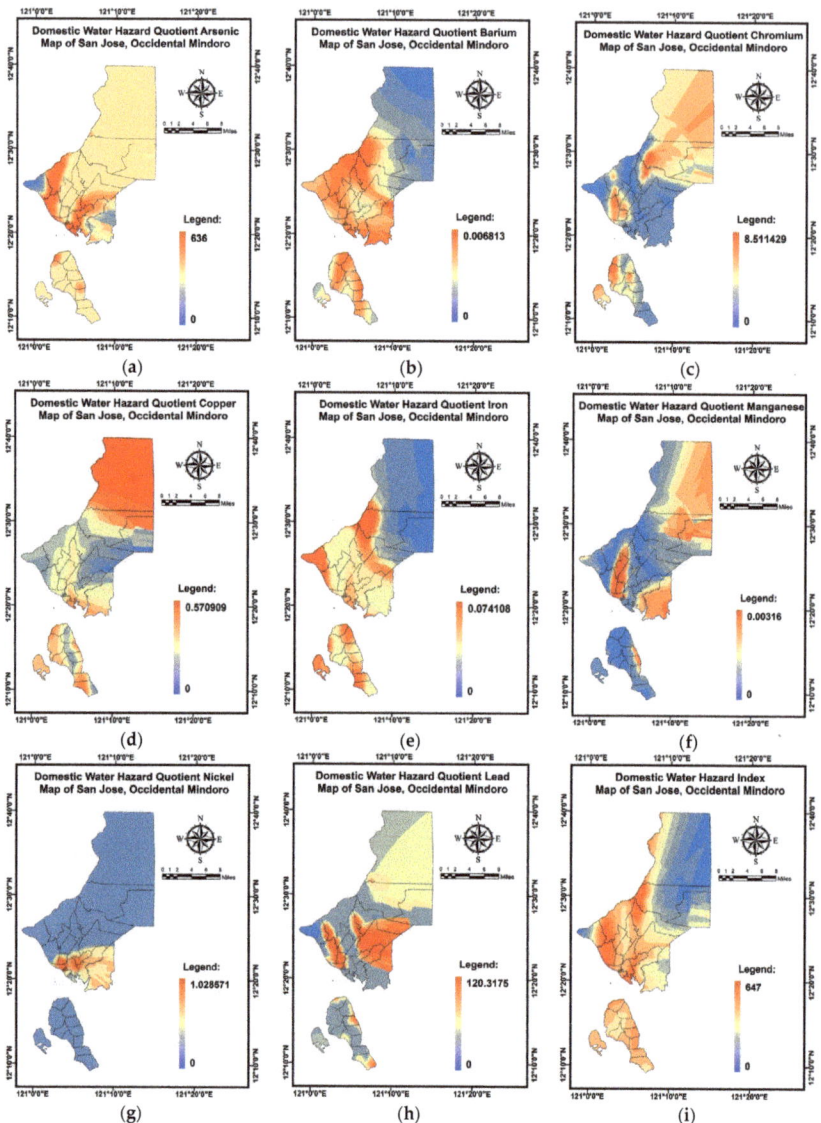

Figure A5. MLGI spatial maps of the (**a**) HQ (As); (**b**) HQ (Ba); (**c**) HQ (Cr); (**d**) HQ (Cu); (**e**) HQ (Fe); (**f**) HQ (Mn); (**g**) HQ (Ni); (**h**) HQ (Pb); (**i**) HI in domestic water.

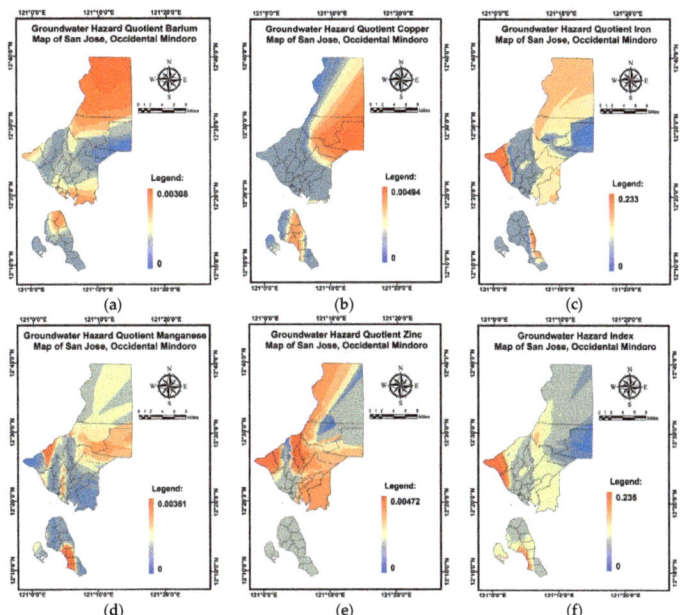

Figure A6. MLGI spatial maps of the (**a**) HQ (Ba); (**b**) HQ (Cu); (**c**) HQ (Fe); (**d**) HQ (Mn); (**e**) HQ (Zn); (**f**) HI in groundwater.

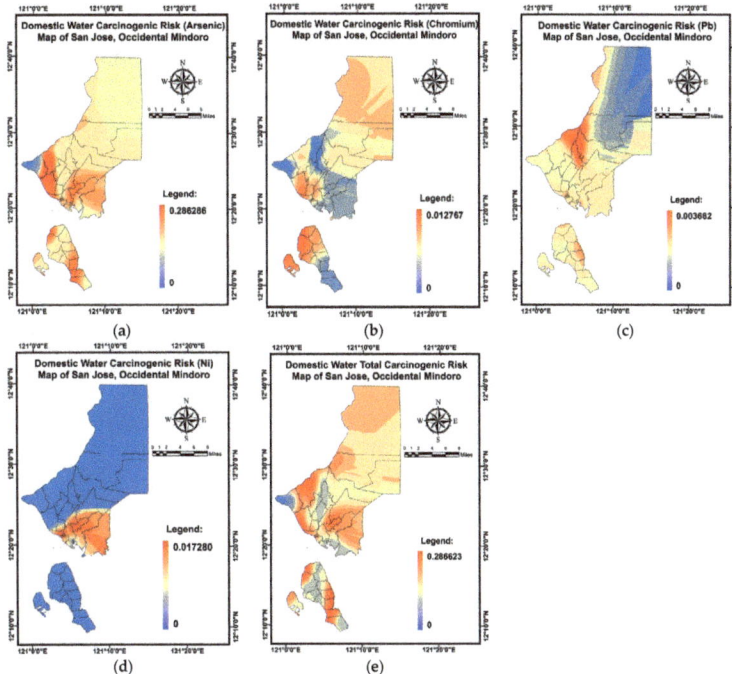

Figure A7. MLGI spatial maps of the (**a**) Cr (As); (**b**) CR (Cr); (**c**) CR (Pb); (**d**) CR (Ni); (**e**) TCR in domestic water.

References

1. Bhattacharya, S.; Das, S.; Das, S.; Kalashetty, M.; Warghat, S.R. An integrated approach for mapping groundwater potential applying geospatial and MIF techniques in the semiarid region. *Environ. Dev. Sustain.* **2021**, *23*, 495–510. [CrossRef]
2. Kumar, V.; Parihar, R.D.; Sharma, A.; Bakshi, P.; Singh Sidhu, G.P.; Bali, A.S.; Karaouzas, I.; Bhardwaj, R.; Thukral, A.K.; Gyasi-Agyei, Y.; et al. Global Evaluation of Heavy Metal Content in Surface Water Bodies: A Meta-Analysis Using Heavy Metal Pollution Indices and Multivariate Statistical Analyses. *Chemosphere* **2019**, *236*, 124364. [CrossRef]
3. Ukah, B.U.; Egbueri, J.C.; Unigwe, C.O.; Ubido, O.E. Extent of heavy metals pollution and health risk assessment of groundwater in a densely populated industrial area, Lagos, Nigeria. *Int. J. Energy Water Resour.* **2019**, *3*, 291–303. [CrossRef]
4. Mirzabeygi, M., Abbasnia, A.; Yunesian, M.; Nodehi, R.N.; Yousefi, N.; Hadi, M.; Mahvi, A.H. Heavy metal contamination and health risk assessment in drinking water of Sistan and Baluchistan, Southeastern Iran. *Hum. Ecol. Risk Assess. Int. J.* **2017**, *23*, 1893–1905. [CrossRef]
5. Mohammadi, A.A.; Zarei, A.; Majidi, S.; Ghaderpoury, A.; Hashempour, Y.; Saghi, M.H.; Alinejad, A.; Yousefi, M.; Hosseingholizadeh, N.; Ghaderpoori, M. Carcinogenic and non-carcinogenic health risk assessment of heavy metals in drinking water of Khorramabad, Iran. *MethodsX* **2019**, *6*, 1642–1651. [CrossRef]
6. Hoang, H.G.; Lin, C.; Tran, H.T.; Chiang, C.F.; Bui, X.T.; Cheruiyot, N.K.; Shern, C.C.; Lee, C.W. Heavy metal contamination trends in surface water and sediments of a river in a highly-industrialized region. *Environ. Technol. Innov.* **2020**, *20*, 101043. [CrossRef]
7. Yeh, G.; Lin, C.; Nguyen, D.H.; Hoang, H.G.; Shern, J.C.; Hsiao, P.J. A five-year investigation of water quality and heavy metal mass flux of an industrially affected river. *Environ. Sci. Pollut. Res.* **2021**, *29*, 12465–12472. [CrossRef] [PubMed]
8. Yousefi, M.; Ghalehaskar, S.; Asghari, F.B.; Ghaderpoury, A.; Dehghani, M.H.; Ghaderpoori, M.; Mohammadi, A.A. Distribution of fluoride contamination in drinking water resources and health risk assessment using geographic information system, northwest Iran. *Regul. Toxicol. Pharmacol.* **2019**, *107*, 104408. [CrossRef]
9. Zhang, Y.; Xu, B.; Guo, Z.; Han, J.; Li, H.; Jin, L.; Chen, F.; Xiong, Y. Human health risk assessment of groundwater arsenic contamination in Jinghui irrigation district, China. *J. Environ. Manag.* **2019**, *237*, 163–169. [CrossRef]
10. Peana, M.; Medici, S.; Dadar, M.; Zoroddu, M.A.; Pelucelli, A.; Chasapis, C.T.; Bjørklund, G. Environmental barium: Potential exposure and health-hazards. *Arch. Toxicol.* **2021**, *95*, 2605–2612. [CrossRef] [PubMed]
11. Miah, M.R.; Ijomone, O.M.; Okoh, C.O.; Ijomone, O.K.; Akingbade, G.T.; Ke, T.; Krum, B.; da Cunha Martins, A., Jr.; Akinyemi, A.; Aranoff, N.; et al. The effects of manganese overexposure on brain health. *Neurochem. Int.* **2020**, *135*, 104688. [CrossRef]
12. Grzeszczak, K.; Kwiatkowski, S.; Kosik-Bogacka, D. The role of Fe, Zn, and Cu in pregnancy. *Biomolecules* **2020**, *10*, 1176. [CrossRef]
13. Long, X.; Liu, F.; Zhou, X.; Pi, J.; Yin, W.; Li, F.; Huang, S.; Ma, F. Estimation of spatial distribution and health risk by arsenic and heavy metals in shallow groundwater around Dongting Lake plain using GIS mapping. *Chemosphere* **2021**, *269*, 128698. [CrossRef] [PubMed]
14. Mahapatra, S.R.; Venugopal, T.; Shanmugasundaram, A.; Giridharan, L.; Jayaprakash, M. Heavy metal index and geographical information system (GIS) approach to study heavy metal contamination: A case study of north Chennai groundwater. *Appl. Water Sci.* **2020**, *10*, 238. [CrossRef]
15. Saleh, H.N.; Panahande, M.; Yousefi, M.; Asghari, F.B.; Oliveri Conti, G.; Talaee, E.; Mohammadi, A.A. Carcinogenic and non-carcinogenic risk assessment of heavy metals in groundwater wells in Neyshabur Plain, Iran. *Biol. Trace Elem. Res.* **2019**, *190*, 251–261. [CrossRef]
16. Gad, M.; El-Safa, A.; Magda, M.; Farouk, M.; Hussein, H.; Alnemari, A.M.; Elsayed, S.; Khalifa, M.M.; Moghanm, F.S.; Eid, E.M.; et al. Integration of water quality indices and multivariate modeling for assessing surface water quality in Qaroun Lake, Egypt. *Water* **2021**, *13*, 2258. [CrossRef]
17. Cüce, H.; Kalıpcı, E.; Ustaoğlu, F.; Kaynar, I.; Baser, V.; Türkmen, M. Multivariate statistical methods and GIS based evaluation of the health risk potential and water quality due to arsenic pollution in the Kızılırmak River. *Int. J. Sediment Res.* **2022**, *37*, 754–765.
18. Yüksel, B.; Ustaoğlu, F.; Arica, E. Impacts of a garbage disposal facility on the water quality of çavuşlu stream in Giresun, Turkey: A health risk assessment study by a validated ICP-MS assay. *Aquat. Sci. Eng.* **2021**, *36*, 181–192. [CrossRef]
19. Bushero, D.M.; Angello, Z.A.; Behailu, B.M. Evaluation of hydrochemistry and identification of pollution hotspots in little Akaki river using integrated water quality index and GIS. *Environ. Chall.* **2022**, *8*, 100587. [CrossRef]
20. Nath, B.K.; Chaliha, C.; Bhuyan, B.; Kalita, E.; Baruah, D.C.; Bhagabati, A.K. GIS mapping-based impact assessment of groundwater contamination by arsenic and other heavy metal contaminants in the Brahmaputra River valley: A water quality assessment study. *J. Clean. Prod.* **2018**, *201*, 1001–1011. [CrossRef]
21. San Jose, Province of Occidental Mindoro. Available online: https://www.philatlas.com/luzon/mimaropa/occidental-mindoro/san-jose.html (accessed on 14 January 2023).
22. Municipality of San Jose, Province of Occidental Mindoro. Available online: https://www.occidentalmindoro.gov.ph/san-jose/ (accessed on 14 January 2023).
23. Executive Summary of San Jose, Province of Occidental Mindoro. Available online: https://coa.gov.ph/download/5184/occidental-mindoro/71786/san-jose-executive-summary-2021-6.pdf (accessed on 14 January 2023).
24. Climate Change-Responsive Integrated River Basin Management and Development Master Plans for the 8 Clustered River Basins. Available online: https://riverbasin.denr.gov.ph/masterplans/8Cluster%20Executive%20Summary/RBCOCluster3ExecutiveSummary.pdf (accessed on 14 January 2023).

25. Municipal Planning and Development Office. *Municipality of San Jose, Province of Occidental Mindoro—Comprehensive Land and Water Use Plan 2017–2030*; Municipal Planning and Development Office: San Jose, Philippines, 2017; Volume 1.
26. de Jesus, K.L.M.; Senoro, D.B.; Dela Cruz, J.C.; Chan, E.B. A Hybrid Neural Network–Particle Swarm Optimization Informed Spatial Interpolation Technique for Groundwater Quality Mapping in a Small Island Province of the Philippines. *Toxics* **2021**, *9*, 273. [CrossRef] [PubMed]
27. Melquiades, F.L.; Appoloni, C. Application of XRF and field portable XRF for environmental analysis. *J. Radioanal. Nucl. Chem.* **2004**, *262*, 533–541. [CrossRef]
28. Zhou, S.; Yuan, Z.; Cheng, Q.; Zhang, Z.; Yang, J. Rapid in situ determination of heavy metal concentrations in polluted water via portable XRF: Using Cu and Pb as example. *Environ. Pollut.* **2018**, *243*, 1325–1333. [CrossRef] [PubMed]
29. Wu, C.M.; Tsai, H.T.; Yang, K.H.; Wen, J.C. How reliable is X-ray fluorescence (XRF) measurement for different metals in soil contamination? *Environ. Forensics* **2012**, *13*, 110–121. [CrossRef]
30. Senoro, D.B.; de Jesus, K.L.M.; Nolos, R.C.; Lamac, M.R.L.; Deseo, K.M.; Tabelin, C.B. In Situ Measurements of Domestic Water Quality and Health Risks by Elevated Concentration of Heavy Metals and Metalloids Using Monte Carlo and MLGI Methods. *Toxics* **2022**, *10*, 342. [CrossRef]
31. Senoro, D.B.; de Jesus, K.L.M.; Mendoza, L.C.; Apostol, E.M.D.; Escalona, K.S.; Chan, E.B. Groundwater quality monitoring using in-situ measurements and hybrid machine learning with empirical Bayesian kriging interpolation method. *Appl. Sci.* **2022**, *12*, 132. [CrossRef]
32. Monjardin, C.E.F.; Senoro, D.B.; Magbanlac, J.J.M.; de Jesus, K.L.M.; Tabelin, C.B.; Natal, P.M. Geo-accumulation index of manganese in soils due to flooding in Boac and Mogpog Rivers, Marinduque, Philippines with mining disaster exposure. *Appl. Sci.* **2022**, *12*, 3527. [CrossRef]
33. Wagh, V.M.; Panaskar, D.B.; Mukate, S.V.; Gaikwad, S.K.; Muley, A.A.; Varade, A.M. Health Risk Assessment of Heavy Metal Contamination in Groundwater of Kadava River Basin, Nashik, India. *Model Earth Syst. Environ.* **2018**, *4*, 969–980. [CrossRef]
34. Kumar, S.; Toppo, S.; Kumar, A.; Tewari, G.; Beck, A.; Bachan, V.; Singh, T.B.N. Assessment of Heavy Metal Pollution in Groundwater of an Industrial Area: A Case Study from Ramgarh, Jharkhand, India. *Int. J. Environ. Anal. Chem.* **2020**, *102*, 7290–7312. [CrossRef]
35. Abdel-Satar, A.M.; Ali, M.H.; Goher, M.E. Indices of water quality and metal pollution of Nile River, Egypt. *Egypt. J. Aquat. Res.* **2017**, *43*, 21–29. [CrossRef]
36. Manan, T.S.B.A.; Beddu, S.; Kamal, N.L.M.; Mohamad, D.; Itam, Z.; Khan, T.; Machmudah, A.; Dutykh, D.; Mohtar, W.H.M.W.; Jusoh, H.; et al. Ecological Risk Indicators for Leached Heavy Metals from Coal Ash Generated at a Malaysian Power Plant. *Sustainability* **2021**, *13*, 10222. [CrossRef]
37. Hoang, H.G.; Lin, C.; Chiang, C.F.; Bui, X.T.; Lukkhasorn, W.; Bui, T.P.T.; Tran, H.T.; Vo, T.D.H.; Le, V.G.; Nghiem, L.D. The individual and synergistic indexes for assessments of heavy metal contamination in global rivers and risk: A review. *Curr. Pollut. Rep.* **2021**, *7*, 247–262. [CrossRef]
38. Senoro, D.B.; Monjardin, C.E.F.; Fetalvero, E.G.; Benjamin, Z.E.C.; Gorospe, A.F.B.; de Jesus, K.L.M.; Ical, M.L.G.; Wong, J.P. Quantitative Assessment and Spatial Analysis of Metals and Metalloids in Soil Using the Geo-Accumulation Index in the Capital Town of Romblon Province, Philippines. *Toxics* **2022**, *10*, 633. [CrossRef]
39. de Jesus, K.L.M.; Senoro, D.B.; Natal, P.; Bonifacio, P. Groundwater Heavy Metal Contamination and Pollution Index in Marinduque Island, Philippines using Empirical Bayesian Kriging Method. *J. Mech. Eng.* **2021**, *10*, 119–141.
40. Eslami, H.; Esmaeili, A.; Razaeian, M.; Salari, M.; Hosseini, A.N.; Mobini, M.; Barani, A. Potentially Toxic Metal Concentration, Spatial Distribution, and Health Risk Assessment in Drinking Groundwater Resources of Southeast Iran. *Geosci. Front.* **2022**, *13*, 101276. [CrossRef]
41. Wu, B.; Zhao, D.Y.; Jia, H.Y.; Zhang, Y.; Zhang, X.X.; Cheng, S.P. Preliminary Risk Assessment of Trace Metal Pollution in Surface Water from Yangtze River in Nanjing Section, China. *Bull. Environ. Contam. Toxicol.* **2009**, *82*, 405–409. [CrossRef]
42. Bortey-Sam, N.; Nakayama, S.M.M.; Ikenaka, Y.; Akoto, O.; Baidoo, E.; Mizukawa, H.; Ishizuka, M. Health Risk Assessment of Heavy Metals and Metalloid in Drinking Water from Communities near Gold Mines in Tarkwa, Ghana. *Environ. Monit. Assess.* **2015**, *187*, 397. [CrossRef] [PubMed]
43. Maleki, A.; Jari, H. Evaluation of Drinking Water Quality and Non-Carcinogenic and Carcinogenic Risk Assessment of Heavy Metals in Rural Areas of Kurdistan, Iran. *Environ. Technol. Innov.* **2021**, *23*, 101668. [CrossRef]
44. Wongsasuluk, P.; Chotpantarat, S.; Siriwong, W.; Robson, M. Heavy Metal Contamination and Human Health Risk Assessment in Drinking Water from Shallow Groundwater Wells in an Agricultural Area in Ubon Ratchathani Province, Thailand. *Environ. Geochem. Health* **2014**, *36*, 169–182. [CrossRef]
45. Karim, Z. Risk Assessment of Dissolved Trace Metals in Drinking Water of Karachi, Pakistan. *Bull. Environ. Contam. Toxicol.* **2011**, *86*, 676–678. [CrossRef]
46. Antoine, J.M.R.; Fung, L.A.H.; Grant, C.N. Assessment of the Potential Health Risks Associated with the Aluminium, Arsenic, Cadmium and Lead Content in Selected Fruits and Vegetables Grown in Jamaica. *Toxicol. Rep.* **2017**, *4*, 181–187. [CrossRef] [PubMed]
47. Oskarsson, A. Barium. In *Handbook on the Toxicology of Metals*; Elsevier: Amsterdam, The Netherlands, 2022; pp. 91–100.
48. Ahmad, W.; Alharthy, R.D.; Zubair, M.; Ahmed, M.; Hameed, A.; Rafique, S. Toxic and Heavy Metals Contamination Assessment in Soil and Water to Evaluate Human Health Risk. *Sci. Rep.* **2021**, *11*, 17006. [CrossRef] [PubMed]

49. Mishra, P.; Pandey, C.M.; Singh, U.; Gupta, A.; Sahu, C.; Keshri, A. Descriptive statistics and normality tests for statistical data. *Ann. Card. Anaesth.* **2019**, *22*, 67. [PubMed]
50. Monjardin, C.E.F.; de Jesus, K.L.M.; Claro, K.S.E.; Paz, D.A.M.; Aguilar, K.L. Projection of water demand and sensitivity analysis of predictors affecting household usage in urban areas using artificial neural network. In Proceedings of the 2020 IEEE 12th International Conference on Humanoid, Nanotechnology, Information Technology, Communication and Control, Environment, and Management (HNICEM), New York, NY, USA, 3–7 December 2020; pp. 1–6.
51. De Jesus, K.L.M.; Senoro, D.B.; Dela Cruz, J.C.; Chan, E.B. Neuro-particle swarm optimization based in-situ prediction model for heavy metals concentration in groundwater and surface water. *Toxics* **2022**, *10*, 95. [CrossRef]
52. Khalid, S.; Shahid, M.; Shah, A.H.; Saeed, F.; Ali, M.; Qaisrani, S.A.; Dumat, C. Heavy metal contamination and exposure risk assessment via drinking groundwater in Vehari, Pakistan. *Environ. Sci. Pollut. Res.* **2020**, *27*, 39852–39864. [CrossRef]
53. Schober, P.; Boer, C.; Schwarte, L.A. Correlation coefficients: Appropriate use and interpretation. *Anesth. Analg.* **2018**, *126*, 1763–1768. [CrossRef]
54. Egbueri, J.C. Groundwater Quality Assessment Using Pollution Index of Groundwater (PIG), Ecological Risk Index (ERI) and Hierarchical Cluster Analysis (HCA): A Case Study. *Groundw. Sustain. Dev.* **2020**, *10*, 100292. [CrossRef]
55. Ghosh, G.C.; Khan, J.H.; Chakraborty, T.K.; Zaman, S.; Kabir, A.H.M.E.; Tanaka, H. Human Health Risk Assessment of Elevated and Variable Iron and Manganese Intake with Arsenic-Safe Groundwater in Jashore, Bangladesh. *Sci. Rep.* **2020**, *10*, 5206. [CrossRef]
56. Sharma, N.; Sodhi, K.K.; Kumar, M.; Singh, D.K. (2021). Heavy metal pollution: Insights into chromium eco-toxicity and recent advancement in its remediation. *Environ. Nanotechnol. Monit. Manag.* **2021**, *15*, 100388.
57. Garcia, A.J.; Okeagu, C.N.; Kaye, A.D.; Abd-Elsayed, A. Metabolism, Pathophysiology, and Clinical Considerations of Iron Overload, a Comprehensive Review. *Essent. Blood Prod. Manag. Anesth. Pract.* **2021**, 289–299. [CrossRef]
58. Das, K.K.; Reddy, R.C.; Bagoji, I.B.; Das, S.; Bagali, S.; Mullur, L.; Khodnapur, J.P.; Biradar, M.S. Primary concept of nickel toxicity—An overview. *J. Basic Clin. Physiol. Pharmacol.* **2019**, *30*, 141–152. [CrossRef] [PubMed]
59. Sachdeva, C.; Thakur, K.; Sharma, A.; Sharma, K.K. Lead: Tiny but mighty poison. *Indian J. Clin. Biochem.* **2018**, *33*, 132–146. [CrossRef]
60. Kianoush, S.; Balali-Mood, M.; Mousavi, S.R.; Shakeri, M.T.; Dadpour, B.; Moradi, V.; Sadeghi, M. Clinical, toxicological, biochemical, and hematologic parameters in lead exposed workers of a car battery industry. *Iranian J. Med. Sci.* **2013**, *38*, 30.
61. Liu, B.; Lv, L.; An, M.; Wang, T.; Li, M.; Yu, Y. Heavy Metals in Marine Food Web from Laizhou Bay, China: Levels, Trophic Magnification, and Health Risk Assessment. *Sci. Total Environ.* **2022**, *841*, 156818. [CrossRef]
62. Alfaifi, H.; El-Sorogy, A.S.; Qaysi, S.; Kahal, A.; Almadani, S.; Alshehri, F.; Zaidi, F.K. Evaluation of heavy metal contamination and groundwater quality along the Red Sea coast, southern Saudi Arabia. *Mar. Pollut. Bull.* **2021**, *163*, 111975. [CrossRef]
63. Sajil Kumar, P.J.; Davis Delson, P.; Thomas Babu, P. Appraisal of heavy metals in groundwater in Chennai city using a HPI model. *Bull. Environ. Contam. Toxicol.* **2012**, *89*, 793–798. [CrossRef]
64. Xu, S.; Li, S.L.; Yue, F.; Udeshani, C.; Chandrajith, R. Natural and anthropogenic controls of groundwater quality in Sri Lanka: Implications for chronic kidney disease of unknown etiology (CKDu). *Water* **2021**, *13*, 2724. [CrossRef]
65. Snousy, M.G.; Morsi, M.S.; Elewa, A.M.; Ahmed, S.A.E.F.; El-Sayed, E. Groundwater vulnerability and trace element dispersion in the Quaternary aquifers along middle Upper Egypt. *Environ. Monit. Assess.* **2020**, *192*, 174. [CrossRef]
66. Sako, A.; Bamba, O.; Gordio, A. Hydrogeochemical processes controlling groundwater quality around Bomboré gold mineralized zone, Central Burkina Faso. *J. Geochem. Explor.* **2016**, *170*, 58–71. [CrossRef]
67. Vasileiou, E.; Papazotos, P.; Dimitrakopoulos, D.; Perraki, M. Expounding the origin of chromium in groundwater of the Sarigkiol basin, Western Macedonia, Greece: A cohesive statistical approach and hydrochemical study. *Environ. Monit. Assess.* **2019**, *191*, 509. [CrossRef]
68. Rezaei, A.; Hassani, H.; Jabbari, N. Evaluation of groundwater quality and assessment of pollution indices for heavy metals in North of Isfahan Province, Iran. *Sustain. Water Resour. Manag.* **2019**, *5*, 491–512. [CrossRef]
69. Çelebi, A.; Şengörür, B.; Kløve, B. Seasonal and spatial variations of metals in Melen Watershed Groundwater, Turkey. *CLEAN–Soil Air Water* **2015**, *43*, 739–745. [CrossRef]
70. Atangana, E.; Oberholster, P.J. Using heavy metal pollution indices to assess water quality of surface and groundwater on catchment levels in South Africa. *J. Afr. Earth Sci.* **2021**, *182*, 104254. [CrossRef]
71. Esmaeili, S.; Asghari Moghaddam, A.; Barzegar, R.; Tziritis, E. Multivariate statistics and hydrogeochemical modeling for source identification of major elements and heavy metals in the groundwater of Qareh-Ziaeddin plain, NW Iran. *Arab. J. Geosci.* **2018**, *11*, 5. [CrossRef]
72. Duggal, V.; Rani, A.; Mehra, R.; Balaram, V. Risk assessment of metals from groundwater in northeast Rajasthan. *J. Geol. Soc. India* **2017**, *90*, 77–84. [CrossRef]
73. Varghese, J.; Jaya, D.S. Metal pollution of groundwater in the vicinity of Valiathura sewage farm in Kerala, South India. *Bull. Environ. Contam. Toxicol.* **2014**, *93*, 694–698. [CrossRef] [PubMed]
74. Karthikeyan1, S.; Arumugam, S.; Muthumanickam, J.; Kulandaisamy, P.; Subramanian, M.; Annadurai, R.; Senapathi, V.; Sekar, S. Causes of heavy metal contamination in groundwater of Tuticorin industrial block, Tamil Nadu, India. *Environ. Sci. Pollut. Res.* **2021**, *28*, 18651–18666. [CrossRef] [PubMed]

75. Hoang, H.G.; Chiang, C.F.; Lin, C.; Wu, C.Y.; Lee, C.W.; Cheruiyot, N.K.; Tran, H.T.; Bui, X.T. Human health risk simulation and assessment of heavy metal contamination in a river affected by industrial activities. *Environ. Pollut.* **2021**, *285*, 117414. [CrossRef]
76. Khairudin, K.; Abu Bakar, N.F.; Ul-Saufie, A.Z.; Abd Wahid, M.Z.A.; Yahaya, M.A.; Mazlan, M.F.; Pin, Y.S.; Osman, M.S. Unravelling Anthropogenic Sources in Kereh River, Malaysia: Analysis of Decadal Spatial-Temporal Evolutions by Employing Multivariate Techniques. *Case Stud. Chem. Environ. Eng.* **2022**, *6*, 100271. [CrossRef]
77. Topal, M.; Arslan Topal, E.I.; Öbek, E. Investigation of Potential Health Risks in Terms of Arsenic in Grapevine Exposed to Gallery Waters of an Abandoned Mining Area in Turkey. *Environ. Technol. Innov.* **2020**, *20*, 101058. [CrossRef]
78. Yang, X.; Duan, J.; Wang, L.; Li, W.; Guan, J.; Beecham, S.; Mulcahy, D. Heavy Metal Pollution and Health Risk Assessment in the Wei River in China. *Environ. Monit. Assess.* **2015**, *187*, 111. [CrossRef] [PubMed]
79. Rahaman, M.S.; Rahman, M.M.; Mise, N.; Sikder, M.T.; Ichihara, G.; Uddin, M.K.; Kurasaki, M.; Ichihara, S. Environmental arsenic exposure and its contribution to human diseases, toxicity mechanism and management. *Environ. Pollut.* **2021**, *289*, 117940. [CrossRef] [PubMed]
80. Izah, S.C.; Inyang, I.R.; Angaye, T.C.; Okowa, I.P. A review of heavy metal concentration and potential health implications of beverages consumed in Nigeria. *Toxics* **2016**, *5*, 1. [CrossRef]
81. Briffa, J.; Sinagra, E.; Blundell, R. Heavy metal pollution in the environment and their toxicological effects on humans. *Heliyon* **2020**, *6*, e04691. [CrossRef] [PubMed]
82. Nickel, W.N.; Steelman, T.J.; Sabath, Z.R.; Potter, B.K. Extra-articular retained missiles; is surveillance of lead levels needed? *Mil. Med.* **2018**, *183*, e107–e113. [CrossRef]
83. Vahidinia, A.; Samiee, F.; Faradmal, J.; Rahmani, A.; Taravati Javad, M.; Leili, M. Mercury, lead, cadmium, and barium levels in human breast milk and factors affecting their concentrations in Hamadan, Iran. *Biol. Trace Elem. Res.* **2019**, *187*, 32–40. [CrossRef] [PubMed]
84. Su, J.F.; Le, D.P.; Liu, C.H.; Lin, J.D.; Xiao, X.J. Critical care management of patients with barium poisoning: A case series. *Chin. Med. J.* **2020**, *133*, 724–725. [CrossRef]
85. Bhoelan, B.S.; Stevering, C.H.; Van Der Boog, A.T.J.; Van der Heyden, M.A.G. Barium toxicity and the role of the potassium inward rectifier current. *Clin. Toxicol.* **2014**, *52*, 584–593. [CrossRef]
86. Yang, F.; Massey, I.Y. Exposure routes and health effects of heavy metals on children. *Biometals* **2019**, *32*, 563–573.
87. Chittick, E.A.; Srebotnjak, T. An analysis of chemicals and other constituents found in produced water from hydraulically fractured wells in California and the challenges for wastewater management. *J. Environ. Manag.* **2017**, *204*, 502–509. [CrossRef]
88. Yuen, H.W.; Becker, W. Iron Toxicity. In *StatPearls*; StatPearls Publishing: Treasure Island, FL, USA, 2022.
89. Alharbi, O.A.; Loni, O.A.; Zaidi, F.K. Hydrochemical assessment of groundwater from shallow aquifers in parts of Wadi Al Hamad, Madinah, Saudi Arabia. *Arab. J. Geosci.* **2017**, *10*, 35. [CrossRef]
90. Rehabilitation/Improvement of Mindoro East Coast Road Project. Available online: https://documents1.worldbank.org/curated/en/709531468095354137/pdf/E14670v50EAP1EA1P079935.pdf (accessed on 11 January 2023).
91. Chotpantarat, S.; Thamrongsrisakul, J. Natural and anthropogenic factors influencing hydrochemical characteristics and heavy metals in groundwater surrounding a gold mine, Thailand. *J. Asian Earth Sci.* **2021**, *211*, 104692. [CrossRef]
92. People Using Safely Managed Drinking Water Services (% of population)—Philippines. Available online: https://data.worldbank.org/indicator/SH.H2O.SMDW.ZS?locations=PH (accessed on 7 February 2023).
93. Chen, X.; Chen, W. GIS-based landslide susceptibility assessment using optimized hybrid machine learning methods. *Catena* **2021**, *196*, 104833. [CrossRef]
94. Bui, D.T.; Van Le, H.; Hoang, N.D. GIS-based spatial prediction of tropical forest fire danger using a new hybrid machine learning method. *Ecol. Inform.* **2018**, *48*, 104–116.
95. Fanos, A.M.; Pradhan, B.; Mansor, S.; Yusoff, Z.M.; Abdullah, A.F.B. A hybrid model using machine learning methods and GIS for potential rockfall source identification from airborne laser scanning data. *Landslides* **2018**, *15*, 1833–1850. [CrossRef]
96. Pham, B.T.; Prakash, I.; Bui, D.T. Spatial prediction of landslides using a hybrid machine learning approach based on random subspace and classification and regression trees. *Geomorphology* **2018**, *303*, 256–270. [CrossRef]
97. Bui, D.T.; Hoang, N.D.; Nguyen, H.; Tran, X.L. Spatial prediction of shallow landslide using Bat algorithm optimized machine learning approach: A case study in Lang Son Province, Vietnam. *Adv. Eng. Inform.* **2019**, *42*, 100978. [CrossRef]
98. Al-Ruzouq, R.; Shanableh, A.; Yilmaz, A.G.; Idris, A.; Mukherjee, S.; Khalil, M.A.; Gibril, M.B.A. Dam site suitability mapping and analysis using an integrated GIS and machine learning approach. *Water* **2019**, *11*, 1880. [CrossRef]
99. Rahman, M.; Islam, M.; Bodrud-Doza, M.; Muhib, M.; Zahid, A.; Shammi, M.; Tareq, S.M.; Kurasaki, M. Spatio-temporal assessment of groundwater quality and human health risk: A case study in Gopalganj, Bangladesh. *Expo. Health* **2018**, *10*, 167–188. [CrossRef]
100. Phan, K.; Phan, S.; Huoy, L.; Suy, B.; Wong, M.H.; Hashim, J.H.; Yasin, M.S.M.; Aljunid, S.M.; Sthiannopkao, S.; Kim, K.W. Assessing mixed trace elements in groundwater and their health risk of residents living in the Mekong River basin of Cambodia. *Environ. Pollut.* **2013**, *182*, 111–119. [CrossRef]
101. Wang, L.; Li, P.; Duan, R.; He, X. Occurrence, controlling factors and health risks of Cr^{6+} in groundwater in the Guanzhong Basin of China. *Expo. Health* **2022**, *14*, 239–251. [CrossRef]
102. Ravindra, K.; Mor, S. Distribution and health risk assessment of arsenic and selected heavy metals in Groundwater of Chandigarh, India. *Environ. Pollut.* **2019**, *250*, 820–830. [CrossRef] [PubMed]

103. Shams, M.; Tavakkoli Nezhad, N.; Dehghan, A.; Alidadi, H.; Paydar, M.; Mohammadi, A.A.; Zarei, A. Heavy metals exposure, carcinogenic and non-carcinogenic human health risks assessment of groundwater around mines in Joghatai, Iran. *Int. J. Environ. Anal. Chem.* **2022**, *102*, 1884–1899. [CrossRef]
104. Al-Jumaily, H.A.; Mohammad, O.A.; Rasheed, B.R. Health Risk Assessment of Heavy Metals in Ground and Tap Water of Chamchamal City-Sulaymaniyah Governorate/Kurdistan Region, Iraq. *Tikrit J. Pure Sci.* **2020**, *25*, 62–70.
105. Al-Hwaiti, M.S.; Brumsack, H.J.; Schnetger, B. Heavy metal contamination and health risk assessment in waste mine water dewatering using phosphate beneficiation processes in Jordan. *Environ. Earth Sci.* **2018**, *77*, 661. [CrossRef]
106. Zhang, W.; Ma, L.; Abuduwaili, J.; Ge, Y.; Issanova, G.; Saparov, G. Distribution characteristics and assessment of heavy metals in the surface water of the Syr Darya River, Kazakhstan. *Pol. J. Environ. Stud.* **2020**, *29*, 979–988. [CrossRef]
107. Lim, H.S.; Lee, J.S.; Chon, H.T.; Sager, M. Heavy metal contamination and health risk assessment in the vicinity of the abandoned Songcheon Au–Ag mine in Korea. *J. Geochem. Explor.* **2008**, *96*, 223–230. [CrossRef]
108. Li, Y.; Ma, L.; Abuduwaili, J.; Li, Y. Spatiotemporal distributions of fluoride and arsenic in rivers with the role of mining industry and related human health risk assessments in Kyrgyzstan. *Expo. Health* **2022**, *14*, 49–62. [CrossRef]
109. Chanpiwat, P.; Lee, B.T.; Kim, K.W.; Sthiannopkao, S. Human health risk assessment for ingestion exposure to groundwater contaminated by naturally occurring mixtures of toxic heavy metals in the Lao PDR. *Environ. Monit. Assess.* **2014**, *186*, 4905–4923. [CrossRef]
110. Kusin, F.M.; Rahman, M.S.A.; Madzin, Z.; Jusop, S.; Mohamat-Yusuff, F.; Ariffin, M. The occurrence and potential ecological risk assessment of bauxite mine-impacted water and sediments in Kuantan, Pahang, Malaysia. *Environ. Sci. Pollut. Res.* **2017**, *24*, 1306–1321. [CrossRef]
111. Tripathee, L.; Kang, S.; Sharma, C.M.; Rupakheti, D.; Paudyal, R.; Huang, J.; Sillanpää, M. Preliminary health risk assessment of potentially toxic metals in surface water of the Himalayan Rivers, Nepal. *Bull. Environ. Contam. Toxicol.* **2016**, *97*, 855–862. [CrossRef] [PubMed]
112. Abeer, N.; Khan, S.A.; Muhammad, S.; Rasool, A.; Ahmad, I. Health risk assessment and provenance of arsenic and heavy metal in drinking water in Islamabad, Pakistan. *Environ. Technol. Innov.* **2020**, *20*, 101171. [CrossRef]
113. Rajmohan, N.; Niyazi, B.A.; Masoud, M.H. Trace metals pollution, distribution and associated health risks in the arid coastal aquifer, Hada Al-Sham and its vicinities, Saudi Arabia. *Chemosphere* **2022**, *297*, 134246. [CrossRef]
114. Nilkarnjanakul, W.; Watchalayann, P.; Chotpantarat, S. Spatial distribution and health risk assessment of As and Pb contamination in the groundwater of Rayong Province, Thailand. *Environ. Res.* **2022**, *204*, 111838. [CrossRef] [PubMed]
115. Mahmoud, M.T.; Hamouda, M.A.; Al Kendi, R.R.; Mohamed, M.M. Health risk assessment of household drinking water in a district in the UAE. *Water* **2018**, *10*, 1726. [CrossRef]
116. Bui Huy, T.; Tuyet-Hanh, T.T.; Johnston, R.; Nguyen-Viet, H. Assessing health risk due to exposure to arsenic in drinking water in Hanam Province, Vietnam. *Int. J. Environ. Res. Public Health* **2014**, *11*, 7575–7591.
117. Benhaddya, M.L.; Halis, Y.; Hamdi-Aïssa, B. Assessment of heavy metals pollution in surface and groundwater systems in Oued Righ region (Algeria) using pollution indices and multivariate statistical techniques. *Afr. J. Aquat. Sci.* **2020**, *45*, 269–284. [CrossRef]
118. Boum-Nkot, S.N.; Nlend, B.; Komba, D.; Ndondo, G.N.; Bello, M.; Fongoh, E.J.; Ntamak-Nida, M.J.; Etame, J. Hydrochemistry and assessment of heavy metal groundwater contamination in an industrialized city of sub-Saharan Africa (Douala, Cameroon). Implication on human health. *HydroResearch* **2023**, *6*, 52–64. [CrossRef]
119. Seleem, E.M.; Mostafa, A.; Mokhtar, M.; Salman, S.A. Risk assessment of heavy metals in drinking water on the human health, Assiut City, and its environs, Egypt. *Arab. J. Geosci.* **2021**, *14*, 427. [CrossRef]
120. Endale, Y.T.; Ambelu, A.; Mees, B.; Du Laing, G. Exposure and health risk assessment from consumption of Pb contaminated water in Addis Ababa, Ethiopia. *Heliyon* **2021**, *7*, e07946. [CrossRef]
121. Opoku, P.A.; Anornu, G.K.; Gibrilla, A.; Owusu-Ansah, E.D.G.J.; Ganyaglo, S.Y.; Egbi, C.D. Spatial distributions and probabilistic risk assessment of exposure to heavy metals in groundwater in a peri-urban settlement: Case study of Atonsu-Kumasi, Ghana. *Groundw. Sustain. Dev.* **2020**, *10*, 100327. [CrossRef]
122. Nyambura, C.; Hashim, N.O.; Chege, M.W.; Tokonami, S.; Omonya, F.W. Cancer and non-cancer health risks from carcinogenic heavy metal exposures in underground water from Kilimambogo, Kenya. *Groundw. Sustain. Dev.* **2020**, *10*, 100315. [CrossRef]
123. Nshimiyimana, F.X.; Faciu, M.E.; El Blidi, S.; El Abidi, A.; Soulaymani, A.; Fekhaoui, M.; Lazar, G. Seasonal influence and risk assessment of heavy metals contamination in groundwater, Arjaat village, Morocco. *Environ. Eng. Manag. J.* **2016**, *15*, 579–587.
124. Ricolfi, L.; Barbieri, M.; Muteto, P.V.; Nigro, A.; Sappa, G.; Vitale, S. Potential toxic elements in groundwater and their health risk assessment in drinking water of Limpopo National Park, Gaza Province, Southern Mozambique. *Environ. Geochem. Health* **2020**, *42*, 2733–2745. [CrossRef]
125. Peleka, J.C.M.; Diop, C.; Foko, R.F.; Daffe, M.L.; Fall, M. Health risk assessment of trace metals in drinking water consumed in Dakar, Senegal. *J. Water Resour. Prot.* **2021**, *13*, 915–930. [CrossRef]
126. Elumalai, V.; Brindha, K.; Lakshmanan, E. Human exposure risk assessment due to heavy metals in groundwater by pollution index and multivariate statistical methods: A case study from South Africa. *Water* **2017**, *9*, 234. [CrossRef]
127. Idriss, I.E.; Abdel-Azim, M.; Karar, K.I.; Osman, S.; Idris, A.M. Isotopic and chemical facies for assessing the shallow water table aquifer quality in Goly Region, White Nile State, Sudan: Focusing on nitrate source apportionment and human health risk. *Toxin Rev.* **2021**, *40*, 764–776. [CrossRef]

128. Ceballos, E.; Dubny, S.; Othax, N.; Zabala, M.E.; Peluso, F. Assessment of human health risk of chromium and nitrate pollution in groundwater and soil of the Matanza-Riachuelo River Basin, Argentina. *Expo. Health* **2021**, *13*, 323–336. [CrossRef]
129. Rupias, O.J.B.; Pereira, S.Y.; de Abreu, A.E.S. Hydrogeochemistry and groundwater quality assessment using the water quality index and heavy-metal pollution index in the alluvial plain of Atibaia river-Campinas/SP, Brazil. *Groundw. Sustain. Dev.* **2021**, *15*, 100661. [CrossRef]
130. Cipriani-Avila, I.; Molinero, J.; Jara-Negrete, E.; Barrado, M.; Arcos, C.; Mafla, S.; Custode, F.; Vilaña, G.; Carpintero, N.; Ochoa-Herrera, V. Heavy metal assessment in drinking waters of Ecuador: Quito, Ibarra and Guayaquil. *J. Water Health* **2020**, *18*, 1050–1064. [CrossRef]
131. Fernández-Macias, J.C.; Ochoa-Martínez, Á.C.; Orta-García, S.T.; Varela-Silva, J.A.; Pérez-Maldonado, I.N. Probabilistic human health risk assessment associated with fluoride and arsenic co-occurrence in drinking water from the metropolitan area of San Luis Potosí, Mexico. *Environ. Monit. Assess.* **2020**, *192*, 712. [CrossRef] [PubMed]
132. Avigliano, E.; Clavijo, C.; Scarabotti, P.; Sánchez, S.; Vegh, S.L.; del Rosso, F.R.; Caffetti, J.D.; Facetti, J.F.; Domanico, A.; Volpedo, A.V. Exposure to 19 elements via water ingestion and dermal contact in several South American environments (La Plata Basin): From Andes and Atlantic Forest to sea front. *Microchem. J.* **2019**, *149*, 103986. [CrossRef]
133. Ccanccapa-Cartagena, A.; Paredes, B.; Vera, C.; Chavez-Gonzales, F.D.; Olson, E.J.; Welp, L.R.; Zyaykina, N.N.; Filley, T.R.; Warsinger, D.M.; Jafvert, C.T. Occurrence and probabilistic health risk assessment (PRA) of dissolved metals in surface water sources in Southern Peru. *Environ. Adv.* **2021**, *5*, 100102. [CrossRef]
134. Podlasek, A.; Jakimiuk, A.; Vaverková, M.D.; Koda, E. Monitoring and assessment of groundwater quality at landfill sites: Selected case studies of Poland and the Czech Republic. *Sustainability* **2021**, *13*, 7769. [CrossRef]
135. Kelepertzis, E. Investigating the sources and potential health risks of environmental contaminants in the soils and drinking waters from the rural clusters in Thiva area (Greece). *Ecotoxicol. Environ. Saf.* **2014**, *100*, 258–265. [CrossRef]
136. Tiwari, A.K.; De Maio, M.; Amanzio, G. Evaluation of metal contamination in the groundwater of the Aosta Valley Region, Italy. *Int. J. Environ. Res.* **2017**, *11*, 291–300. [CrossRef]
137. Dąbrowska, D.; Witkowski, A.J. Groundwater and Human Health Risk Assessment in the Vicinity of a Municipal Waste Landfill in Tychy, Poland. *Appl. Sci.* **2022**, *12*, 12898. [CrossRef]
138. Dippong, T.; Hoaghia, M.A.; Mihali, C.; Cical, E.; Calugaru, M. Human health risk assessment of some bottled waters from Romania. *Environ. Pollut.* **2020**, *267*, 115409. [CrossRef]
139. Ulniković, V.P.; Kurilić, S.M. Heavy metal and metalloid contamination and health risk assessment in spring water on the territory of Belgrade City, Serbia. *Environ. Geochem. Health* **2020**, *42*, 3731–3751. [CrossRef]
140. Dippong, T.; Resz, M.A. Quality and Health Risk Assessment of Groundwaters in the Protected Area of Tisa River Basin. *Int. J. Environ. Res. Public Health* **2022**, *19*, 14898. [CrossRef]
141. Kumar, S.; Islam, A.R.M.T.; Islam, H.T.; Hasanuzzaman, M.; Ongoma, V.; Khan, R.; Mallick, J. Water resources pollution associated with risks of heavy metals from Vatukoula Goldmine region, Fiji. *J. Environ. Manag.* **2021**, *293*, 112868. [CrossRef] [PubMed]
142. Doaemo, W.; Betasolo, M.; Montenegro, J.F.; Pizzigoni, S.; Kvashuk, A.; Femeena, P.V.; Mohan, M. Evaluating the Impacts of Environmental and Anthropogenic Factors on Water Quality in the Bumbu River Watershed, Papua New Guinea. *Water* **2023**, *15*, 489. [CrossRef]

Disclaimer/Publisher's Note: The statements, opinions and data contained in all publications are solely those of the individual author(s) and contributor(s) and not of MDPI and/or the editor(s). MDPI and/or the editor(s) disclaim responsibility for any injury to people or property resulting from any ideas, methods, instructions or products referred to in the content.

Article

Pollution Characteristics and Risk Assessments of Mercury in Jiutai, a County Region Thriving on Coal Mining in Northeastern China

Yuliang Xiao [1,†], Gang Zhang [1,2,†], Jiaxu Guo [1], Zhe Zhang [1], Hongyi Wang [1], Yang Wang [1], Zhaojun Wang [1,2], Hailong Yuan [3,*] and Dan Cui [4,5,6,*]

[1] School of Environment, Northeast Normal University, Changchun 130117, China
[2] State Environmental Protection Key Laboratory of Wetland Ecology and Vegetation Restoration, Changchun 130117, China
[3] Institute of Ice and Snow Economy, Jilin International Studies University, Changchun 130117, China
[4] School of English, Beijing Foreign Studies University, Beijing 100089, China
[5] Comparative Literature and Cross-Culture Discipline, Institute of International Language and Culture, Jilin International Studies University, Changchun 130117, China
[6] College of Foreign Languages, Yanbian University, Yanji 133002, China
* Correspondence: yuanhailongxuany@126.com (H.Y.); summerelephant@126.com (D.C.); Tel.: +86-13604433833 (D.C.)
† Y.X. and G.Z. are co-first authors of the article.

Citation: Xiao, Y.; Zhang, G.; Guo, J.; Zhang, Z.; Wang, H.; Wang, Y.; Wang, Z.; Yuan, H.; Cui, D. Pollution Characteristics and Risk Assessments of Mercury in Jiutai, a County Region Thriving on Coal Mining in Northeastern China. *Sustainability* **2022**, *14*, 10366. https://doi.org/10.3390/su141610366

Academic Editor: Said Muhammad

Received: 6 July 2022
Accepted: 10 August 2022
Published: 19 August 2022

Publisher's Note: MDPI stays neutral with regard to jurisdictional claims in published maps and institutional affiliations.

Copyright: © 2022 by the authors. Licensee MDPI, Basel, Switzerland. This article is an open access article distributed under the terms and conditions of the Creative Commons Attribution (CC BY) license (https://creativecommons.org/licenses/by/4.0/).

Abstract: Among human activities, coal mining and the combustion of fossil fuels are important sources of mercury in the environment. Research on mercury pollution in coal mining areas and surrounding cities, especially in densely populated areas, has always been at the forefront of this research field. In order to study the characteristics of environmental mercury pollution in small and medium-sized coal mining areas and surrounding towns in China, this study selected the main urban area of Jiutai District, a typical mining town in the Changchun City circle industrial base in northeast China, as the research object. In this study, the geo-accumulation index (I_{geo}) was used to study the soil mercury pollution degree in Jiutai District, the potential ecological risk index (Er) was used to evaluate the potential ecological risk of soil mercury in the study area, and the human exposure risk assessment model was used to evaluate the non-carcinogenic risk of soil mercury to the human body. The results showed that 32% of the soil samples in the study area had a higher mercury content than the regional soil background value of Jilin Province (0.04 mg·kg^{-1}). According to the I_{geo}, 19% of the sample sites in the study area were polluted (index > 0). In general, the soil mercury pollution level in Jiutai District is low, and the polluted areas are mainly concentrated in the northeast of the study area. The Er of the soil mercury in the study area ranged from 7.2 to 522.0, with 32% of the sampling sites having a moderate or above potential ecological risk (Er > 40), and the potential ecological risk level of the soil mercury was higher in the northeast of the study area. The non-carcinogenic risk index (*HQ*) and total non-carcinogenic risk value (*HI*) of the soil mercury were all far less than 1, indicating that soil mercury pollution in the study area did not harm the health of local adults. The oral ingestion of soil mercury is the main form of human exposure to mercury.

Keywords: mercury; coal mine; urban soil; pollution characteristics; risk assessments; county

1. Introduction

Mercury (Hg) is a global pollutant [1], with the characteristics of persistent pollution, concealment, easy migration, high bioenrichment, and high biotoxicity [2]. Mercury is unevenly distributed in the natural environment. Its distribution, transformation, and migration are closely related to the geological environment, soil parent rocks, vegetation, and human activities [3]. Coal mines and their surrounding environment frequently have high background mercury levels, and mercury concentrations in regional soil rise as coal mining output increases [4,5]. The average mercury content of Chinese coal is 0.20 mg·kg^{-1} [6], and the average mercury content of the coal in northeast China is 0.158 mg·kg^{-1}, both higher than the worldwide average (0.10 mg·kg^{-1}) [7,8]. In the process of coal mining, large-scale excavation and stripping decompose the sulfide in coal when it meets groundwater or is exposed to air, resulting in a large amount of acidic mine wastewater with strong leachability. About 83.8 wt.% mercury is discharged from coal into the environment through acidic wastewater [9]. The accumulative amount of coal gangue is increasing significantly due to coal mining activities, becoming one of the largest industrial residues [10]. We found that Jilin Province produced 2349.3 million tons of standard coal in 2020, among which the output of coal gangue accounted for about 10–15% of the coal output [11]. Most of the coal gangue mined in Jiutai District in the early stages of the industry was stored in the open air, and now most of it is successively used in the production of coal gangue bricks. In the process of the spontaneous combustion and weathering of coal gangue, mercury can be continuously enriched in the surrounding soil. Research shows that the content of mercury in the soil around coal gangue after spontaneous combustion and weathering can reach 1.980 mg·kg^{-1}, and the enrichment coefficient is up to 22 times higher than the concentration of mercury in the background soil [12,13]. After spontaneous combustion and leaching, the mercury in coal gangue particles is released into the surrounding environment; migrates, transforms, and accumulates in the ecosystem; and finally enters into the human body through various routes, affecting human health [14]. The soot, dust, and waste water generated during coal processing cause the mercury in the coal to enter the environment, and the combustion of coal-preparation byproducts may cause secondary emissions, resulting in regional environmental mercury pollution [15].

Towns that prospered from coal mining accumulated mercury in the soil as the coal was mined [5]. With the development of towns, the number and density of urban populations have gradually increased, and this has put urban populations at increased risk of exposure to mercury pollution [16] that directly or indirectly affects the local residents' welfare, natural resources storage, economic development, cultural communication, tourism, etc. Such typical environmental pollution has always been a concern of Westerners of all walks of life. Taking London at the end of the 18th and beginning of the 19th centuries as an example, we can see that the coal pollution issue was of great concern, as made clear by environmentalists, ecologists, governors, administrators, and even literary figures such as the English romantic poets William Blake and William Wordsworth. The poet John Keats was particularly concerned by this issue, since members of his family developed tuberculosis due in large part to the frequent burning of coal.

Coal mines are usually characterized by a long mining cycle, a wide development range, and a strong environmental impact. In addition, a large number of coal-related industries gather around coal mining areas, which leads to more serious regional mercury pollution than in ordinary towns. Therefore, it is necessary to understand the current situation regarding the level of mercury accumulation in the soil environments of coal mining towns. According to previous research results, the soil mercury in coal mining towns mainly comes from coal mining and the coal industry [17]. However, with the comprehensive development of cities and towns, the sources of soil mercury pollution have increased, and it is believed to mainly originate from the following urban sources [18–21]:

1. Atmospheric mercury from fossil fuel combustion eventually returns to the soil through wet and dry deposition—e.g., coal-fired thermal power plants and coal-fired power for industrial mercury release, oil smelting, and automobile emissions.
2. The raw materials of non-ferrous metal smelting and the cement industry contain mercury, which is activated in the production process so that it migrates and transforms, causing mercury pollution to the air, water, and soil.
3. The production and use of items containing mercury such as fluorescent lamps, dental amalgams, thermometers, pressure gauges, and electronic products.
4. Emissions caused by waste treatment and incineration processes: various waste treatment facilities and landfills also produce mercury pollution.

Although much attention has been paid to the risk of heavy metal pollution in coal mining areas, less is currently known regarding the level and spatial distribution of soil mercury pollution in urban areas after coal mining in China [22,23]. This information is critical, considering the continuous exposure of urban residents to the accumulated mercury in urban soil [24]. Such exposure may take the form of oral ingestion, inhalation, or dermal contact. At present, Chinese researchers have investigated the mercury content levels in the soil of Huainan [25], Qianxi [26], and Lianyuan [27], as well as some of the counties in Shanxi Province [28] and other cities in China. On an international scale, the status of soil mercury accumulation in the Rohini OCP NK Area of India [17], Pszczyna County Poland [29], and Ostrava in the Czech Republic have [30] been examined. However, a comprehensive risk assessment of the mercury in the soils of coal mining cities at the county scale in China is still lacking [31]. Therefore, this study focuses on the mercury content level in urban soil and a risk assessment of mercury's negative impact on urban residents' health after the depletion of coal resources in coal mining cities.

The main objective of this study was to assess the risks of heavy metal mercury pollution in the soil of Jiutai District, a town in northeastern China that has prospered because of coal mining. It is believed that long-term coal mining causes soil mercury pollution, and, despite the adjustments to the energy structures that have contributed to the formation of mercury pollution, it will remain in the soil for a long period of time, possibly impacting human living environments [32]. Therefore, we chose to discuss the nature of such impacts in this study. Surface soil samples were collected in Jiutai District, and their heavy metal mercury and atmospheric mercury concentrations were determined. The ecological risks of the heavy metal mercury in the soil were assessed by calculating the geo-accumulation index (I_{geo}) [33] and the potential ecological risk index (Er) [34]. The associated health risks were then assessed by determining the levels of metal exposure through oral ingestion, inhalation, and dermal contact and then calculating the non-carcinogenic risk index (HQ) [35] and the total non-carcinogenic risk value (HI) [36]. The results of this study will contribute to a more comprehensive understanding of the risks arising from human exposure to heavy metal mercury in Chinese coal mining towns.

2. Materials and Methods

2.1. Study Area

The study area of Jiutai District is located in the northeast of China and the central part of Jilin Province. It is under the jurisdiction of Changchun, lying 47 km to the northeast of the main city of Changchun and located between E125°25′ and 126°30′ and N43°51′ and 44°32′. Jiutai District covers an area of 3375.27 km^2, and its main urban area is 41 km^2. As of 2021, Jiutai District had a permanent population of 570,000, with a population density of about 198 people/km^2 [37,38]. Jiutai District has a mid-temperate continental monsoon climate, with an annual average temperature of 4.7 °C and annual average precipitation of 658.5 mm. The dominant wind direction in the city is southwest [19]. The green space coverage area of the whole district is 610.15 hm^2, and the green coverage rate is 20.93% [39].

After the residents of Yingcheng discovered coal mines in 1868, Jiutai gradually developed from an ordinary town into a typical mining town, and it has now become one of the important mineral bases of coal resources in Jilin Province [40]. The coal mining industry has long supported the economic development of the city and affected its industrial layout. Jiutai District's main urban area is the center of the administration, economy, culture, and transportation for the coal mining area. There are large coal processing enterprises, and the mining and extraction points are mainly distributed in the mining area and workers' residential area. To develop the coal industry in the early years after its founding in China, there was a rapid rise in coal mining in Jiutai District, which became home to several key state-owned coal mines; this greatly promoted the development of Jiutai District as well as the formation of the industrial landscape. The city's coal mines were rich in this early period, and Jiutai District could not only meet the demands of the industry in the local area, but could also supply a large amount of coal to Changchun and Jilin Province, taking up a decisive position in China's coal mining industry. Then, with the exhaustion of the coal resources in Yingcheng, the main pillar industry in Jiutai District gradually changed from the coal industry to the manufacturing industry. According to the Jiutai District Mineral Resources Plan (2016–2020), Jiutai District still accounts for 78% of Changchun's coal production capacity.

At present, Jiutai District is a sub-central city with good-quality living infrastructure. It is a comprehensive industrial city with leading industries such as food processing, building materials, machinery, chemicals, and coal. The northern part of Jiutai District belongs to the old city, with large residential areas and farmland protected by the state, mainly for commercial and residential functions. The southern part of the city belongs to the new urban area, which is dominated by residential buildings, commercial services, administrative offices, culture, and entertainment, with a small number of industrial and logistics parks. In the eastern part of the city, the former Yingcheng mining area has been optimized to form the present Jiutai industrial concentration area, but the problems of soil pollution and geological environment left over from the historical mines have not been completely solved. There are a few industrial parks and residential areas in the west.

2.2. Sampling and Analysis Processes

According to the spatial pattern of Jiutai District, the main urban area was selected as the research scope, since it includes both built-up areas and suburban areas. These are composed of the early development zones and new urban areas of Jiutai District, which comprehensively reflect the basic characteristics of the typical urban core area and peripheral expansion areas. A total of 100 sampling units (1 km × 1 km) were selected by the random grid distribution method. Sampling points were randomly distributed in different grids. Finally, the locations of sampling points were plotted according to the latitude and longitude of each sampling point. In this study, we collected data from 100 sample points, and the location of each sample point is depicted in Figure 1. Based on the urban regional planning of Jiutai District, and for the convenience of subsequent research on the relationship between wind direction and urban soil mercury concentration, taking the intersection of Xiqiao Park and Nanshan Park and Changtong Road as the center, the main urban area of Jiutai District was divided into four regions: southeast, northeast, southwest, and northwest.

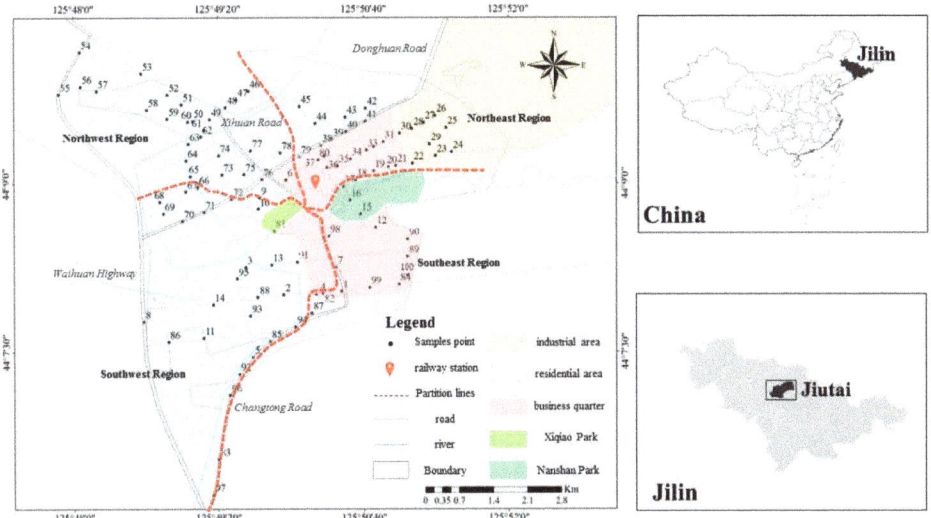

Figure 1. Study area and sampling points.

On meteorologically typical days, each sample was collected using the five-point mixed sampling method. Five equal-volume samples were collected in the range of 5 m × 5 m and uniformly mixed into one sample. In addition, all of the samples were collected from the 0~10 cm surface soil layers, and any plants, dead branches, fallen leaves, rocks, sand grains, and so on were removed. The samples were sealed and stored in polyethylene bags, with each sample weighing approximately 500 g. For the soil samples obtained in areas without hardened sampling points, the dust removed within a certain area around the sampling points was taken as the test sample. Following the completion of the sample collection process, the samples were stored in a cool place until the analysis was completed.

The total atmospheric mercury content was monitored by a Zeeman LUMEX RA-915+ effect mercury analyzer. All of the sampling points were divided into two layers in the vertical direction of the air within a range of 0 cm and 100 cm, and an in-vehicle mercury meter (Zeeman LUMEX RA-915+) was used for the monitoring process according to the sample point sequence number. The instrument was self-calibrated before the measurements, and the data were obtained every 10 s. A total of 120 s–180 s were measured, and 12–18 monitoring data points were obtained at each sample point, with a data accuracy of 1 ng·m^{-3}. Blank samples were taken from urban clean and open green space, and corresponding background values were measured as a reference. Air-dried soil samples of 50~200 mg were weighed out, and the total mercury content of the soil was determined in the one-way optical path sample pool of the LUMEX RA-915+ mercury meter combined with the UMA solid sample test unit. The air transport speed of the instrument was 4 L·min^{-1}. The minimum detection limit of solid samples determined by the instrument was 0.5 ng·g^{-1}. For each soil sample, 4–7 parallel determinations were carried out to eliminate abnormal data. In the experimental process, GBW07424 standard material of the soil composition in the Songnen Plain (GSS-10) was used to construct standard curves.

2.3. Ecological Risk Assessment Processes

2.3.1. Geo-Accumulation Index (I_{geo})

The I_{geo} is used to assess the contamination levels of a specific metal in soil by evaluating the metal enrichment above the baseline or background values. The I_{geo} was calculated according to Equation (1), as follows [41,42]:

$$I_{geo} = log_2 \left[\frac{C_i}{K \times B_i} \right] \quad (1)$$

where I_{geo} represents the geo-accumulation index; C_i is the measured concentration of mercury (mg·kg^{-1}); K represents a modified index (typically 1.5) to account for variations in background values that may be due to differences in rock characteristics at different locations; and B_i denotes the municipal soil background value (0.040 mg·kg^{-1}). The classification of the I_{geo} is shown in Table 1.

Table 1. The classification of the I_{geo}.

Geo-Accumulation Index (I_{geo})	Grading	Degree of Contamination
$5 < I_{geo} \leq 10$	6	Extremely strong
$4 < I_{geo} \leq 5$	5	Strong/extremely strong
$3 < I_{geo} \leq 4$	4	Strong
$2 < I_{geo} \leq 3$	3	Medium-strong
$1 < I_{geo} \leq 2$	2	Medium
$0 < I_{geo} \leq 1$	1	Slight/medium
$I_{geo} \leq 0$	0	Non-pollution

2.3.2. Potential Ecological Risk Index (Er)

The Er was used to evaluate the potential degree of ecological harm posed by the mercury in the soil and atmosphere. The Er was calculated according to Equation (2), as follows [43,44]:

$$\text{Er} = \text{Tr} \cdot \frac{C_i}{C_0} \tag{2}$$

where Er is the potential ecological harm coefficient of the mercury; Tr represents the toxicity coefficient of the mercury, which was set at 40; C_i is the measured value of mercury content; and C_0 is the background value of Hg, which was set at 0.040 mg·kg^{-1}. The relationship between the potential ecological risk coefficient and hazard degree of Er was graded as shown in Table 2.

Table 2. Criteria for potential ecological risk analysis.

Er	<40	40–80	80–160	160–320	>320
Potential Ecological Risk Index	Slight	Medium	Strong	Very strong	Extremely strong

2.4. Health Risk Assessments

2.4.1. Exposure Assessments

This study referred to the environmental health risk assessment method recommended by the United States Environment Program (USEPA) for soil mercury health risk assessment [45]. Soil mercury mainly enters the human body through oral ingestion, inhalation, and dermal contact [46]. The human exposure risk assessment model was used to calculate the long-term daily exposure dose of non-carcinogenic risk for the above three exposure pathways. The formulas are as follows [47,48]:

$$ADI_{inh} = \frac{C \times IR_{inh} \times EF \times ED}{BW \times AT \times PEF} \tag{3}$$

$$ADI_{oral} = \frac{C \times IR_{oral} \times EF \times ED}{BW \times AT} \times 10^{-6} \tag{4}$$

$$ADI_{dermal} = \frac{C \times SA \times SL \times ABS \times EF \times ED}{BW \times AT} \times 10^{-6} \tag{5}$$

In Equations (3)–(5), ADI_{inh}, ADI_{oral}, and ADI_{dermal} represent the average daily exposure dose (mg·(kg·d)$^{-1}$) of inhalation, oral intake, and skin contact, respectively; C represents the measured concentration of mercury in soil (mg·kg^{-1}); IR_{inh} represents respiratory intake (m^3·d^{-1}), with a reference value of 20 m^3·d^{-1}; EF represents exposure frequency (d·a^{-1}), with a reference value of 350 d·a^{-1}; ED stands for exposure years (a), with a reference value of 25 a; BW represents body weight (kg), with a reference value of 55.9 kg; AT represents exposure period (d), with a reference value of 365 × ED (d); PEF stands for particulate matter emission factor (m^3·kg^{-1}), with a reference value of 1.32 × 10^9 m^3·kg^{-1}; IR_{oral} represents oral intake (mg·d^{-1}), with a reference value of 114 mg·d^{-1}; SA represents the exposed skin area (cm^2·d^{-1}), with a reference value of 5000 cm^2·d^{-1}; SL is skin adhesion (mg·(cm^2)$^{-1}$), with a reference value of 1 mg·(cm^2)$^{-1}$; and ABS stands for skin absorption factor, with a reference value of 0.001.

2.4.2. Non-Carcinogenic Risk Assessments

Mercury is a non-carcinogenic heavy metal, so the non-carcinogenic risk index (HQ) was used to evaluate the non-carcinogenic risk. The non-carcinogenic risk index was calculated according to Equation (6), as follows [35]:

$$HQ = \frac{ADIi}{RfDi} \tag{6}$$

where ADI_i is the average daily exposure dose $(mg·(kg·d)^{-1})$; RfD_i is the reference dose $(mg·(kg·d)^{-1})$; $RfD_{dermal} = 2.4 \times 10^{-5}$ mg·(kg·d)$^{-1}$; $RfD_{oral} = 3.0 \times 10^{-4}$ mg·(kg·d)$^{-1}$; and $RfD_{inh} = 1.07 \times 10^{-4}$ mg·(kg·d)$^{-1}$.

The total non-carcinogenic risk value of mercury for the different pathways is indicated by HI [36]:

$$HI = \sum_i^n HQ \quad (7)$$

When HI or $HQ > 1$, there is a potential non-carcinogenic risk, and when HI or $HQ < 1$, the risk is considered small and can be ignored [49,50].

3. Results

3.1. Mercury Concentrations in the Environment

3.1.1. Mercury in Surface Soils

According to the statistics, the average mercury content in the soil samples measured in this study was 0.0484 ± 0.0633 mg·kg^{-1}, and the concentration range was 0.00720~0.522 mg·kg^{-1}. It can be seen that the mercury content in Jiutai District varied widely, and the mercury content levels in the four regions are shown in Table 3. Of the 100 samples, 32% presented soil mercury contents that exceeded the regional soil background value of Jilin Province (0.04 mg·kg^{-1}), and the maximum level was 13.05 times higher than the background value. It can be seen that the soil in different regions of Jiutai District was polluted by mercury to different degrees.

Table 3. Soil mercury levels in different regions.

Region	Number of Soil Samples (n)	Average (mg·kg^{-1})	Range (mg·kg^{-1})	Standard Deviation (mg·kg^{-1})	Coefficient of Variation (%)
Southeast	18	0.0358	0.0142–0.125	0.0266	0.742
Northeast	29	0.0837	0.00940–0.522	0.103	1.23
Southwest	26	0.0377	0.00720–0.123	0.0260	0.690
Northwest	27	0.0289	0.00810–0.0704	0.0156	0.539
All Regions	100	0.0484	0.00720–0.522	0.0633	1.30

3.1.2. Atmospheric Mercury in Urban Areas

The atmospheric mercury concentration levels in the northeast, northwest, southeast, and southwest regions of Jiutai District of Changchun are shown in Figure 2. The mean atmospheric mercury contents at 0 cm and 100 cm were 3.2 ± 4.1 ng·m^{-3} and 4.5 ± 5.2 ng·m^{-3}, respectively, and the concentrations ranged from 0 to 21.7 ng·m^{-3} and 0 to 23.3 ng·m^{-3}, respectively. The atmospheric mercury content at 0 cm and 100 cm in blank samples was the same, with an average of 1.2 ± 1.5 ng·m^{-3}. The results showed that 59% of the samples had a higher atmospheric mercury content at 0 cm than the blank control, and the highest mercury content was 18.1 times that of the blank control. Sixty percent of the samples had a higher atmospheric mercury content at 100 cm than the blank control, and the highest mercury content was 19.4 times that of the blank control. The atmospheric mercury content in the Jiutai District urban area varied widely. It can also be seen from Figure 2 that there were differences in the atmospheric mercury content between the different regions of the study area. The atmospheric mercury concentration at 0 cm in the northwest region was high, and that at 100 cm in the southwest region was also high.

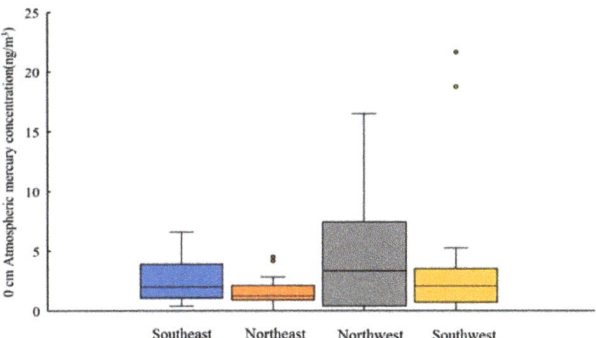

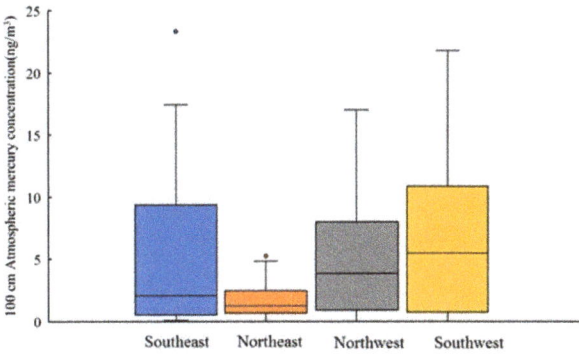

Figure 2. Box chart of the mercury content levels in the four regions.

3.2. Risk Assessment

3.2.1. Ecological Risk Assessment of Soil Mercury Pollution

Using the I_{geo} method as the evaluation standard, 19% of the sample sites were contaminated with mercury (index > 0). The pollution level of sample 30 was strong; that of sample 41 was medium-strong; that of samples 36, 23, 80, 22, 17, 76, and 79 was medium; that of samples 31, 57, 26, 84, 74, 96, 91, 40, 58, and 49 was slight-medium; and the other samples were judged to be non-polluted.

In general, the mercury pollution in Jiutai District was not high. According to the evaluation of the soil mercury accumulation index method, all the samples with moderate or above soil mercury pollution were concentrated in the areas near the industrial center of Jiutai District (the former Yingcheng mining area), Jiutai Railway Station, and on both sides of the main traffic circle of the city. The samples with mild soil mercury pollution were distributed near logistics parks, business centers, automobile markets, and driving schools and were mainly affected by mercury emission from traffic.

It was found that the Er of the soil mercury in the study area ranged from 7.2 to 522.0, and 32% of the samples presented medium or above potential ecological risk (Er > 40). Among these, some samples demonstrated an extremely strong potential ecological risk (Er up to 522 > 320). The possible reason for this was that the sample site was close to the industrial concentration area of Jiutai District (the former Yingcheng mining area), and early mining activities and current industrial activities have increased the soil mercury concentration, resulting in extremely strong and very strong potential ecological risk scores for the soil mercury in this area. In addition, 11% of the samples posed a strong potential ecological risk, 21% posed a medium potential ecological risk, and the rest posed a slight potential ecological risk. In general, this study found that the mercury pollution level of the soil in Jiutai District presented a slight potential ecological risk.

3.2.2. Health Risk Assessment of Soil Mercury

In terms of average *HI*, the regions of the town were ranked as follows: northeast (0.580×10^{-2}) > northwest (0.260×10^{-2}) > southeast (0.250×10^{-2}) > southwest (0.200×10^{-2}). The northeast region contains concentrated industrial parks, experienced frequent early coal mining activities, and is located downwind of the city, so the total health risk value is high. The Jiutai District's new urban area is located in the southwest and is primarily made up of residential areas and low-pollution high-tech industrial parks. The environmental quality is high, the soil mercury pollution level is low, and it is located upwind of the city, so the total health risk value is low. Vehicle testing companies, driving schools, and the outer ring road of the city are distributed. In the northwest region. There is a large traffic flow and a large amount of automobile exhaust emissions, which have a significant impact on the soil mercury content, so the total health risk value is also high. In general, the total non-carcinogenic health risk value in Jiutai District was determined to be far less than 1. Currently, the urban soil mercury does not harm the health of local adults.

In this study, by referring to the environmental health risk assessment method recommended by the USEPA for soil mercury health risk assessment [45], we determined the non-carcinogenic risk posed by the three main soil mercury exposure pathways as follows: oral ingestion (average 0.320×10^{-2}) > dermal contact (average 0.170×10^{-3}) >> inhalation (average 0.118×10^{-6}). Therefore, oral ingestion through food is the main route of soil mercury exposure, and the contribution of respiratory exposure is very small compared with exposure through ingestion.

4. Discussion

4.1. Mercury Distribution in Urban Environment

4.1.1. Spatial Distribution Characteristics of Soil Mercury Concentration

In order to intuitively display the spatial distribution characteristics of the soil mercury in the study area, we used ArcGIS 10.7 for geostatistical analysis. The Kriging interpolation method and the inverse distance weight method were used for the analysis, and the method with the best performance was selected to determine the research result. The spatial distribution map of the soil mercury content in the study area is shown in Figure 3. The spatial differences in the soil mercury content in the study area are obvious. The surface soil mercury content in the northeast of the study area is significantly higher than that in the southwest. The northeast of the research area is close to the industrial concentration area of Jiutai District (the former mining area of Yingcheng), and after long-term underground mining in the Yingcheng Coal Mine, a large area of coal mining subsidence and a coal gangue mountain was formed. It has been found that mercury is enriched in the soil by the weathering and leaching of coal gangue that has been piled up for a long time. Most of the mercury that enters the soil is bound by organic matter and remains there for a long time. Mercury accumulated in the soil can be slowly released into the surface water and other media over long periods, resulting in regional environmental mercury pollution [51,52]. In addition, this area now contains a concentration of Jiutai District industries, and the discharge of industrial mercury-containing wastes, coal combustion, and coal transportation have increased the accumulation of mercury in the surface soil. Due to the rapid urbanization in recent years, new urban areas in the south have also been developed. With the increase in the built-up area and the decrease in the population density in Jiutai District, the spatial distribution of soil mercury in the city has become more diffuse. A small number of soil mercury enrichment areas have also appeared in the new southern urban area, mainly distributed in large residential areas with high environmental quality and high-tech parks with low pollution and high output, and the soil mercury content is relatively low.

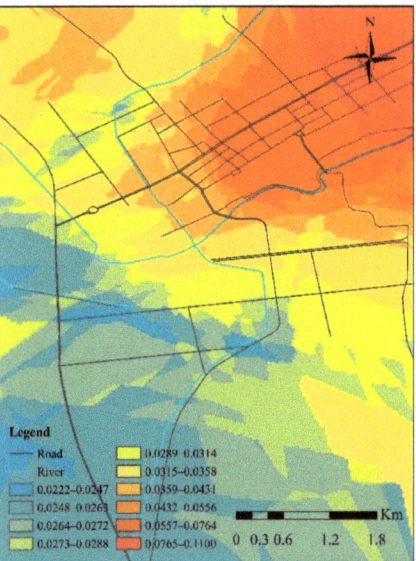

Figure 3. Spatial distribution of soil mercury concentration (mg·kg^{-1}).

Higher mercury levels were also found in the commercial and northern residential regions of the study area. The commercial district has a large flow of people and vehicles, and the traffic contributes a lot to mercury emissions. In addition, the garbage output is large, and the total amount of mercury-containing waste is relatively large, so the mercury content in the surface soil is relatively high. Furthermore, Jiutai Railway Station is located in this area. The main cargo transportation categories of Jiutai Railway Station include grain, coal, fertilizer, small amounts of mineral materials, and building materials. Regional soil mercury arises from fly ash and coal dust deposited by coal and other minerals in the process of transportation. However, the northern residential area belongs to the old urban area, which used to contain many large and medium-sized heavy industries, such as thermal power plants and cement plants. Now, most of these enterprises have moved out, but the area still contains enterprises such as Jiutai Rongxiang Heating Supply and Xinhua Heating Supply. Studies have shown that the mercury content in the soil around coal-fired enterprises is much higher than the world average mercury concentration of 0.03 mg·kg^{-1}. This is directly impacted by the mercury-containing flue gas emitted by enterprises, and the soil mercury concentration around coal-fired enterprises was found to decrease as the distance from the enterprise increased [27], which is consistent with the results of this study.

A comparison between the soil mercury content in Jiutai District and other typical mining towns worldwide is shown in Table 4. The six coal mines of Zaozhuang, Yuancang, Daihe, Gaoyang, Longfeng, and Yingcheng have all experienced the extensive exploitation of resources. Yingcheng, Yuancang, Daihe, and Longfeng are now closed due to resource exhaustion, whereas the Zaozhuang and Gaoyang mines are still in operation. Like the Gaoyang coal mine, the others are small and medium-sized coal mines. Rohini OCP NK Area in India, Pszczyna County Poland, and Ostrava in the Czech Republic are large and medium-sized coal mines. Through the comparison, it was found that the soil mercury content in the urban regions of the mining areas was higher than the local soil mercury background value, indicating that coal mining activities contribute to local soil mercury levels. By comparing the soil mercury content in the mining towns where the coal mine has ceased operations with that in the mining towns where the coal mine has been exploited, we found that the former did not present a lower level, indicating that early mining activities may have a long-term influence on the soil mercury content in mining towns. The background value of the soil mercury content in Jiutai District is higher than that in Zaozhuang, Huaibei, Xiaoyi, Fushun, and Pszczyna County, but the mean value of the soil mercury content is lower than that of these five cities. This indicates that Jiutai District is exposed to low exogenous mercury pollution, and it may be the case that coal mining activities had a small impact on the soil mercury content in the main urban area of Jiutai District.

Table 4. Soil mercury content in urban areas of different coal mining towns.

Town	Range (mg·kg^{-1})	Average (mg·kg^{-1})	Background (mg·kg^{-1})	Maximum Coal Yield (Mt·a^{-1})	Operating Time	Reference
Zaozhuang	-	0.040	0.010	1.92	1880–now	[53]
Huangshi (Yuancang Coal Mine)	0.012~1.823	0.137	0.041	0.600	1949–2014	[54]
Huaibei (Daihe Coal Mine)	0.008~0.154	0.067	0.014	1.40	1965–2017	[55]
Xiaoyi (Gaoyang Coal Mine)	0.073~0.100	0.087	0.025	6.00	1965–now	[56]
Fushun (Longfeng Coal Mine)	N.D~0.646	0.102	0.037	1.02	1907–1999	[57]
Jiutai District (Yingcheng Coal Mine)	0.007~0.522	0.048	0.040	1.97	1880–2001	This study
Rohini OCP, NK Area, India	-	0.090	0.050	3.00	2007–now	[17]
Pszczyna County, Poland	0.020~0.460	0.070	0.010	2.00	1792–now	[29]
Ostrava, Czech Republic	0.080~1.310	-	0.070	9.00	1830–now	[30]

4.1.2. Spatial Distribution Characteristics of Atmospheric Mercury Concentration

The spatial distribution of the atmospheric mercury concentration at 0 cm and 100 cm in Jiutai District is shown in Figure 4. The Pearson correlation analysis of the mercury content in the atmosphere at 0 cm and 100 cm showed that there was a positive correlation between the two (R = 0.645), and the data showed that only 38% of the sampling points at 0 cm had a higher atmospheric mercury concentration than at 100 cm. The atmospheric mercury concentration at 100 cm was generally higher than that at 0 cm, which may have been caused by secondary dust [58]. As most sampling points were located around main traffic lines with large traffic flows, the secondary dust phenomena were more pronounced, resulting in a higher atmospheric mercury content at 100 cm than at ground level. In addition, the low level of atmospheric mercury at 0 cm may have been due to the adsorption of surface atmospheric mercury by vegetation and soil [59].

The areas with a high atmospheric mercury content at 0 cm were mainly located around heating enterprises, logistics parks, driving schools, and automobile markets. The waste gas and waste residue discharged from coal burning in heating enterprises contain different concentrations of mercury, which easily enters into the environmental medium. The areas with a high atmospheric mercury content at 100 cm were mainly located near logistics parks, commercial and residential areas, and main traffic lines and stations. The atmospheric mercury concentration level at 100 cm near logistics parks, commercial districts, and main traffic arteries and stations was relatively high, which was mainly due to heavy traffic flows. Studies have shown that traffic exhaust fumes are the main source of soil mercury on both sides of urban streets [60]. The main possible causes of high atmospheric mercury concentrations in residential areas are domestic waste and the use of mercury-containing products.

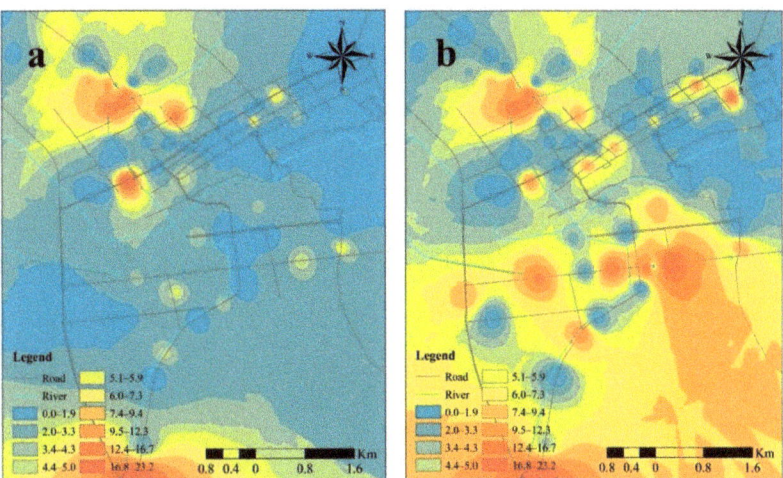

Figure 4. Distribution map of the atmospheric mercury content at: (**a**) 0 cm; (**b**) 100 cm.

4.1.3. Relationship between Soil Mercury Concentration and Atmospheric Mercury Concentration

Through mapping the atmospheric mercury concentration at 0 cm and 100 cm and the soil mercury concentration on a scatter diagram, as shown in Figure 5, we found no significant linear correlation between the mercury concentration in the atmosphere and the soil mercury content. In this study, the soil mercury levels were less affected by atmospheric deposition. This may have been due to the fact that atmospheric mercury has a stronger ability to migrate over long distances. In addition, in this study, the monitoring was conducted on cloudy days with less solar radiation and a low oxidation level of gaseous elemental mercury, thus improving the migration capacity of atmospheric mercury [61].

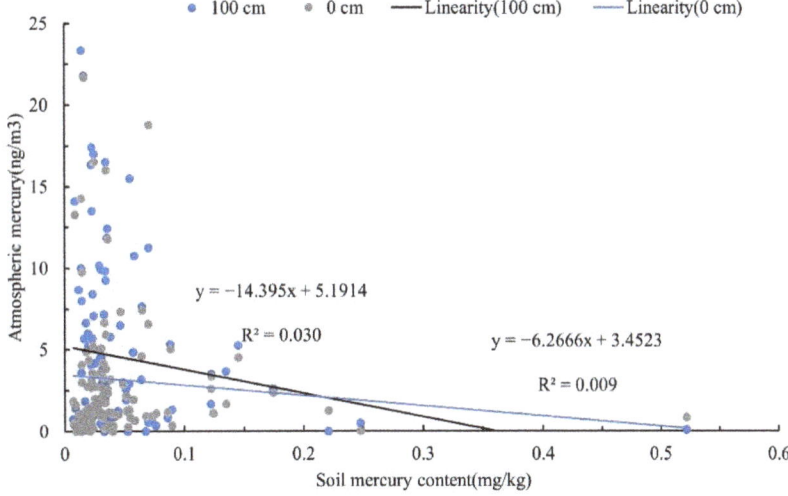

Figure 5. Scatter plot of atmospheric mercury concentration and soil mercury concentration.

4.1.4. Relationship between Soil Mercury Concentration and Wind Direction

A box diagram and variance analysis were used to analyze the influence of four wind directions (northeast, southeast, northwest, and southwest) on the soil mercury concentration in the main urban area of Jiutai District, as shown in Figure 6.

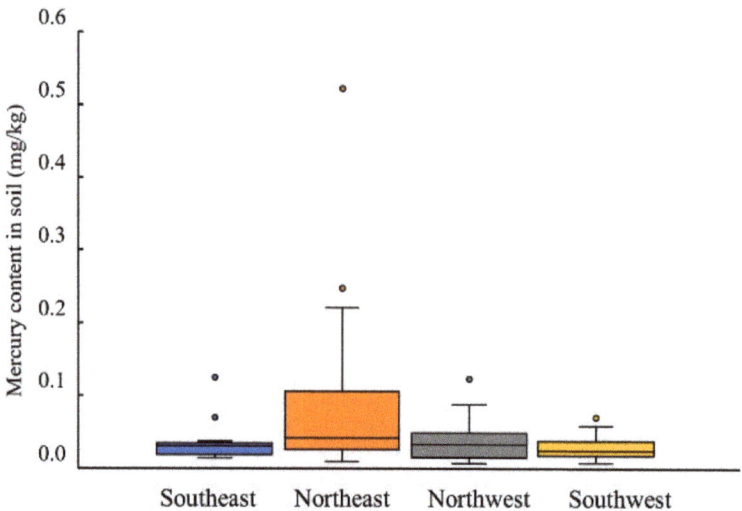

Figure 6. Box chart of the mercury content levels in the different wind directions.

As can be seen from Figure 6, there were differences in soil mercury concentration between the different wind directions, with the highest mercury concentration in the northeast direction. The results of one-way ANOVA showed that there was a significant difference in soil mercury concentration between the northeast and southwest directions ($p = 0.001 < 0.05$). The soil mercury concentration in the northeast direction was significantly higher than that in southwest direction. The results of one-way ANOVA showed that there was no significant difference in soil mercury concentration between the northwest and southeast directions ($p = 0.913 > 0.05$), while for the west and east directions we found a significant difference in mercury concentration ($p = 0.012 < 0.05$). The soil mercury concentration in the eastern direction was significantly higher than that in the western direction. We also found that there was a significant difference in mercury concentration between the south direction and the north direction ($p = 0.018 < 0.05$). The soil mercury concentration in the northern region was significantly higher than that in the southern region.

Therefore, there is a direct relationship between soil mercury pollution and wind direction in Jiutai District. The sampling time of this study ranged from the end of October to the beginning of November. According to the urban industrial layout of Jiutai District, its heating enterprises are mainly distributed in the northeast section of the main urban area, and the pollutants generated by coal-fired heating in winter are mainly concentrated in this area. At this time, the prevailing southwesterly wind in Jiutai District led to the migration of mercury-containing air pollutants from the southwest to the northeast. Atmospheric mercury enters regional soils through snowfall and rainfall. The water-soluble Hg^{2+} that had accumulated on the soil surface directly or indirectly entered the soil through leaching [62], resulting in the soil mercury concentration on the downwind surface being significantly higher than that on the upwind surface.

4.1.5. Influence of Mining Area on Mercury Content in Urban Soil

As a coal mining town, Jiutai District is relatively concentrated in coal resources. Since large-scale mining began, many polluting enterprises, such as thermal power, coal burning, and coal washing, have developed around the coal industry and are clustered and distributed around the town, affecting the mercury content and spatial distribution in urban soil to a certain extent [27]. This study mainly considered the influence of Yingcheng Coal Mine in the northeast of Jiutai District's main urban area. The results showed that the soil mercury was enriched in the coal mine and its surrounding area in the early stages of coal mining [63]. Based on the investigation of soil mercury pollution sources in Jiutai District, combined with the experimental analysis results of this study and soil mercury pollution studies in other coal mining towns, the sources and locations of soil mercury in the urban environment of a typical coal mining town were schematized, as shown in Figure 7. The piling of coal gangue, the transportation of coal, the processing of coal by-products, and the industrial activities of coal enterprises built on the edge of coal mines are the main sources of soil mercury in early-stage coal mining. Long-term coal mining led to the establishment of industrial structures in Jiutai District that are dominated by coal mining and processing and caused a spatial distribution pattern of mercury discharge enterprises such as concentrators, smelters, and coal-fired power plants interspersed among the urban areas. Urban soil mercury shows a spatial distribution characteristic of gradually decreasing pollution severity from the source center to the surrounding areas. As the coal resource was depleted, the Jiutai District's major pillar industry also gradually shifted from the coal industry to processing and manufacturing products containing mercury, waste incineration, transportation, and industrial activities, which have become the main pollution sources of mercury in the soil. The soil mercury from the early mining activities of Yingcheng Coal Mine in Jiutai District has little influence on the present urban soil environment. In conclusion, through this study, it was found that there are few areas of soil mercury enrichment in Jiutai District's main urban area, and the early coal mining activities in Yingcheng Coal Mine had little impact on the soil environment in the main urban area.

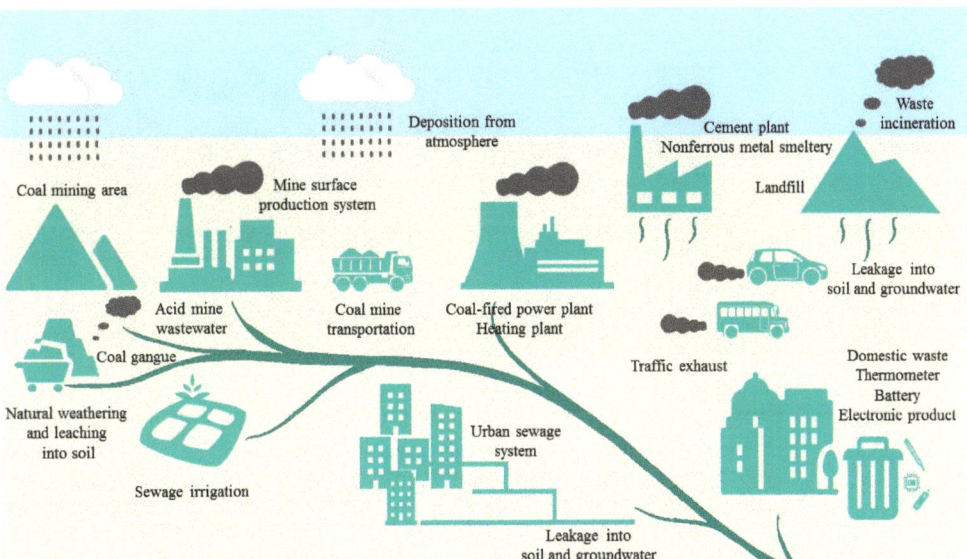

Figure 7. Sources and locations of environmental elemental mercury in urban production and life in typical mining area.

4.2. Risk Assessment of Soil Mercury Pollution in Jiutai District

4.2.1. Ecological Risk Assessments of the Soil Mercury Pollution

A descending chart of the I_{geo} is shown in Figure 8. The distribution of the soil mercury I_{geo} is shown in Figure 9.

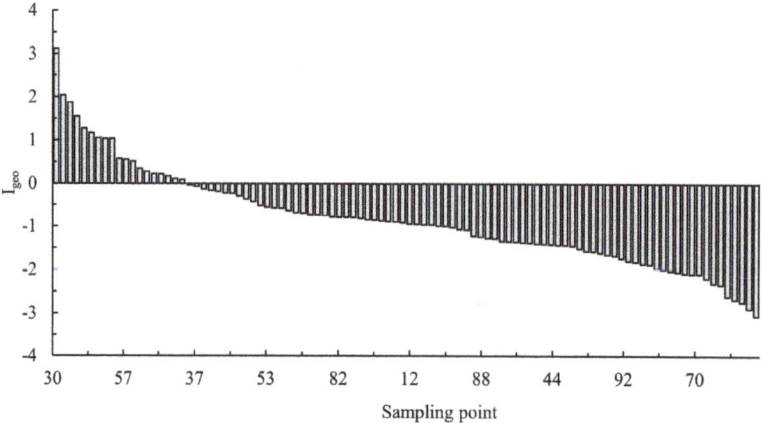

Figure 8. Descending chart of the I_{geo}.

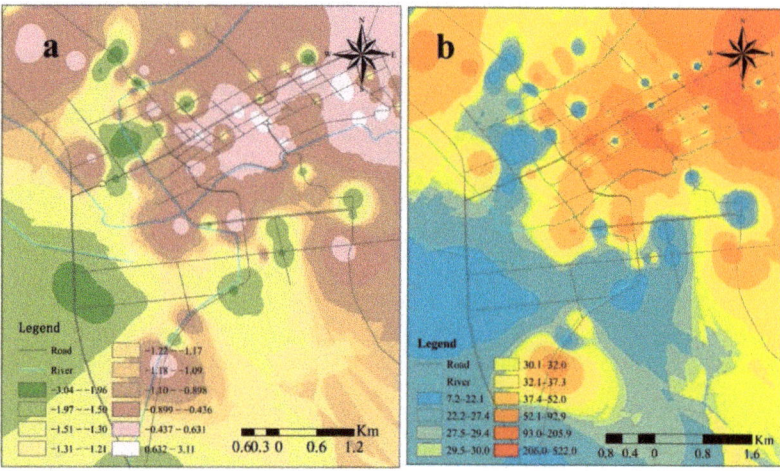

Figure 9. Characteristic distribution of soil mercury pollution: (**a**) distribution of the soil mercury I_{geo} in Jiutai District; (**b**) distribution of the soil mercury Er in Jiutai District.

Based on the results of the I_{geo}, only 19% of the sample sites in the study area were polluted (index > 0), and only one of the sample sites was heavily polluted. Eight percent of the sample sites were moderately polluted, ten percent of the sample sites were mildly polluted, and the rest of the sample sites were not polluted. In general, the evaluation of the soil mercury accumulation index in Jiutai District showed that the soil mercury pollution level was low and was mainly concentrated in certain areas. As shown in Figure 9a, the polluted areas were mainly concentrated in the northeast region of Jiutai District, which is consistent with the research results of the relationship between soil mercury concentration and wind direction. The northeast region is close to the Jiutai industrial concentration area, indicating that industrial activities caused obvious mercury pollution to the surrounding soil. Different degrees of mercury pollution also existed in the northern old city of Jiutai District, and the polluted areas were mainly concentrated in the commercial area, both sides of the main traffic lines, and around major traffic stations (such as Jiutai Railway Station and Jiutai Highway Passenger Station), indicating that the contribution of traffic to urban soil mercury pollution was significant.

The spatial distribution results of the potential ecological risk levels of soil mercury pollution in the study area are shown in Figure 9b. The Er of soil mercury ranged from 7.2 to 522.0, and its spatial distribution showed significant variation. Figure 9b shows that the northeast of the study area had the highest soil mercury overall. Based on the analysis of urban industrial distribution in Jiutai District, it can be seen that the early mining activities and current industrial activities in the northeast region are close to the industrial concentration area of Jiutai District (the former Yingcheng Coal Mining area), and these early mining activities and current industrial activities increased the soil mercury concentration. Therefore, the regional soil mercury potential ecological risk was high. Another section of the study area with a high potential risk of soil mercury was located near Jiutai Railway Station. This area is the main commercial traffic area of the city, and mercury emission from traffic has a great impact on soil mercury content. Furthermore, there is a large flow of people in this area, which poses a high potential ecological risk, so attention should be paid to timely mercury pollution control.

4.2.2. Assessment of the Potential Health Risks of Soil Mercury Contamination

In order to clarify the sources and paths of the soil mercury exposure risk to residents in coal mining towns, we drew a schematic diagram of human mercury exposure in coal mining towns by investigating the soil mercury exposure risk faced by the population in typical mining cities and combining this with the experimental results of our study, as shown in Figure 10. Generally, urban residents in typical mining areas are exposed to mercury by the inhalation of mercury-containing exhaust gases and dust from coal combustion, industrial activities, garbage incineration, and transportation; skin contact with mercury-containing soil, mercury-containing waste, and polluted water; and the consumption of vegetables, grains, meat, drinking water, etc.

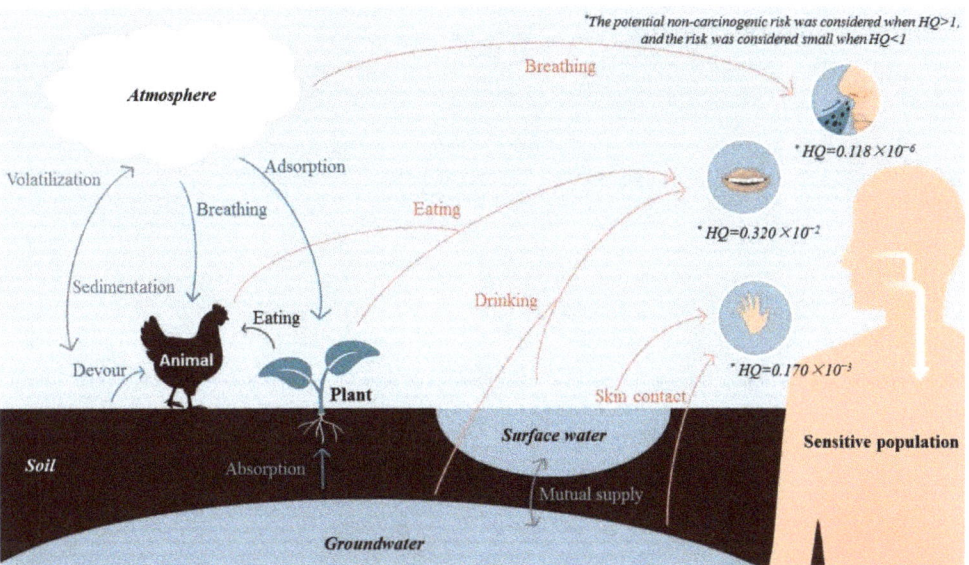

Figure 10. Schematic diagram of human mercury exposure pathways.

In this study, a radar map of the non-carcinogenic risk levels in the southeast, southwest, northeast, and northwest regions of the study area was plotted for the three exposure pathways of oral ingestion, inhalation, and dermal contact, as shown in Figure 11.

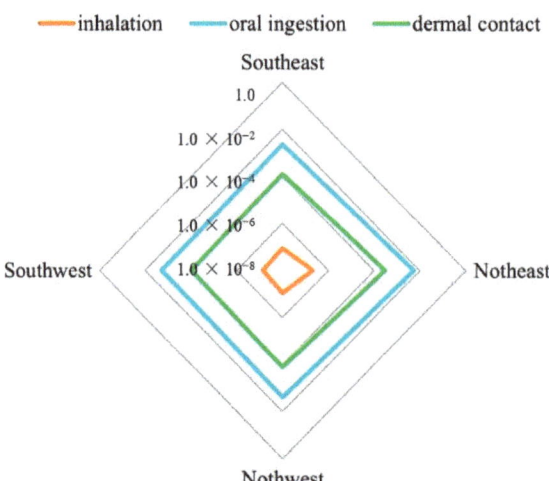

Figure 11. Radar map of soil mercury non-carcinogenic risk levels for three exposure pathways.

It can be seen from Figure 11 that the levels of non-carcinogenic risk for the three exposure pathways were ranked as follows: oral ingestion > dermal contact >> inhalation, with the oral ingestion of soil mercury being the main exposure route. Related studies have also focused on the risks to human health posed by mercury exposure through air, water, and other media, including the consumption of animal or plant products [64]; in the latter case, exposure may occur due to the intake of vegetables, while the bioaccumulation effect in animals can also lead to a large amount of soil mercury exposure through the intake of animal products. In the north of Jiutai District, there are a large number of basic cultivated land areas protected by the state. The northeast area of Jiutai District is now an industrial concentration area. Long-term coal mining and industrial activities cause different levels of farmland soil mercury pollution, which means that the output of agricultural and sideline products in the region may pose a potential environmental health risk. Many studies have shown that the health risk of mercury in farmland soil near abandoned coal mines is high. For example, mercury in the farmland soil of the Xiaoyi industrial and mining area, a typical coal mining city, exceeded the local background value by more than three times, and mercury accumulation in its agricultural products was also found [56]. Vegetables in the vegetable fields of Panji Coal Mine in Huainan were found to be contaminated to varying degrees. Although studies have shown that local soil mercury poses a low non-carcinogenic risk to the human body through vegetable intake, the soil in mining areas is often contaminated by multiple heavy metals, thus posing higher health risks to humans [65]. In general, the health risks posed by the three routes of soil mercury exposure and the total health risks were far less than 1, showing that in the study area, soil mercury pollution does not harm the health of local adults. However, further work should be carried out to improve the environment, promote the construction of an urban ecological civilization, and manage the existing soil pollution in the town.

5. Conclusions

The average soil mercury concentration in the Jiutai District urban area was 0.0484 ± 0.0633 mg·kg^{-1}, and the range was 0.00720~0.522 mg·kg^{-1}. The mercury content of 32% of the soil samples exceeded the regional soil background value of Jilin Province (0.04 mg·kg^{-1}). There was a significant positive correlation between the atmospheric mercury content at 0 cm and that at 100 cm (R = 0.645), and the atmospheric mercury concentration at 100 cm was generally higher than that at 0 cm. According to the I_{geo}, 19% of the sample sites in the study area were polluted (index > 0). In general, the soil mercury pollution level in Jiutai District was low, and the polluted areas were mainly concentrated in the northeast of Jiutai District's main urban area. The Er of the soil mercury in the study area ranged from 7.2 to 522.0, and 32% of the sampling sites presented a moderate or above potential ecological risk (Er > 40). The potential ecological risk level of the soil mercury was higher in the northeast of the study area. In terms of non-carcinogenic risk, the three main soil mercury exposure pathways were ranked as follows: oral ingestion (average 0.320×10^{-2}) > dermal contact (average 0.170×10^{-3}) >> inhalation (average 0.118×10^{-6}). The oral ingestion of soil is the main route of human mercury exposure. In addition, the health risk and total health risk values of the three types of soil mercury exposure were far less than 1, indicating that the soil mercury pollution in the study area poses no harm to the health of local adults.

Author Contributions: Conceptualization, Y.X. and G.Z.; methodology, Y.X., H.W., Z.Z. and Y.W.; software, Y.X. and H.W.; validation, Y.W.; formal analysis, Y.X. and Z.W.; data curation, J.G. and G.Z.; writing—original draft preparation, Y.X. and G.Z.; writing—review and editing, H.Y. and D.C.; visualization, H.Y.; supervision and project administration, G.Z. and D.C.; funding acquisition, G.Z. and D.C. All authors have read and agreed to the published version of the manuscript.

Funding: Dan Cui as Postdoc of Beijing Foreign Studies University, this research was funded by the Chinese National Natural Science Foundation (31230012, 31770520); the Key Social Development Project of Jilin Science and Technology Department, China (20190303068SF); and the Chinese Postdoctoral Science Foundation (2021M700496).

Institutional Review Board Statement: Not applicable.

Informed Consent Statement: Informed consent was obtained from all subjects involved in the study.

Data Availability Statement: The data presented in this study are available on request from the corresponding author.

Acknowledgments: We are grateful to the Key Laboratory of Vegetation Ecology of the Ministry of Education for their help and support.

Conflicts of Interest: The authors declare that they have no known competing financial interests or personal relationships that could have influenced the work reported in this paper.

References

1. Driscoll, C.T.; Mason, R.; Chan, H.M.; Jacob, D.J.; Pirrone, N. Mercury as a Global Pollutant: Sources, Pathways, and Effects. *Environ. Sci. Technol.* **2013**, *47*, 4967–4983. [CrossRef] [PubMed]
2. Wang, J.; Ma, L.Q.; Letcher, R.; Bradford, S.A.; Feng, X.; Rinklebe, J. Biogeochemical cycle of mercury and controlling technologies: Publications in critical reviews in environmental science & technology in the period of 2017–2021. *Crit. Rev. Environ. Sci. Technol.* **2022**, 1–6. [CrossRef]
3. Feng, X.; Chen, J.; Fu, X.; Hu, H.; Li, P.; Qiu, G.; Yan, H.; Yin, R.; Zhang, H.; Zhu, W. Progresses on Environmental Geochemistry of Mercury. *Bull. Mineral. Petrol. Geochem.* **2013**, *32*, 503–530. (In Chinese)
4. Osipova, N.A.; Tkacheva, E.V.; Arbuzov, S.I.; Yazikov, E.G.; Matveenko, I.A. Mercury in Coals and Soils from Coal-Mining Regions. *Solid Fuel Chem.* **2019**, *53*, 411–417. [CrossRef]
5. Wu, Q.; Wang, S.; Li, G.; Liang, S.; Lin, C.-J.; Wang, Y.; Cai, S.; Liu, K.; Hao, J. Temporal Trend and Spatial Distribution of Speciated Atmospheric Mercury Emissions in China During 1978–2014. *Environ. Sci. Technol.* **2016**, *50*, 13428–13435. [CrossRef] [PubMed]
6. Tian, H.Z.; Lu, L.; Hao, J.M.; Gao, J.J.; Cheng, K.; Liu, K.Y.; Qiu, P.P.; Zhu, C.Y. A Review of Key Hazardous Trace Elements in Chinese Coals: Abundance, Occurrence, Behavior during Coal Combustion and Their Environmental Impacts. *Energy Fuels* **2013**, *27*, 601–614. [CrossRef]
7. Yudovich, Y.E.; Ketris, M. Mercury in coal: A review: Part 1. Geochemistry. *Int. J. Coal Geol.* **2005**, *62*, 107–134. [CrossRef]
8. Nriagu, J.O.; Pacyna, J.M. Quantitative assessment of worldwide contamination of air, water and soils by trace metals. *Nature* **1988**, *333*, 134–139. [CrossRef]
9. Feng, X.; Hong, Y. Modes of occurrence of mercury in coals from Guizhou, People's Republic of China. *Fuel* **1999**, *78*, 1181–1188. [CrossRef]
10. Li, J.; Wang, J. Comprehensive utilization and environmental risks of coal gangue: A review. *J. Clean. Prod.* **2019**, *239*, 117946. [CrossRef]
11. Jilin Provincial Bureau of Statistics. *Jilin Statistical Yearbook*; Changchun Audio and Video Electronic Publishing House: Changchun, China, 2021.

12. Song, W.; He, T. Environmental effects of mercury in coal mine spoils in Panxian of Guizhou Province. *Chin. J. Ecol.* **2009**, *28*, 1589–1593. (In Chinese) [CrossRef]
13. Wang, X.; Tang, Y.; Ma, T.; Chen, Y. Research progress onmigration and transformation behavior of hazardous mineral matter in coals during coal combustion. *Coal Sci. Technol.* **2021**, *49*, 207–219. (In Chinese)
14. Miller, C.L.; Watson, D.B.; Lester, B.P.; Lowe, K.A.; Pierce, E.M.; Liang, L. Characterization of soils from an industrial complex contaminated with elemental mercury. *Environ. Res.* **2013**, *125*, 20–29. [CrossRef] [PubMed]
15. Ouyang, D.; Liu, K.; Wu, Q.; Wang, S.; Tang, Y.; Li, Z.; Liu, T.; Han, L.; Cui, Y.; Li, G.; et al. Effect of the Coal Preparation Process on Mercury Flows and Emissions in Coal Combustion Systems. *Environ. Sci. Technol.* **2021**, *55*, 13687–13696. [CrossRef]
16. Li, R.; Yuan, Y.; Li, C.; Sun, W.; Yang, M.; Wang, X. Environmental Health and Ecological Risk Assessment of Soil Heavy Metal Pollution in the Coastal Cities of Estuarine Bay—A Case Study of Hangzhou Bay, China. *Toxics* **2020**, *8*, 75. [CrossRef]
17. Raj, D.; Kumar, A.; Maiti, S.K. Evaluation of toxic metal(loid)s concentration in soils around an open-cast coal mine (Eastern India). *Environ. Earth Sci.* **2019**, *78*, 645. [CrossRef]
18. Zhang, L.; Wong, M.H. Environmental mercury contamination in China: Sources and impacts. *Environ. Int.* **2007**, *33*, 108–121. [CrossRef]
19. Fang, F.; Wang, Q.; Li, J. Urban environmental mercury in Changchun, a metropolitan city in Northeastern China: Source, cycle, and fate. *Sci. Total Environ.* **2004**, *330*, 159–170. [CrossRef]
20. Wang, Z.; Wang, L.; Xiao, T.; Yu, T.; Li, X.; Tang, Z.; Zhang, G. Pollution Characteristics and Risk Assessments of Mercury in the Soil of the Main Urban Regions in a Typical Chinese Industrial City: Changchun. *Pol. J. Environ. Stud.* **2021**, *30*, 3829–3841. [CrossRef]
21. Rodrigues, S.; Pereira, M.; Duarte, A.; Ajmone-Marsan, F.; Davidson, C.; Grčman, H.; Hossack, I.; Hursthouse, A.; Ljung, K.; Martini, C.; et al. Mercury in urban soils: A comparison of local spatial variability in six European cities. *Sci. Total Environ.* **2006**, *368*, 926–936. [CrossRef]
22. Fei, X.; Lou, Z.; Xiao, R.; Ren, Z.; Lv, X. Contamination assessment and source apportionment of heavy metals in agricultural soil through the synthesis of PMF and GeogDetector models. *Sci. Total Environ.* **2020**, *747*, 141293. [CrossRef] [PubMed]
23. Bi, X.; Zhang, M.; Wu, Y.; Fu, Z.; Sun, G.; Shang, L.; Li, Z.; Wang, P. Distribution patterns and sources of heavy metals in soils from an industry undeveloped city in Southern China. *Ecotoxicol. Environ. Saf.* **2020**, *205*, 111115. [CrossRef] [PubMed]
24. Khan, S.; Munir, S.; Sajjad, M.; Li, G. Urban park soil contamination by potentially harmful elements and human health risk in Peshawar City, Khyber Pakhtunkhwa, Pakistan. *J. Geochem. Explor.* **2016**, *165*, 102–110. [CrossRef]
25. Liu, W.; Li, Y.; Li, J.; Tang, Z. Distribution and temporal trend of heavy metals in soils from typical areas in a coal-mining city. *Environ. Pollut. Control.* **2021**, *43*, 984–989. (In Chinese)
26. Wu, X.; Huang, X.; Quan, W.; Hu, J.; Qin, F.; Tang, F. Chemical Forms and Risk Assessment of Heavy Metals in Soils and Selected Hypertolerant Plants around a Coal Mining Area in Western Guizhou Province. *Bull. Soil Water Conserv.* **2018**, *38*, 313–321. (In Chinese)
27. Liang, J.; Feng, C.; Zeng, G.; Gao, X.; Zhong, M.; Li, X.; Li, X.; He, X.; Fang, Y. Spatial distribution and source identification of heavy metals in surface soils in a typical coal mine city, Lianyuan, China. *Environ. Pollut.* **2017**, *225*, 681–690. [CrossRef]
28. Pan, L.; Ma, J.; Hu, Y.; Su, B.; Fang, G.; Wang, Y.; Wang, Z.; Wang, L.; Xiang, B. Assessments of levels, potential ecological risk, and human health risk of heavy metals in the soils from a typical county in Shanxi Province, China. *Environ. Sci. Pollut. Res.* **2016**, *23*, 19330–19340. [CrossRef]
29. Loska, K.; Wiechuła, D.; Korus, I. Metal contamination of farming soils affected by industry. *Environ. Int.* **2004**, *30*, 159–165. [CrossRef]
30. Doležalová Weissmannová, H.; Mihočová, S.; Chovanec, P.; Pavlovský, J. Potential ecological risk and human health risk assessment of heavy metal pollution in industrial affected soils by coal mining and metallurgy in Ostrava, Czech Republic. *Int. J. Environ. Res. Public Health* **2019**, *16*, 4495. [CrossRef]
31. Yin, W.; Lu, Y.; Li, J.-H.; Chen, C.-X.; Zhang, C.; Dong, F. Distribution Characteristics and Pollution Assessment of Mercury in Urban Soils of Guangzhou. *J. Soil Sci.* **2009**, *40*, 1185–1188. (In Chinese)
32. Yao, Y.; Fang, F.; Wu, J.; Zhu, Z.; Lin, Y.; Zhang, D.; Zhu, H. Concentrations and pollution assessment of mercury in surfaces dust at urban streets and elementary school campuses of Huainan City. *Environ. Sci. Technol.* **2017**, *37*, 844–852. (In Chinese)
33. Christophoridis, C.; Evgenakis, E.; Bourliva, A.; Papadopoulou, L.; Fytianos, K. Concentration, fractionation, and ecological risk assessment of heavy metals and phosphorus in surface sediments from lakes in N. Greece. *Environ. Geochem. Heal.* **2020**, *42*, 2747–2769. [CrossRef] [PubMed]
34. Liu, K.; Li, C.; Tang, S.; Shang, G.; Yu, F.; Li, Y. Heavy metal concentration, potential ecological risk assessment and enzyme activity in soils affected by a lead-zinc tailing spill in Guangxi, China. *Chemosphere* **2020**, *251*, 126415. [CrossRef] [PubMed]
35. *OSWER 9355*; 4-24 Supplemental Guidance for Developing Soil Screening Levels for Superfund Sites. Environmental Protection Agency: Washington, DC, USA, 2001.
36. US EPA. *Risk-Based Concentration Table*; Environmental Protection Agency: Philadelphia, PA, USA, 2000.
37. Xu, J.; Zhang, B.; Wang, Z.; Guo, Y.; Song, K. Chromium Accumulation in Soils for Different Land Uses and Spatial Distribution of Chromium in Jiutai City of Jilin Province. *J. Soil Water Conserv.* **2006**, *20*, 36–39. (In Chinese)
38. Xu, Z.; Liu, W.; Ding, R.; Wang, J. Perlite in Yingcheng Coal Mine, Jiutai: Characteristics, Formation and Reservoir Significance, Journal of Jilin University. *Earth Sci. Ed.* **2007**, *37*, 1139–1145. (In Chinese)

39. Zhu, J.; Li, G.; You, B. Application of overlying strata damage theory in evaluation on foundation stability: Taking Yingcheng coal mine in Jiutai of Jilin as an example. *World Geol.* **2012**, *31*, 584–588. (In Chinese)
40. Changchun Jiutai District Chorography Compilation Committee. *Jiutai Yearbook*; Fang Zhi Publishing House: Changchun, China, 2021.
41. Sabouhi, M.; Ali-Taleshi, M.S.; Bourliva, A.; Nejadkoorki, F.; Squizzato, S. Insights into the anthropogenic load and occupational health risk of heavy metals in floor dust of selected workplaces in an industrial city of Iran. *Sci. Total Environ.* **2020**, *744*, 140762. [CrossRef]
42. Meng, X.X.; Li, S.Z. *Background Value of Soil Elements in Jilin Province*; Science Press: Beijing, China, 1995; p. 64. (In Chinese)
43. Maanan, M.; Saddik, M.; Maanan, M.; Chaibi, M.; Assobhei, O.; Zourarah, B. Environmental and ecological risk assessment of heavy metals in sediments of Nador lagoon, Morocco. *Ecol. Indic.* **2015**, *48*, 616–626. [CrossRef]
44. Xu, Z.; Ni, S.; Tuo, X.; Zhang, C. Calculation of Heavy Metals' Toxicity Coefficient in the Evaluation of Potential Ecological Risk Index. *Environ. Sci. Technol.* **2008**, *31*, 112–115. (In Chinese)
45. Means, B. *Risk-Assessment Guidance for Superfund Volume 1: Human Health Evaluation Manual (Part A)*; EPA-540/1-89/002; Environmental Protection Agency: Washington, DC, USA, 1989.
46. Ying, L.; Shaogang, L.; Xiaoyang, C. Assessment of heavy metal pollution and human health risk in urban soils of a coal mining city in East China. *Hum. Ecol. Risk Assess. Int. J.* **2016**, *22*, 1359–1374. [CrossRef]
47. Du, Y.; Gao, B.; Zhou, H.; Ju, X.; Hao, H.; Yin, S. Health Risk Assessment of Heavy Metals in Road Dusts in Urban Parks of Beijing, China. *Procedia Environ. Sci.* **2013**, *18*, 299–309. [CrossRef]
48. Vilavert, L.; Nadal, M.; Schuhmacher, M.; Domingo, J.L. Concentrations of Metals in Soils in the Neighborhood of a Hazardous Waste Incinerator: Assessment of the Temporal Trends. *Biol. Trace Element Res.* **2012**, *149*, 435–442. [CrossRef] [PubMed]
49. Hassett-Sipple, B.; Swartout, J.; Mahaffey, K.R.; Rice, G.E.; Schoeny, R. *Mercury Study Report to Congress: Health Effects of Mercury and Mercury Com-Pounds*; US EPA EPA-452/R-97-007; Environmental Protection Agency: Washington, DC, USA, 1997.
50. US EPA. *Supplemental Guidance for Developing Soil Screeninglevels for Superfund Sites*; Olfflice of Soild Waste and Emergency Response: Washington, DC, USA, 2001; pp. 4–24.
51. Wang, S.; Luo, K.; Wang, X.; Sun, Y. Estimate of sulfur, arsenic, mercury, fluorine emissions due to spontaneous combustion of coal gangue: An important part of Chinese emission inventories. *Environ. Pollut.* **2016**, *209*, 107–113. [CrossRef] [PubMed]
52. Liang, Y.; Liang, H.; Zhu, S. Mercury emission from spontaneously ignited coal gangue hill in Wuda coalfield, Inner Mongolia, China. *Fuel* **2016**, *182*, 525–530. [CrossRef]
53. Wei, Q.; Chen, W.; Jin, L. Health risk assessment and source analysis of heavy metal elements in PM2.5 in Zaozhuang city. *China Powder Sci. Technol.* **2020**, *26*, 69–78. (In Chinese)
54. Yang, Y.; Liu, S.; Yang, Y.; Li, L.; Liu, S.; Kang, Y.; Fei, X.; Gao, Y.; Gao, B. Heavy metals in peri-urban soil of Huangshi: Their distribution, risk assessment and source identification. *Geophys. Geochem. Explor.* **2021**, *45*, 1147–1156. (In Chinese)
55. Zheng, L.; Liu, G.; Wang, L.; Chou, C.-L. Composition and quality of coals in the Huaibei Coalfield, Anhui, China. *J. Geochem. Explor.* **2008**, *97*, 59–68. [CrossRef]
56. Tian, L. The status and evaluation of environment in the Xiaoyi city of Shanxi province. *Shanxi Archit.* **2011**, *37*, 179–180. (In Chinese)
57. Xie, Y.; Tang, W. Investigation & Assessment for Heavy metal in Soil of Fushun Typical Areas. *Environ. Sci. Manag.* **2009**, *34*, 38–40+78. (In Chinese)
58. Kim, K.-H.; Kim, M.-Y. The effects of anthropogenic sources on temporal distribution characteristics of total gaseous mercury in Korea. *Atmos. Environ.* **2000**, *34*, 3337–3347. [CrossRef]
59. Leonard, T.L.; Taylor, G.E., Jr.; Gustin, M.S.; Fernandez, G.C. Mercury and plants in contaminated soils: 1. Uptake, partitioning, and emission to the atmosphere. *Environ. Toxicol. Chem.* **1998**, *17*, 2063–2071. [CrossRef]
60. Lei, L.; Yu, D.; Chen, Y.; Song, W.; Liang, D.; Wang, Z. Spatial distribution and sources of heavy metals in soils of Jinghui Irrigated Area of Shaanxi, China. *Trans. Chin. Soc. Agric. Eng.* **2014**, *30*, 88–96. (In Chinese)
61. Lin, C.-J.; Pehkonen, S. The chemistry of atmospheric mercury: A review. *Atmos. Environ.* **1999**, *33*, 2067–2079. [CrossRef]
62. Manta, D.S.; Angelone, M.; Bellanca, A.; Neri, R.; Sprovieri, M. Heavy metals in urban soils: A case study from the city of Palermo (Sicily), Italy. *Sci. Total Environ.* **2002**, *300*, 229–243. [CrossRef] [PubMed]
63. Li, K.; Gu, Y.; Li, M.; Zhao, L.; Ding, J.; Lun, Z.; Tian, W. Spatial analysis, source identification and risk assessment of heavy metals in a coal mining area in Henan, Central China. *Int. Biodeterior. Biodegrad.* **2018**, *128*, 148–154. [CrossRef]
64. Xu, Y.; Lu, F.; Dong, L.; Yang, J.; Nai, C.; Liu, Y. Human health effects of heavy metals around an incineration plant through multi-media exposure. *Acta Sci. Circumst.* **2016**, *36*, 3464–3471. (In Chinese)
65. Liu, X.; Zheng, L.; Chen, X.; Yang, T.; Chen, Y.; Cheng, H. Study on the heavy metal pollution characteristics of agricultural soiland their accumulation characteristics in wheat in Panji mining area, Huainan. *Environ. Pollut. Prevention.* **2019**, *41*, 959–964. (In Chinese)

Article

Sequestration of Lead Ion in Aqueous Solution onto Chemically Pretreated *Pycnanthus angolensis* Seed Husk: Implications for Wastewater Treatment

Arinze Longinus Ezugwu [1,*], Hillary Onyeka Abugu [1], Ifeanyi Adolphus Ucheana [2], Samson Ifeanyi Eze [1], Johnbosco C. Egbueri [3], Victor Sunday Aigbodion [4] and Kovo Godfrey Akpomie [1,5]

[1] Department of Pure and Industrial Chemistry, University of Nigeria, Nsukka 410105, Enugu State, Nigeria; hillary.abugu@unn.edu.ng (H.O.A.); eze.samson@unn.edu.ng (S.I.E.); kovo.akpomie@unn.edu.ng (K.G.A.)
[2] Central Science Laboratory, University of Nigeria, Nsukka 410105, Enugu State, Nigeria; ifeanyi.ucheana@unn.edu.ng
[3] Department of Geology, Chukwuemeka Odumegwu Ojukwu University, Awka 432107, Uli Anambra State, Nigeria; johnboscoegbueri@gmail.com
[4] Faculty of Engineering and Built Environment, University of Johannesburg, Johannesburg 2092, South Africa; victor.aigbodion@unn.edu.ng
[5] Department of Chemistry, University of Free State, Bloemfontein 9301, South Africa
* Correspondence: arinze.ezugwu.192194@unn.edu.ng

Citation: Ezugwu, A.L.; Abugu, H.O.; Ucheana, I.A.; Eze, S.I.; Egbueri, J.C.; Aigbodion, V.S.; Akpomie, K.G. Sequestration of Lead Ion in Aqueous Solution onto Chemically Pretreated *Pycnanthus angolensis* Seed Husk: Implications for Wastewater Treatment. *Sustainability* **2023**, *15*, 15446. https://doi.org/10.3390/su152115446

Academic Editor: Said Muhammad

Received: 26 September 2023
Revised: 23 October 2023
Accepted: 25 October 2023
Published: 30 October 2023

Copyright: © 2023 by the authors. Licensee MDPI, Basel, Switzerland. This article is an open access article distributed under the terms and conditions of the Creative Commons Attribution (CC BY) license (https://creativecommons.org/licenses/by/4.0/).

Abstract: This novel study investigated and proposes the use of *Pycnanthus angolensis* seed husk for the sequestration of Pb(II) from contaminated solutions, with the aim of contributing to the urgent need for accessibility to quality water, sustainable management of water and the environment in line with the Sustainable Development Goals (SDGs). The activated *Pycnanthus angolensis* seed husk was developed by modifying the pure sample (P-PA) with ethylene-glycol (E-PA) and Iso-butanol (I-PA). Infrared spectroscopy (FTIR), scanning electron microscopy (SEM), the Brunauer-Emmett-Teller (BET) analyzer, thermogravimetric analyzer (TGA), and X-ray diffractometer (XRD) were used to characterize the adsorbents before and after adsorption. The batch adsorption studies carried out revealed the highest adsorption of Pb(II) at pH 6 and 180 min for all the adsorbents. The functional groups, as well as the shifts in peaks after modification, were confirmed using FTIR analysis. In addition, SEM images show a heterogeneous, rough surface with sufficient cavities of the adsorbent after modification. The physiochemical characteristics indicated that BET pore volume and pore diameter increased for E-PA and I-PA compared to P-PA. The experimental data obtained indicated that Langmuir and pseudo-first-order (PFO) best described the isotherm and kinetic models, respectively. The adsorption mechanism revealed that the adsorption of Pb(II) was controlled mainly by pore filling, while electrostatic interaction, surface complexation, and ionic exchange also occurred minimally. The thermodynamic parameters, $\Delta H°$ and $\Delta G°$, suggest an endothermic and spontaneous adsorption process, respectively. The findings in this study indicate that *Pycnanthus angolensis* seed husks offer cost-effective and sustainable solutions that are readily accessible for wastewater treatment.

Keywords: biosorbent; adsorption mechanism; lead adsorption; error function; thermodynamics; *Pycnanthus angolensis*; chemical-treated biomass

1. Introduction

Recent reports have shown that heavy metals, even at a minimal concentration, could be detrimental to life forms [1–3]. Due to their highly toxic nature and their inability to undergo degradation, they have become a global concern and a major target for most scientists [4–6]. These metals have also been found in living cells [1]. Within living organisms, they tend to accumulate, binding to nucleic acids and proteins, thereby disrupting essential cellular functions and causing severe health consequences [1,7,8].

Lead (Pb) contamination has been listed as a major global concern due to its toxicity and negative impact on the human system [9–11]. It enters the environment primarily via the metal processing industries, chemical industries, mining, paint, and battery production processes [12–14]. Although there is a minimal level of Pb(II) in the surface water, when water is used without being purified, it accumulates in the environment [15]. More so, when substances contaminated with Pb(II) enter the food chain and are consumed at concentrations above the acceptable limit by plants or animals, it is detrimental to their health [16–18].

Among the different technologies used for wastewater treatment, adsorption has proven to be a reliable and environmentally friendly method [19–21]. Recent research has explored the use of agricultural and biological waste materials as adsorbents for removing metal ions from contaminated solutions. Notable examples include black cumin seed [22], *Terminalia mantaly* [23], fruit peels from banana, granadilla, and orange [24], peanut shells and compost [25], Ajwa Date Pits [26], and endocarp waste of Gayo Coffee [27]. These materials offer cost-effective and sustainable solutions that are readily accessible in large quantities [28,29]. Other reported adsorbents employed for the adsorption of contaminants are *Gigantochloa* bamboo [30], *Nepenthes rafflesiana* pitcher (NP), *Nepenthes rafflesiana* leaves [31], and modified waste shrimp shell (MSS) [32].

Previous reports showed that different functional groups on biomaterial surfaces are the driving factor in the removal of contaminants [33]. To improve its adsorption efficiency, researchers have used different modification methods (thermal, acid, base, and organic reagents) to address it. These modifications result in an increase in the biosorbents' pore cavities and enhance the number of oxygen-containing functional groups on the surface of the material, thereby improving its adsorption efficiency [34]. While many chemicals have been used for the modification of adsorbents [35–37], research is still scanty on the use of organic reagents to increase adsorption efficiency.

The *Pycnanthus angolensis* seed husk is widely called African nutmeg because of its close resemblance to nutmeg species seeds. It comes from a tropical plant in the *Myristicaceae* family and has been used to treat a wide range of health problems in Africa [38]. The fruits, which possess an oblong shape, consist of seeds abundant in oil content that are enclosed within a rigid outer shell [39]. The tree exhibits a substantial size, characterized by a diminutive crown consisting of branches positioned perpendicularly to the trunk [40]. In Nigeria, it is used in the treatment of chest pain, malaria, and bacterial infections [41]. This is because the seed contains bioactive compounds like quinone-terpenoids, lignans, isoflavonoids, kombic acid, tocochromanols, vitamins, and other compounds found in essential oils [38]. The *Pycnanthus angolensis* seed husk contains predominantly fat, carbohydrate, protein, and lignocellulosic materials (cellulose, hemicellose, and lignin), which are suitable materials for adsorption [29,31]. The structure of lignocellulose, tocochotrienol derivatives, and myristoleic acid contains sufficient (OH), (CO), (COOH), and (C=C) functional groups that could necessitate an effective removal process, thus necessitating this investigation for its potential as an adsorbent [16,42]. However, while this seed has been employed for medical applications [42] and for anti-malaria analysis [39], there has been no reported work on the use of *Pycnanthus angolensis* seed husk for the removal of contaminants in aqueous solutions.

Therefore, the goal of this work is to use *Pycnanthus angolensis* seed husk as an adsorbent to remove Pb(II) from aqueous solutions, activate some of the adsorbents with ethylene glycol and iso-butanol to improve its adsorption capacity and compare the adsorbents. The research also targets to characterize the prepared adsorbents using infrared spectroscopy (FTIR), scanning electron microscopy (SEM), the Brunauer-Emmett-Teller (BET) analyzer, thermogravimetric analyzer (TGA), and X-ray diffractometer (XRD). Finally, the research hopes to investigate the potential of the adsorbents for the sequestration of Pb(II) from wastewater. Lead contamination in aqueous environments poses a significant threat to both human health and ecosystems, demanding immediate attention. Traditional wastewater treatment methods often rely on non-renewable resources and can be energy-intensive, presenting sustainability challenges. Thus, the justification for this research lies at the intersection of pressing environmental concerns [43,44], the need for sustainable solutions [45], and the promise of innovative and eco-friendly materials. With these critical issues in view, this study proposes the sequestration of lead ions using innovative and eco-friendly chemically modified *Pycnanthus angolensis* seed husk—a novel approach with multifaceted sustainability implications. The novelty of this present research centers around the application of this natural, renewable, and biodegradable material, the *Pycnanthus angolensis* seed husk, as an efficient adsorbent for lead ions.

2. Materials and Method

2.1. Adsorbent Preparation

The appropriate amount of sample required for the analysis was collected in line with the stipulated guidelines set by the Convention on International Trade in Endangered Species of Wild Fauna and Flora and was identified in the Plant Science and Biotechnology Department, University of Nigeria, Nsukka.

The seed husk was washed with clean water, sun-dried for 72 h, and subsequently pulverized to achieve a very fine particle size. The dried residue was sieved to get a homogenous size (diameter less than 1.0 to 250 μm). 100 g of the powdered seed husk was treated with 250 mL of iso-butanol for 48 h to get the iso-butanol-modified *Pycnanthus angolensis* (I-PA). 250 mL of ethylene glycol was also used to treat another 100 g of the powdered seed husk for 48 h, and the filtrate was labeled (E-PA) while another sample of powered, unmodified seed husk was labeled (P-PA). The samples were washed with de-ionized water, and the filtrate was tested several times until it became neutral. The residues were dried in an oven at 105 °C, allowed to cool, and stored in an airtight container. The dried samples were characterized to ascertain the impacts of organic solvent on the *Pycnanthus angolensis* seed husk on adsorption. A graphical illustration of this procedure is presented in Figure 1.

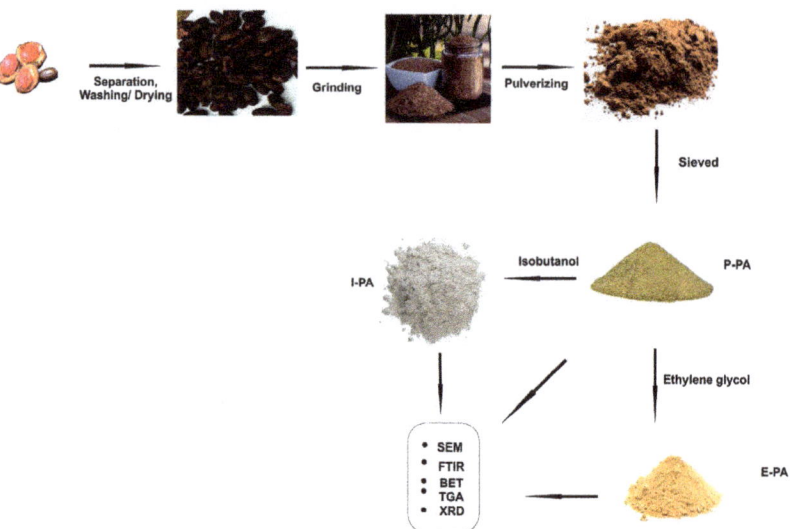

Figure 1. Flow chart of the adsorbent preparation procedure.

2.2. Characterization of Adsorbents

As shown in Figure 2, the samples were characterized using Fourier transform infrared spectroscopy (FTIR), scanning electron microscopy (SEM), the Brunauer-Emmett-Teller (BET) analyzer, the thermogravimetric analyzer (TGA), and an X-ray diffractometer (XRD). Fourier-transformed infrared spectroscopy (Perkin Elmer Spectrum 65 FT-IR spectrophotometer, Rodgau, Germany) was used to determine the functional groups present in the biosorbents, and the spectra were recorded in the range of 500–4000 cm^{-1}. The surface structure of the sorbent under different magnifications was obtained using scanning electron microscopy (SEM). The surface area, pore diameter, and pore sizes were investigated using the Brunauer-Emmett-Teller (BET) analyzer (JW-DA: 76502057en, Beijing, China). The Rigaku D/Max-llC X-ray diffractometer (Applied Rigaku Technologies, Cedar Park, TX, USA) was used to measure the XRD spectra of the modified adsorbents to determine their crystallinity, and the TGA was used to measure the thermal stability of the adsorbents.

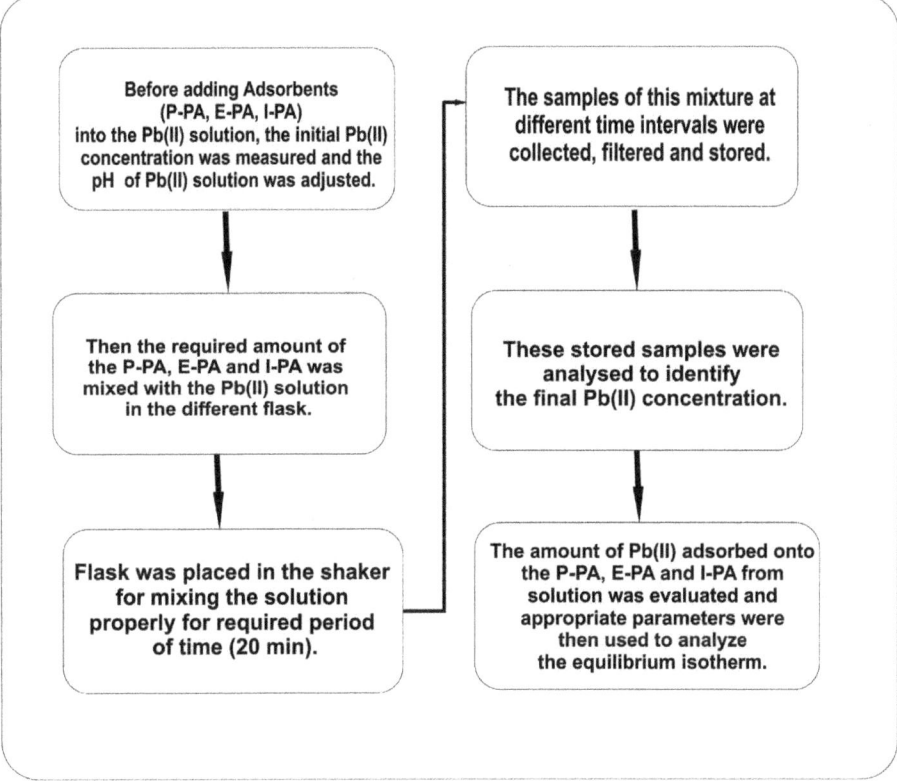

Figure 2. Experimental batch adsorption process.

2.3. Adsorbate Preparation and Models

2.3.1. Preparation and Dilution of Pb(NO$_3$)$_2$ Stock Solutions

A stock solution of Pb with a concentration of 1000 mg/L was made by dissolving 1.598 g of Pb(NO$_3$)$_2$ salt in a small amount of de-ionized water in a beaker and stirring it with a glass rod until the salt was completely dissociated. It was appropriately transferred into a 1 L volumetric flask and made up to the mark with de-ionized water. Several concentrations of synthetic wastewater containing lead (II) ions were later prepared using serial dilution.

2.3.2. Batch Adsorption Studies

In this work, the influence of the reaction conditions on the adsorption capacity of the prepared adsorbents for the uptake of lead was ascertained using the batch technique. 10 mL of the prepared adsorbate at different known concentrations were charged individually into five different beakers. I g of the adsorbents were accurately weighed and put into each of the 250 mL beakers containing the adsorbate at a constant pH of 5 and room temperature. The beakers were put in a magnetic flask shaker at 100 rpm for 20 min to ensure adsorption efficiency was attained. After equilibrium was attained, these mixtures were filtered at various time intervals, and the filtrate was stored. The final concentration of Pb(II) was analyzed with an atomic absorption spectrometer. The quantity of Pb(II) ion uptake onto the P-PA, E-PA, and I-PA from mixtures was estimated, and the appropriate variables were employed to determine the isotherm and kinetic models. Figure 2 shows the breakdown of the experimental batch adsorption process.

2.3.3. Isotherm Models

Five isotherm models were used to evaluate the adsorption efficiency of the modified sorbents with the target metal ions [46]. This study was done using the linearized forms of the various models, such as Langmuir, Freundlich, and Temkin, the Dubinin–Radushkevich models, and Flory–Huggins models [47]. However, Langmuir and Freundlich models are mostly used for the adsorption of contaminants from aqueous solutions [48]. The summary of the parameters used for the isotherm and kinetic plots is presented in Table 1.

Table 1. Summary of parameters used for the Isotherm and Kinetic model linear plots.

Model	Plot	Equations
Langmuir	$\frac{C_e}{q_e}$ vs. C_e	(1)
Freundlich	$\ln q_e$ vs. $\ln C_e$	(2)
Tempkin	q_e vs. $\ln C_e$	(3)
	$B = \frac{RT}{b_T}$	(4)
D-R	$\ln q_e$ vs. ε^2	(5)
Florry Huggins	$\log(1-\theta)$ vs. $\log\left(\frac{\theta}{C_o}\right)$	(6)
PFO	$\log(q_{e,cal} - q_t)$ vs. t	(7)
PSO	$\frac{t}{q_t}$ vs. t	(8)
ID	q_t vs. $t^{0.5}$	(9)

2.3.4. Langmuir Isotherm Model

Langmuir predicts the monolayer adsorption mechanism of metal ions onto the sorbent surfaces [25]. The linearized equation of the Langmuir model (Equation (1)) was employed to evaluate the isotherm variables.

$$\frac{C_e}{q_e} = \frac{1}{q_m K_L} + \frac{1}{q_m} C_e \qquad (1)$$

q_e (mg/g) signifies maximum adsorption, C_e (mg/L) denotes the optimum concentration of the metal ion, K_L (L/mg) denotes constant, and q_m (mg/g) is the theoretical optimum adsorption capacity.

2.3.5. Freundlich Model

The Freundlich isotherm (Equation (2)) describes multilayer adsorption mechanisms in which the metal ion interacts with the heterogeneous surface, giving divergent adsorption energy on the entire surface [49].

$$q_e = \frac{1}{n_F} \ln C_e + \ln K_F \qquad (2)$$

where q_e = equilibrium adsorption of metal ions onto the functionalized adsorbent, K_F = Freundlich constant, and n_F = intensity of the adsorbents.

2.3.6. Temkin Isotherm Model

Temkin isotherm model (Equation (3)) suggests that the adsorbate/adsorbent interaction indirectly affects the adsorption process [50]. This model only accounts for the irregular range of metal ion concentrations [51]. The linearized expressions used in this study are shown in Equations (3) and (4).

$$q_e = B \ln A_T + B \ln C_e \tag{3}$$

where

$$B = \frac{RT}{b_T} \tag{4}$$

where T equals temperature (K), R = universal gas constant, b_T is the constant connected to adsorption heat (KJ/mol), and A = constant (L/g).

2.3.7. Dubinin-Radushkevich Isotherm Model (D-R)

The D-R model describes the relationship between the adsorption curve and the surface area of the biosorbents [52]. Additionally, this isotherm model predicts the multilayer interaction using Van Der Waal's force, which is used in the physical adsorption procedure [53]. The Dubinin-R linear equation is generally represented as follows:

$$\ln q_e = \ln q_m - B\varepsilon^2 \tag{5}$$

where B denotes constant, while q_m (mg/g) signifies theoretical equilibrium.

2.3.8. Flory-Huggins Isotherm

Flory-Huggins model gives insight into the extent of surface coverage of the sorbate interactions on the sorbent [54]. The linearized equation used in this study is shown in Equation (6).

$$\ln\left(\frac{\theta}{C_o}\right) = \ln K_{FH} + n \ln(1 - \theta) \tag{6}$$

where θ = degree of surface coverage, n = number of sorbates entrapped in the adsorption areas, and K_{FH} are Flory–Huggin's constants (Lmol^{-1}).

2.4. Kinetics Study

In this study, pseudo-first-order (PFO), pseudo-second-order (PSO), and intra-particle diffusion models (ID) were employed to investigate the adsorption rates. The linearized expression of the PFO, PSO, and intra-particle diffusion model equations are shown in Equations (7)–(9), respectively [30,31,55,56].

$$\log(q_{e,cal} - q_t) = \log q_{e,cal} - k_1 t \tag{7}$$

$$\frac{t}{q_t} = \frac{1}{k_2 q_{e,cal}} + \frac{t}{q_{e,cal}} \tag{8}$$

$$q_t = k_3 t^{0.5} + C \tag{9}$$

where q_t (mg/g) denotes adsorption efficiency in a specific time t (min), K_1 (min^{-1}), K_2 (mg/g min), and K_3 (g/mg min$^{-1/2}$) describe the rate constants of the PFO, PSO, and ID models, respectively, and C = intercept.

2.5. Thermodynamic Equilibrium

Thermodynamic equilibrium was estimated to study the process of metal ion uptake, its feasibility, and spontaneity. The following equations were used in the study [57].

$$\Delta G° = -RT \ln K_C \tag{10}$$

$$\ln K_C = -\left(\frac{\Delta H°}{R}\right) + \left(\frac{\Delta S°}{R}\right) \tag{11}$$

where the values for K_C were estimated using Equation (12)

$$K_C = \frac{q_e}{C_e} \tag{12}$$

where $\Delta G°$ = change in standard Gibbs free energy, R is a universal gas constant 8.314 Jmol^{-1} K^{-1}, and T = Temperature, $\Delta H°$, $\Delta G°$, and $\Delta S°$ denotes change in enthalpy, free energy, entropy, respectively, and $_{FH}$ = adsorption equilibrium constant.

2.6. Removal Efficiency and Uptake Capacity

The percentage of the metal ions removed and the adsorption capacity of the adsorbents for Pb(II) were estimated using Equations (13) and (14), respectively.

$$\text{Percentage removal efficiency} = \frac{100(C_o - C_e)}{C_o} \tag{13}$$

$$q_e = \frac{(C_o - C_e)V}{m} \tag{14}$$

where C_o = the initial metal ion (mg/L), C_e (mg/L) = equilibrium concentration, q_e (mg/g) = maximum adsorption capacity, V (L) denotes volume, and m (g) = adsorbent mass.

Error Functions

To assess how well the kinetic models fit the experimental data, two error functions were examined. The Hybrid Fractional Error Function (HYBRID) was used to improve the fit of the square of errors function at low concentration levels. The Marquardt's Percent Standard Deviation (MPSD) error function is similar to a geometric mean error distribution that has been adjusted based on the system's degree of freedom [58]. The HYBRID is presented in Equation (15)

$$\text{HYBRID} = \frac{100}{n-p} \sum \left[\frac{\left(q_{e,exp} - q_{e,cal}\right)}{q_{e,exp}}\right]^2 \tag{15}$$

MPSD error function is presented in Equation (16).

$$\text{MPSD} = 100 \sqrt{\frac{1}{(n-p)} \sum \left(\frac{q_{e,exp} - q_{e,cal}}{q_{e,exp.}}\right)^2} \tag{16}$$

where $q_{e,exp}$ = experimental adsorption capacity, $q_{e,cal}$ = calculated adsorption capacity, n = number of experimental data points, and p = the number of each parameter.

3. Results and Discussion

3.1. Characterization of Adsorbents

3.1.1. Fourier Transform Infra-Red Spectroscopy (FTIR)

Figure 3A–F illustrate the FTIR spectra of the biosorbents and the functional groups present in the adsorbent before and after the adsorption process. As presented in Figure 3A, the broad peak between 1250–750 cm^{-1} is attributed to the C–O bending vibration of carboxylic groups, alcohol, ketones, and aldehydes [59]. The bending peak at 1543 and 1464 cm^{-1} was due to the C–O bending vibration of carboxylate (–COO–) and the N–H of the amide 1 group [60]. The band between the range of 1700–1733 and 1650 cm^{-1} suggests conjugated carbonyl (C=O) and (C=C). 2063 cm^{-1} denotes carbon triple bond carbon or

CN). The peaks between 2923 and 2852 cm^{-1} range correspond to the stretching vibration of the C–H aliphatic. The broad peak at 3433 cm^{-1} correlates with the stretching vibration of the OH$^-$ and N–H groups of the cellulose, hemicellulose, and lignin materials [61]. The oxygen-rich functional groups (–OH), (–C=O), and (–COOH) that account for the binding of the metal ion with the adsorbent are rich in both cellulosic and hemicellulose materials [62].

On activation, there was a sudden disappearance of 1650 cm^{-1} attributed to C=C and the C-O bending vibration of carboxylate (–COO–)in the ethylene glycol modified (E-PA) adsorbent (Figure 3B), as well as observable shifts in the wave numbers of the activated adsorbent before and after adsorption. Some of the observed shifts are 3433 to 3505, 2923 to 2905, 2063 to 2099, 1733 to 1734 and 1464 to 1462 cm^{-1} for E-PA, and 3433 to 3432, 2063 to 2086, and 1543 to 1556 cm^{-1} for I-PA (Figure 3C). The observed shifts in peaks in the FTIR spectra after modification can be attributed to the interactions between functional groups of the adsorbent and the Pb(II) ion. This modification ensured that there were no interferences with undesirable compounds during the adsorption process and increased the pore cavities and surface area of the adsorbent [63].

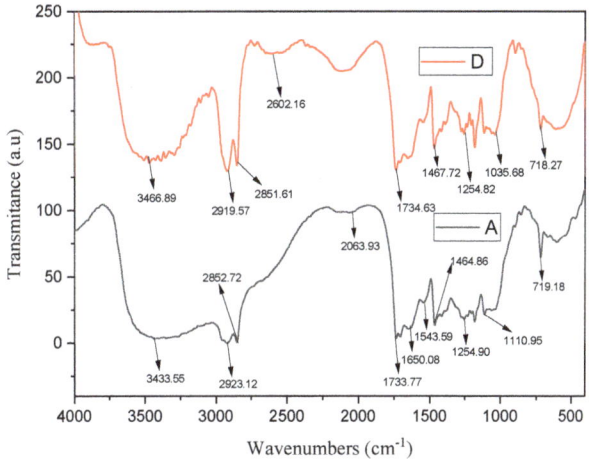

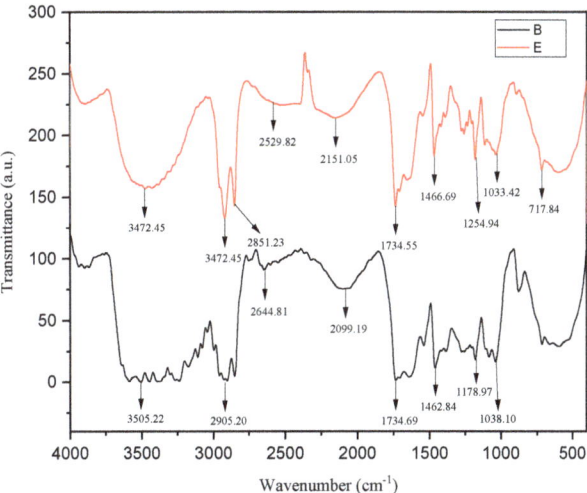

Figure 3. *Cont.*

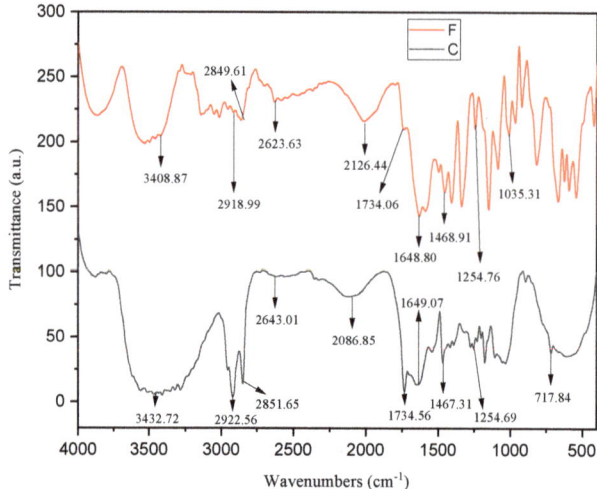

Figure 3. FTIR spectra of P-PA (A), Pretreated E-PA (B), I-PA (C), and spent P-PA (D), spent E-PA (E), spent I-PA (F) of *Pycnanthus angolesis* seed husk.

The spectra of the spent adsorbents confirm the participation of (–OH, –NH, C–O, and C=O) functional groups in the adsorption of Pb(II) ion and changes in the broadband of some peaks after Pb(II) adsorption, indicating the adsorption process of the target metal binding on the sorbent surface [26].

3.1.2. SEM Analysis

The surface morphologies of the adsorbents before modification, after modification, and after adsorption under different magnifications are presented in (Figure 4). The pure adsorbent (see Figure 4A) shows shallow pore structures. However, after modification (Figure 4B,C), the outer surface showed a heterogeneous, rough, uneven display of its surface with sufficient cavities, which will possibly be favorable for the adsorption of Pb(II) ion. The rough, sufficient, and larger cavities observed could be attributed to the impacts of the activation on the sorbent [25].

After adsorption, due to the presence of Pb(II) impurities, the SEM image revealed a homogenous and filled surface (Figure 4E,F) [64]. Similar observations have been reported by other researchers [23,65]. Shafiq et al. [65] developed a biosorbent from *Eucalyptus camdulensis* biochar. Before adsorption, the external EU-biochar surface was rough and had significant pore structures, while after adsorption processes, the adsorbent surface consisted of minute particles and brighter zones.

3.1.3. X-ray Diffraction

The XRD analyses were employed to understand and give insight into the crystal nature of the modified biosorbents. The results obtained for the I-PA and E-PA are shown in Figure 5. E-PA (see Figure 5a) spectra showed five diffraction peaks at 2θ = 25°, 30°, 36°, 44° and 62° attributed to the reflective planes of (110), (111), (121), (200), (311), respectively, while I-PA (see Figure 5b) reveals similar cellulosic diffraction at 2θ = 30°, 35°, 40°, 55.5°, 65.2° corresponding to (110), (111), (121), (200), and (311) planes of reflection, respectively. The sharp peaks obtained in these diffractions suggest a distinct alignment of a perfect crystalline structure [66]. Akpomie and Conradie [55] obtained similar cellulose diffraction peaks for a natural organic-silver nanocomposite for oil removal in water. The sharp peaks obtained in these spectra could be attributed to the influence of the chemical modification on *Pycnanthus angolesis* seed husk.

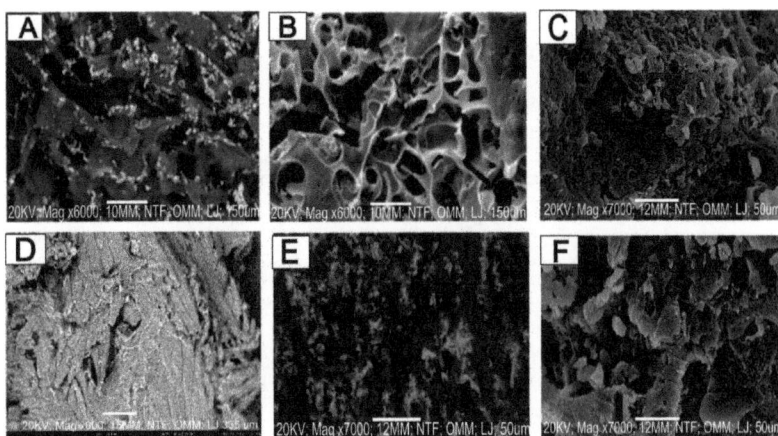

Figure 4. SEM images of (P-PA (**A**), E-PA (**B**), I-PA (**C**), spent P-PA (**D**), spent E-PA (**E**), spent I-PA (**F**)).

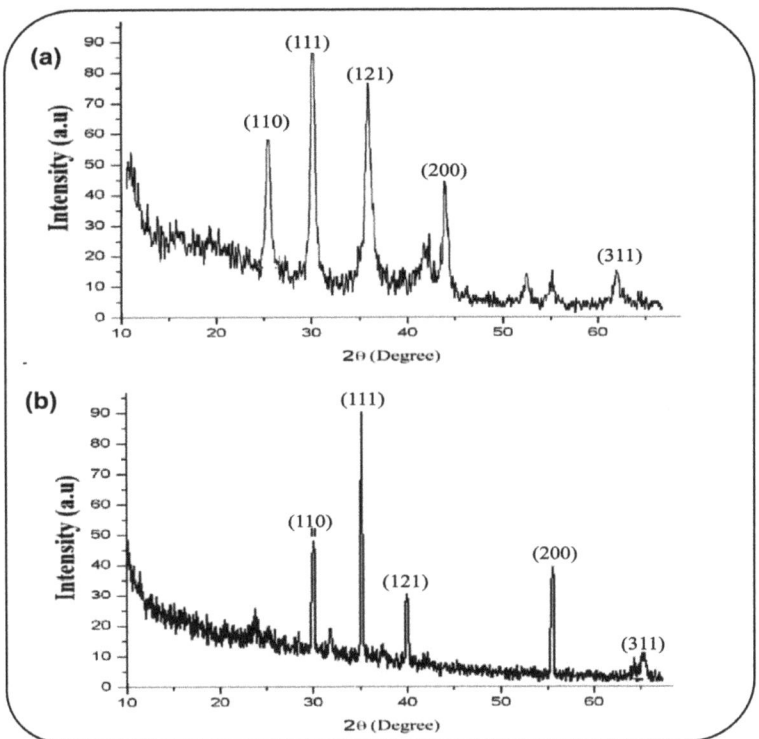

Figure 5. XRD spectra of E-PA (**a**) and I-PA (**b**) modified *Pycnanthus angolensis* seed husk.

3.1.4. Brunauer-Emmett-Teller (BET) Analysis

The physicochemical features of the pure *Pycnanthus angolesis* seed husk (P-PA) and the modified (E-PA and I-PA) were presented in Table 2. From the data generated, the surface area and pore volume of the pure biosorbent are 65.200 m^2/g and 0.055 cm^3/g, respectively. The surface area of the ethylene-glycol-activated adsorbent is 66.400 m^2/g, with a pore volume of 0.6355 cm^3/g, while 60.500 m^2/g and 0.6355 cm^3/g were obtained for the surface area and pore volume of the Iso-butanol activated adsorbent, respectively.

Table 2. Physiochemical features of pure (P-PA) and modified (E-PA and I-PA) *Pycnanthus angolensis* seed husk adsorbent.

Adsorbent Type	Surface Area (m^2/g)	Pore Volume (cm^3/g)	Pore Diameter (Å)
P-PA	65.200	0.605 24.540	24.540
E-PA	66.400	0.635	28.340
I-PA	60.500	0.635	30.340

The surface area obtained for the iso-butanol and ethylene glycol adsorbents is higher than some of the previously reported works, such as the acid-modified onion skin at 10.62 m^2/g [67], garlic waste at 5.62 m^2/g [68], and fennel seed at 3.668 m^2/g [16]. Comparatively, the activated adsorbents showed higher pore sizes than the pure adsorbent. The high pore size obtained could be attributed to the dissolution of hemicelluloses and pectin from the cell wall of the sorbent [69]. The high surface area and pore diameter obtained indicate an abundant adsorption site on the adsorbent surfaces [70]. Table (see Table 2) shows an increase in pore size, pore volume, and surface area of the treated adsorbent compared to the pure sample (P-PA), suggesting that the ethylene glycol and iso-butanol (E-PA and I-PA) activated adsorbent could be more efficient in the adsorption process.

3.2. Thermal Stability

Figure 6 shows the thermal stability of the pure ethylene glycol and iso-butanol-modified *Pycnanthus angolesis* seed husk (P-PA, E-PA, and I-PA). The result showed that the samples were stable up to 500 °C with three stages of degradation. An initial mass loss of −0.035 mg was observed in all the samples below 275, 255, and 270 °C for the P-PA, E-PA, and I-PA, respectively.

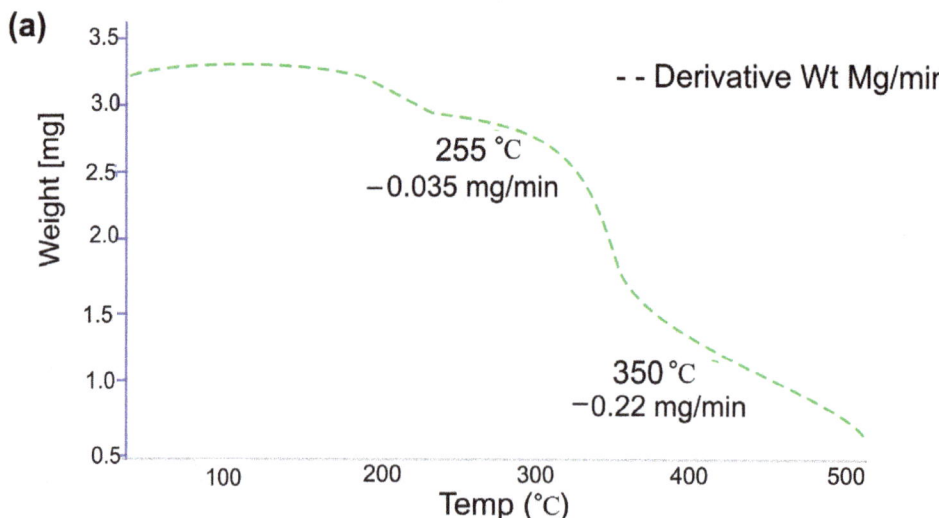

Figure 6. *Cont.*

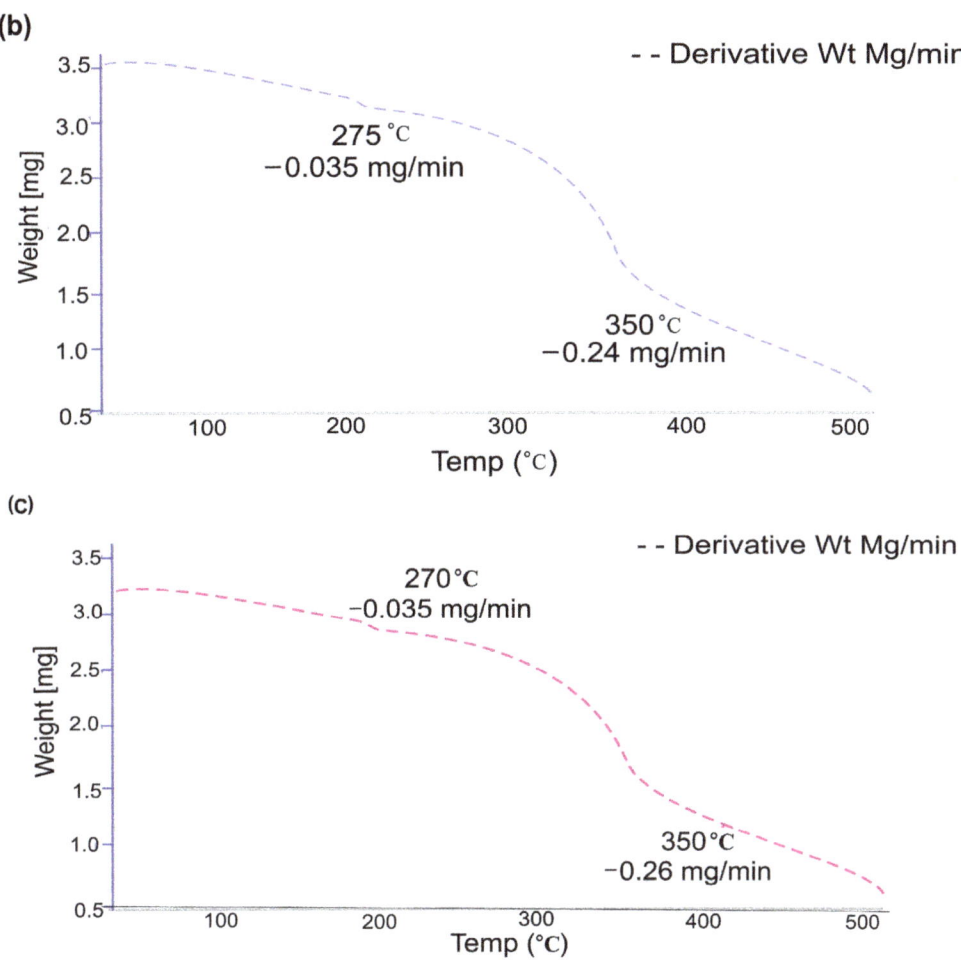

Figure 6. TGA thermogram of P-PA (**a**), E-PA (**b**) and I-PA (**c**).

This initial weight loss is linked to the loss of water from the adsorbents. Another mass loss of 0.24, 0.22, and 0.26 mg was observed on the TGA curve from 255 to 350 °C for P-PA, E-PA, and I-PA, respectively, due to the breakdown of hemicellulose and cellulose and the subsequent formation of a carbonaceous residue, which decomposes until it reaches 500 °C for all the adsorbent. This observation agrees with the reports of Eze et al. [23] on the TGA/DTA analysis of *Terminalia mantaly* seed husk biosorbent for the sequestration of ions from an aqueous solution.

3.2.1. Effect of pH

In adsorption studies, the pH study is an important parameter because it affects both the oxidation state of the contaminants in the aqueous system and the surface property of the adsorbent [71]. Its impact in this study was determined at pH 2, 4, 6, 8, and 10 on 10 mg/L at 298 K, as presented in Figure 7 for the P-PA and E-PA and I-PA, respectively. From the graph, there was an observable rapid increase in Pb(II) ion adsorption with an increase in pH value from 2 to 4 in the pure adsorbent before attaining equilibrium at pH 6 (see Figure 7). A spontaneous increase at lower pH was also observed for the modified adsorbents up to pH 6 (Figure 7).

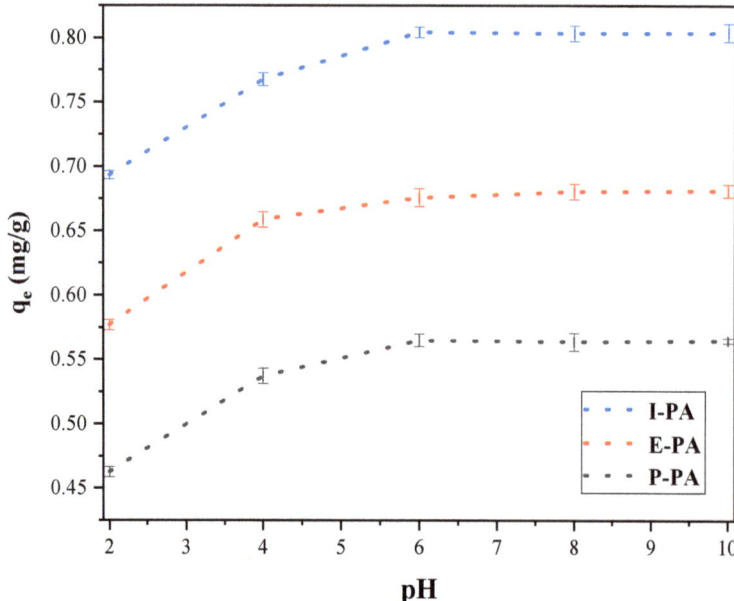

Figure 7. Effect of pH on Pb(II) onto P-PA, E-PA, and I-PA at 10 mg/L at 298 K reaction condition.

This is due to the activation of the surface functional groups, a decrease in hydrogen ion concentration, and electrostatic desirability, which promote metal ion adsorption [72]. The optimum adsorption efficiency of Pb(II) ion was achieved at pH 6 for all the adsorbents. At this point, the adsorption of Pb(II) ions remained constant. This is because, at a higher pH, the adsorption efficiency decreases as a result of the formation of soluble hydroxyl complexes and insoluble hydroxide precipitation [73]. Similar observations were also reported by Wang et al. [74] and Radha et al. [75].

3.2.2. Effect of Contact Time

For the pure adsorbent (Figure 8), the adsorption occurs relatively rapidly within the first 120 min and remains constant after 180 min. I-PA (Figure 8) showed rapid adsorption until maximum adsorption removal was attained at 180 min and decreased gradually after 200 min, while E-PA increased gradually from 50 to 120 min before attaining equilibrium at 180 min. The rapid increase in adsorption could be attributed to sufficient adsorption sites, making Pb(II) interact easily with the site [76]. In this study, the maximum adsorption time obtained was faster than in the study of Radha et al. [75], which used a chitosan-derived copolymeric blend for Pb(II) ion and Cd^{2+} ion removal.

This result is corroborated using the SEM image (see Figure 4), which shows that after modification, the outer surface showed a heterogeneous, rough, uneven display of its surface with sufficient cavities, which will possibly be favorable for the uptake of Pb(II) ion [77]. After equilibrium was achieved, further increases in time resulted in no significant change in the Pb(II) adsorption, showing that the binding sites are already saturated for the adsorption of Pb(II) ion [4]. Similar results were observed for the uptake of Pb(II) ion using biosorbent derived from the endocarp waste of Gayo coffee [27].

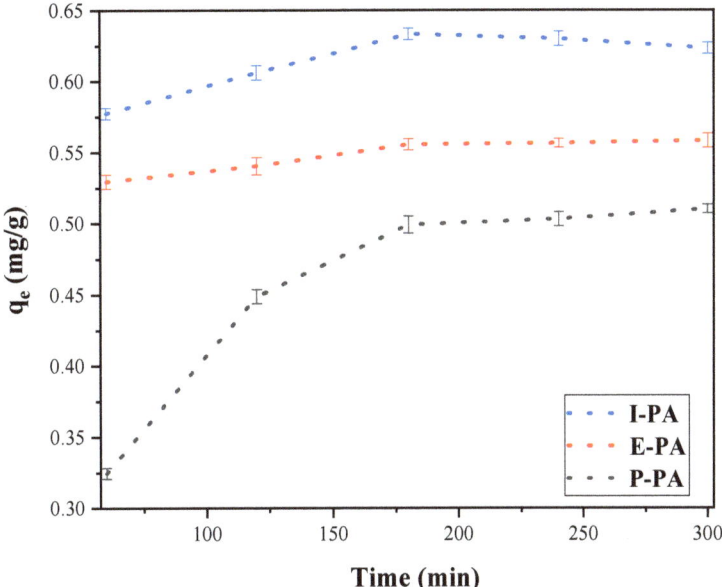

Figure 8. Effect of contact time on Pb(II) onto P-PA, E-PA, and I-PA.

3.2.3. Effect of Initial Metal Concentration

Figure 9 illustrates the impacts of metal ion concentration on the adsorption capacity of the prepared adsorbents for the uptake of lead. The sequestration of Pb(II) ion onto the P-PA, E-PA, and I-PA increased rapidly with an increase in the initial concentration of Pb(II) ion in the aqueous system. The reason is that at the initial adsorption stage, there is a higher usage of the active sites or enormous availability of unused active sites on the adsorbents [78]. Previous studies have reported a similar rapid increase in adsorption capacity with an increase in initial metal concentration [79].

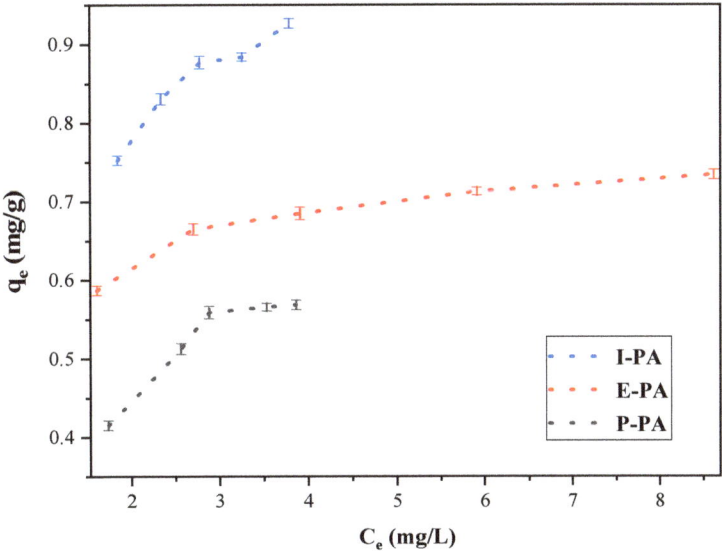

Figure 9. Effect of concentration on Pb(II) onto P-PA, E-PA, and I-PA.

3.3. Isotherm Studies

The isotherm models give insight into the adsorption mechanism, adsorbent-adsorbate interaction, and adsorbent materials [80]. The coefficient of regression (R^2) obtained from the plot of each model was used to select the isotherm model that correlates better with the adsorption of Pb(II). The parameters and estimated adsorption isotherm are shown in Table 3.

Table 3. Estimated values obtained for different Parameters of the isotherm models.

Isotherm Model	Parameters	P-PA	E-PA	I-PA
Langmuir	q_m (mg/g)	0.714	0.581	0.523
	K_L (L/g)	0.866	3.685	−20.141
	R_L	0.025	0.0059	−0.011
	R^2	0.985	0.992	0.987
Freundlich	K_F ((mg/g)/(mg/L) n)	0.483	0.443	0.414
	n	14.854	7.622	5.328
	R^2	0.628	0.674	0.749
Temkin	A_t (L/g)	473.212	1.387	3.301
	B	0.073	0.161	0.197
	b_t (kJ/mol)	34.047	15.415	12.601
	R^2	0.788	0.733	0.704
D-R	q_m	0.803	0.785	0.8074
	E	−30.30	−36.523	−25.649
	R^2	0.905	0.551	0.805
Florry Huggins	K_{FH}	0.014	0.001	0.002
	n_{FH}	−1.734	−3.034	−2.677
	R^2	0.410	0.733	0.501

Langmuir Isotherm model suggests a monolayer adsorption mechanism and the possible formation of a homogenous surface on the sorbent. The regression coefficients (R^2 = 0.985, 0.992, and 0.987) obtained from this model for P-PA, E-PA, and I-PA adsorbents, respectively, are higher than the ones obtained from other models, suggesting a perfect description of the experimental data. The perfect fits obtained suggest that the adsorption surfaces are homogenous, and the adsorption process follows a monolayer adsorption mechanism [81]. The maximum adsorption capacities (q_m) estimated were 0.714, 0.581, and 0.523 mg/g for the P-PA, E-PA, and I-PA, respectively. The Freundlich isotherm suggests a multilayer adsorption mechanism and a likely heterogeneous adsorption surface [55]. The values of the correlation coefficient (R^2 = 0.628, 0.674, and 0.749) obtained for the pure adsorbent and both modified adsorbents are lower than those obtained in the Langmuir isotherm. This implies that the Freundlich isotherm does not correlate best with the removal of Pb(II) ion onto the sorbents. However, the n values obtained for the adsorption study are <10, indicating a favorable removal of Pb(II) ions by the adsorbents [56]. Temkin models give insight into adsorbent and adsorbate interactions. The lower R^2 values (0.788, 0.733, and 0.704) obtained are lower than the ones obtained for the Langmuir model and thus do not correlate better with the uptake of Pb(II) ion onto the adsorbents. The Dubinin–Radushkevich isotherm is used to evaluate the nature of the adsorption process (a physical or chemical reaction) [82]. The lower regression coefficient values (R^2 = 0.905, 0.551, and 0.805) and (R^2 = 0.410, 0.733, and 0.501) obtained for the Dubinin–Radushkevich Isotherm and Flory Huggins models for P-PA, E-PA, and I-PA, respectively, denote that these models do not correlate best in describing the process of Pb(II) ion removal.

3.4. Kinetics Studies

The kinetic data obtained from the studies of PFO, PSO, and ID were used to study the adsorption mechanism of the Pb(II) ion on the adsorbents. The estimated kinetic parameters are presented in Table 3. The comparison of the kinetic models shows that

PFO best describes the kinetic adsorption of Pb(II) ion onto the adsorbents than PSO due to the degree of closeness between the $q_{e,Exp}$ and the $q_{e,cal}$ suggesting a physical adsorption process (formation of a weaker bond), which could be the reason for the lower adsorption capacity.

The lesser values obtained in HYBRID and MPSD error function analysis for the PFO (Table 4) also supported this observation. The intra-particle plot for pure and Iso-butanol *Pycnanthus angolensis* seed husk did not pass through the origin, suggesting that film and intra-article diffusion simultaneously controlled the rate-limiting step [83]. This observation is similar to the adsorption of Pb(II) ions onto naturally aged and virgin microporous materials [84].

Table 4. Kinetic parameters for the uptake of Pb(II) ion on the pure and modified adsorbents.

Kinetic Models	Parameter	P-PA	E-PA	I-PA
$q_{e,Exp}$		0.526	0.534	0.565
Pseudo-first-order				
	K_1 (min^{-1})	−0.007	−0.008	−0.002
	$q_{e,cal}$ (mg/g)	0.344	0.387	0.153
	R^2	0.763	0.897	0.885
HYBRID		5.133	3.23	17.79
MPSD		36.172	32.201	49.337
Pseudo-second-order				
	$q_{e,cal}$ (mg/g)	1.139	1.098	2.517
	K_2 (L/mg min)	−0.023	−0.021	0.022
	h (mg/L min)	−0.029	−0.024	0.142
	R^2	0.942	0.954	0.972
HYBRID		34.332	29.651	397.862
MPSD		58.137	56.302	107.238
Intra-particle diffusion				
	K_3 (mg/g min$^{-1/2}$)	0.125	0.340	0.094
	C	0.555	−1.829	0.952
	R^2	0.823	0.861	0.719

3.5. Thermodynamic Studies

Table 5 lists the estimated thermodynamic parameters for the adsorption of Pb(II) ions onto the P-PA, E-PA, and I-PA. The equations used in evaluating these parameters were established (see Equations (10) and (11)). The $\Delta G°$ values for the sequestration of Pb(II) ion unto the adsorbents were negative for all adsorbents, showing that the adsorption process was spontaneous (i.e., occurred without the input of external energy).

Table 5. Calculated values of Thermodynamic parameters.

Adsorbents	T (K)	$\Delta G°$ (kJ/mol)	$\Delta H°$ (kJ/mol)	$\Delta S°$ (J/mol K)
P-PA	313	−8.83	−33.91	−81.91
	323	−6.06		
	333	−7.32		
E-PA	313	−1.04	32.82	78.48
	323	−5.90		
	333	−7.47		
I-PA	313	−8.56	73.85	206.60
	323	−8.08		
	333	−4.45		

For E-PA, it was observed that Gibbs free energy decreased as the temperature increased, indicating that the removal efficiency was enhanced at high temperatures [85]. The negative value of $\Delta H°$ obtained for the pure adsorbent depicts exothermic adsorption, while upon impregnation, the endothermic nature of the adsorption process was indicated by the positive values of $\Delta H°$ [86]. In addition, the increased random interaction observed during the adsorption process of the Pb(II) ion onto the modified adsorbents was indicated by the positive ΔS° against the negative value ΔS° obtained for the pure adsorbents [87]. Ghanim et al. [79] observed a similar result using acid-modified poultry litter-derived hydrochar.

3.6. Adsorption Mechanism

The determination of the adsorption mechanism is of considerable significance in understanding the adsorption process. This process relies on various factors, including the morphology and functional groups that exist on the surface of the adsorbent, as well as the size and charge characteristics of the contaminants [16]. *Pycnanthus angolensis* seed husks consist primarily of lignocellulosic materials, which consist of hemicellulose, cellulose, and lignin. Lignocellulosic materials possess a significant abundance of oxygen-rich functional groups, including hydroxyl (–OH), carbonyl (–CO), and carboxyl (–COOH), which makes them very suitable for adsorption processes [88]. The mechanism that controls the adsorption process is illustrated in Figure 10. To assess the mechanism of adsorption that controls the sequestration of Pb(II) from aqueous solution unto the *Pycnanthus angolensis* seed husk, spectroscopic techniques (SEM and FTIR) were employed. Deduction from the SEM analysis showed that the adsorption of Pb(II) unto the surface of the adsorbents was necessitated due to the large pore structures observed, which served as a binding site for Pb(II) and were further corroborated by the assessment of the kinetic model that suggested physisorption. In addition, the shifts and disappearance of the broad band observed in the FTIR spectra (see Figure 3B,C) suggest the likelihood of a chemical adsorption mechanism. The charged ion on some of the functional groups (OH, N–H, COOH, C=O) identified using the FTIR on the adsorbents may interact with the ion of the sorbate via the interaction between the negatively charged oxygen and the positively charged metal ion (electrostatic interaction), ion exchange, and the formation of a complex with the oxygen-carrying group on the adsorbents. Therefore, it is assumed that, with the functional groups present on the adsorbent's surface and the presence of hydrogen bonding and cation interaction, the removal of Pb(II) from the aqueous solution is probable. This observation is similar to the report of Mabungela et al. [16] on the adsorption of Pb(II) onto fennel seeds and Wang et al. [74] on the Pb(II) adsorption using pectin/activated carbon-based porous microspheres.

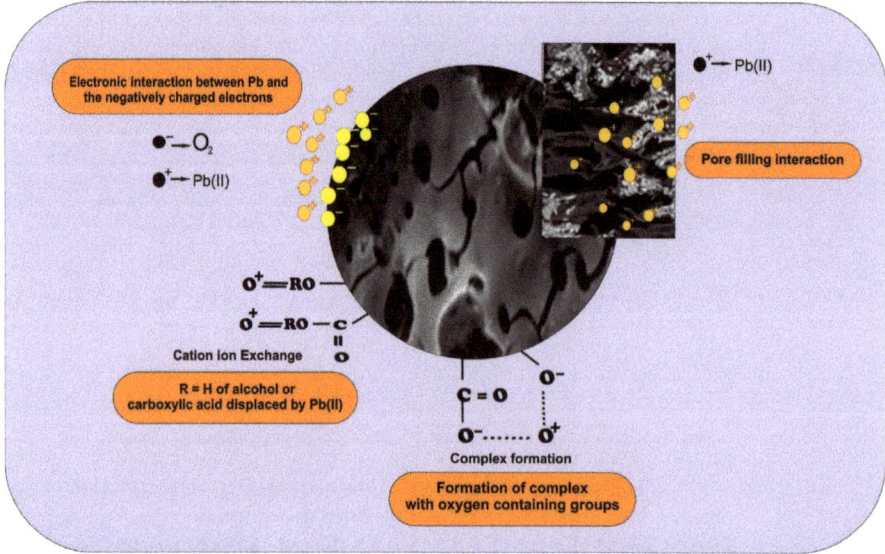

Figure 10. Schematic mechanism of Pb(II) adsorption onto pure and modified *Pycnanthus angolensis* seed husk.

3.7. Comparing the Performance of Pycnanthus angolensis Seed Husk with Other Adsorbents

The results of this study were compared to those of previous studies that used different adsorbents to study the adsorption of Pb(II). A comparison of these results is presented in Table 6. Pretreated *Pycnanthus angolensis* showed lower adsorption capacity compared to some of the adsorbents previously reported. However, despite the lower adsorption capacity observed, it showed a higher percentage removal efficiency of 96.75, 96.44, and 95.91% compared to most biosorbents previously reported (Table 5). Therefore, it is still recommended for the treatment of wastewater and for the use of other forms of modifying agents (acid, base, or thermal) to enhance its absorption capacity.

Table 6. Comparison of properties of *Pycnanthus angolensis* seed husk with other adsorbents used for the adsorption of Pb(II).

Biosorbent	q_m (mg/g)	Isotherm Model	Optimal pH	% Removal Efficiency	Reference
Chemically pretreated *Pycnanthus angolensis* Seed Husk	0.714	Langmuir	6	96.75	This study
Acid-treated fennel seed	18.79	Langmuir	8	-	[16]
Banana peel	2.18	Langmuir	5	-	[89]
Bagasse fly ash	2.50	Langmuir	5	95.00	[90]
Black sapote seeds	5.50	Langmuir	9	96.60	[91]
Melocanna baccifera (Poaceae)	10.66	Langmuir	6	81.69	[92]
Orange Peel Cellulose	50.10	Langmuir	4	98.33	[29]
Dyed shells of groundnut	0.106	Langmuir	-	-	[93]
Soya bean	0.72	Langmuir	4	70.42	[94]

3.8. Research Implications for Wastewater Treatment and Contributions to Sustainability

The need for proficient wastewater treatment methods has been driven by the dearth of safe drinking water as well as other environmental consequences. Adsorption techniques are gaining popularity as a practical method for detoxifying contaminated water. Despite a wide range of adsorbents being discussed in the literature, biosorption is a viable alternative method for eliminating contaminants from water [95]. It has a number of benefits, including ease of use, good adsorption capacity, high recoverability, and the ability to be modified [96]. A growing number of biosorbents have been developed and enhanced using environmentally friendly, secure, and cost-effective processes adhering to the ideals of sustainable chemistry [97]. Biosorbents offer interesting and intriguing properties for the removal of contaminants. Additionally, the sorbents exhibit excellent adsorption qualities and capacities following a variety of activations that enhance the rate of the adsorption process. In light of these, the use of modified *Pycnanthus angolensis* seed husk in the sequestration of Pb(II) from wastewater, as well as its kinetics, isotherm, and mechanism of adsorption, were explored.

This research paper holds significant implications for sustainability in the context of wastewater treatment and environmental protection. The results demonstrated the potential of the use of modified *Pycnanthus angolensis* seed husk for applications in real-life scenarios for the regeneration of biosorbents. Below are some of the key ways in which this research aligns with sustainability principles:

Pollution mitigation: This research addresses the pressing issue of lead pollution in water, a matter of significant environmental concern in recent times. Lead contamination has detrimental effects on human health and aquatic ecosystems. By developing an effective method for lead removal from wastewater, this research contributes to reducing water pollution and protecting the environment.

Resource efficiency: The utilization of the *Pycnanthus angolensis* seed husk as a natural adsorbent for lead ion removal represents a sustainable approach, as it would potentially reduce the reliance on synthetic and non-renewable materials in metal removal from aqueous solutions, such as in wastewater treatment process. This method thereby promotes resource efficiency and minimizes the ecological footprint of conventional treatments.

Renewable and biodegradable material: *Pycnanthus angolensis* seed husk is a renewable and biodegradable resource, offering the choice for commitment to sustainable practices of using environmentally friendly alternatives to synthetic adsorbents. Studies have reported that synthetic adsorbents often pose disposal challenges and environmental harm.

Low-cost solution: The findings of this research offer a cost-effective solution for lead removal from aqueous media and for wastewater treatment, especially for many resource-constrained developing countries and regions. The cost-effectiveness of the *Pycnanthus angolensis* seed husk method enhances its applicability and aligns with sustainability goals of affordability and accessibility.

Reduced energy consumption: The proposed lead adsorption method also contributes to sustainability, as it requires less energy compared to traditional treatment processes. This reduction in energy consumption will not only lower the operational costs but also reduce the carbon footprint associated with wastewater treatment.

Social and environmental impact: The potential positive impacts of adopting this method on local communities and ecosystems cannot be overstated. As more lead contamination cases are addressed in affected areas, the research would help in improving the quality of life of residents and wildlife.

Regulatory compliance: The present research complies with environmental regulations and standards pertaining to wastewater quality. This alignment with established standards underscores commitment to responsible and sustainable research practices.

Long-term benefits: The *Pycnanthus angolensis* seed husk method offers the prospect of long-term sustainability by utilizing a renewable resource and promoting environmentally responsible practices in wastewater treatment. The potential for lasting benefits is central to the sustainability of the proposed approach.

In summary, the research findings have indicated that modified *Pycnanthus angolensis* seed husk can be used to remove Pb(II) from aqueous solutions such as wastewater. This research paper significantly contributes to the needed remediation strategies towards achieving the Sustainable Development Goal (SDG) of ensuring access to clean, safe water and promoting sustainable wastewater treatment practices and water management.

4. Conclusions

In conclusion, the pure adsorbent was successfully modified using ethylene glycol (E-PA) and iso-butanol (I-PA) in order to enhance its adsorption capacity. The FTIR spectra confirm the participation of OH, –NH, C–O, C=O, and C–H functional groups in the uptake of Pb(II) ion and shifts in the broadband of some peaks after Pb(II) ion removal, indicating effective binding of the target metal on the sorbent surface. XRD confirms the influence of the modification on the *Pycnanthus angolesis* seed husk. SEM images of modified adsorbents (E-PA and I-PA) under different magnifications exhibited abundant cracked surfaces, enhanced pore structures, and cavities that can aid in the entrapment of Pb(II) ion compared to the pure adsorbent (P-PA). This result is corroborated by the higher pore size recorded for the E-PA and I-PA compared to the pure adsorbent (P-PA) in BET analysis. TGA revealed that the samples recorded three steps of degradation and were stable up to 500 °C. The linear plot of the isotherm and kinetic models shows that the uptake of Pb(II) ion on the adsorbents correlates better with the Langmuir model, which indicates a monolayer adsorption mechanism and possible formation of a homogenous surface of the sorbent. The kinetics of Pb(II) ion uptake fit best with the pseudo-first-order model and were corroborated by the relatively lower values estimated in the error function for all adsorbents. Thermodynamics revealed a spontaneous and endothermic adsorption process. At the same time, the adsorption mechanism revealed that the uptake of Pb(II) onto the P-PA, E-PA, and I-PA mainly involves electrostatic interaction, surface complexation, physical adsorption, and ionic exchange. In addition, the maximum adsorption efficiency was attained at 180 min at pH 6, while the adsorption increases as the concentration increases for all adsorbents. The result demonstrated the potential of the use of *Pycnanthus angolensis* seed husk for applications in real-life scenarios for the regeneration of biosorbents. The use of *Pycnanthus angolensis* seed husk significantly contributes to the needed remediation strategies to further the Sustainable Development Goal (SDG) of ensuring access to clean water and promoting sustainable water management.

Author Contributions: A.L.E.: conceptualization, experiments, H.O.A. and J.C.E.: manuscript writing, data analysis, I.A.U.: AAS, FTIR Interpretation, assistance, and proofreading. S.I.E.: data analysis, proofreading. V.S.A.: resources, data analysis, reviewing. K.G.A.: Literature Writing and XRD, TGA Interpretation. All authors contributed to the revisions of the manuscript and approved it for publication. All authors have read and agreed to the published version of the manuscript.

Funding: The authors did not receive any funding.

Institutional Review Board Statement: Not applicable.

Informed Consent Statement: No human or other animals were used in the study.

Data Availability Statement: All data generated or analyzed are available within the manuscript.

Conflicts of Interest: The authors declare that there is no conflict of interest.

References

1. Tsvetanova, L.; Barbov, B.; Rusew, R.; Delcheva, Z.; Shivachev, B. Equilibrium Isotherms and Kinetic Effects during the Adsorption of Pb(II) on Titanosilicates Compared with Natural Zeolite Clinoptilolite. *Water* **2022**, *14*, 2152. [CrossRef]
2. Guo, Z.; Zhan, R.; Shi, Y.; Zhu, D.; Pan, J.; Chao, Y.; Wang, Y.; Wang, J. Innovative and Green Utilization of Zinc-Bearing Dust by Hydrogen Reduction: Recovery of Zinc and Lead, and Synergetic Preparation of Fe/C Micro-Electrolysis Materials. *Chem. Eng. J.* **2023**, *456*, 141157. [CrossRef]
3. Li, H.; Zhao, S.; Zhang, W.; Du, H.; Yang, X.; Peng, Y.; Han, D.; Wang, B.; Li, Z. Efficient Esterification over Hierarchical Zr-Beta Zeolite Synthesized via Liquid-State Ion-Exchange Strategy. *Fuel* **2023**, *342*, 127786. [CrossRef]
4. Adebisi, E.U.; Adebisi, A.A.; Okechukwu, V.U.; Omokpariola, D.O.; Umeh, T.C. Adsorption potential of Lead and Nickel ions by *Chrysophyllum albidum* G.Don (African Star Apple) seed shells. *World Sci. News* **2022**, *166*, 132–145.
5. Liu, W.; Zheng, J.; Ou, X.; Liu, X.; Song, Y.; Tian, C.; Rong, W.; Shi, Z.; Dang, Z.; Lin, Z. Effective Extraction of Cr(VI) from Hazardous Gypsum Sludge via Controlling the Phase Transformation and Chromium Species. *Environ. Sci. Technol.* **2018**, *52*, 13336–13342. [CrossRef] [PubMed]
6. Hu, F.; Qiu, L.; Xiang, Y.; Wei, S.; Sun, H.; Hu, H.; Weng, X.; Mao, L.; Ming, Z. Spatial Network and Driving Factors of Low-Carbon Patent Applications in China from a Public Health Perspective. *Front. Public Health* **2023**, *11*, 1121860. [CrossRef]
7. Zhang, N.; Lu, D.; Sheng, H.; Xia, J.; Kan, P.; Yao, Z.; Chen, H.; Li, G.; Zhu, D.Z.; Liu, H. Constructed Wetlands as Hotspots of Antibiotic Resistance Genes and Pathogens: Evidence from Metagenomic Analysis in Chinese Rural Areas. *J. Hazard. Mater.* **2023**, *447*, 130778. [CrossRef]
8. Li, Y.; Li, Y.; Shan, Y.; Tong, H.; Zhao, J. Effect of Magnesium Ions on the Mechanical Properties of Soil Reinforced by Microbially Induced Carbonate Precipitation. *J. Mater. Civ. Eng.* **2023**, *35*, 04023413. [CrossRef]
9. Mao, S.-X.; Song, J.; Zhu, W.; Ji, M.; Pang, J.; Dang, D.-B.; Pang, J. Heterogeneous Oxidative Desulfurization of Fuels Using Amphiphilic Mesoporous Phosphomolybdate-Based Poly(Ionic Liquid) over a Wide Temperature Range. *Fuel* **2023**, *352*, 128982. [CrossRef]
10. Sun, X.; Chen, Z.; Sun, Z.; Wu, S.; Guo, K.; Dong, Z.; Peng, Y. High-Efficiency Utilization of Waste Shield Slurry: A Geopolymeric Flocculation-Filtration-Solidification Method. *Constr. Build. Mater.* **2023**, *387*, 131569. [CrossRef]
11. Zhao, Y.; Li, Q.; Cui, Q.; Ni, S.-Q. Nitrogen Recovery through Fermentative Dissimilatory Nitrate Reduction to Ammonium (DNRA): Carbon Source Comparison and Metabolic Pathway. *Chem. Eng. J.* **2022**, *441*, 135938. [CrossRef]
12. Hayeeye, F.; Qian, Y.; Sattar, M.; Chinpa, W.; Sirichote, O. Adsorption of Pb^{2+} Ions from Aqueous Solutions by Gelatin/Activated Carbon Composite Bead Form. *Adsorpt. Sci. Technol.* **2017**, *36*, 355–371. [CrossRef]
13. Wang, Z.; Chen, C.; Liu, H.; Hrynshpan, D.; Savitskaya, T.; Chen, J.; Chen, J. Enhanced Denitrification Performance of Alcaligenes Sp. TB by Pd Stimulating to Produce Membrane Adaptation Mechanism Coupled with Nanoscale Zero-Valent Iron. *Sci. Total Environ.* **2020**, *708*, 135063. [CrossRef]
14. Zheng, Y.; Liu, Y.; Guo, X.; Chen, Z.; Zhang, W.; Wang, Y.; Tang, X.; Zhang, Y.; Zhao, Y. Sulfur-Doped g-C3N4/RGO Porous Nanosheets for Highly Efficient Photocatalytic Degradation of Refractory Contaminants. *J. Mater. Sci. Technol.* **2020**, *41*, 117–126. [CrossRef]
15. Rind, I.K.; Sarı, A.; Tuzen, M.; Lanjwani, M.F.; Karaman, I.; Saleh, T.A. Influential biosorption of lead from aquatic solution using Escherichia coli/carbon nanofibers. *Environ. Nanotechnol. Monit. Manag.* **2023**, *19*, 100776. [CrossRef]
16. Mabungela, N.; Shooto, N.D.; Mtunzi, F.M.; Naidoo, E.B. The Adsorption of Copper, Lead Metal Ions, and Methylene Blue Dye from Aqueous Solution by Pure and Treated Fennel Seeds. *Adsorpt. Sci. Technol.* **2022**, *2022*, 5787690. [CrossRef]

17. Bai, B.; Rao, D.R.V.P.; Chang, T.; Guo, Z. A Nonlinear Attachment-Detachment Model with Adsorption Hysteresis for Suspension-Colloidal Transport in Porous Media. *J. Hydrol.* **2019**, *578*, 124080. [CrossRef]
18. Dong, Y.; Yuan, H.; Ge, D.; Zhu, N. A Novel Conditioning Approach for Amelioration of Sludge Dewaterability Using Activated Carbon Strengthening Electrochemical Oxidation and Realized Mechanism. *Water Res.* **2022**, *220*, 118704. [CrossRef]
19. Hu, J.; Zhao, L.; Luo, J.; Gong, H.; Zhu, N. A Sustainable Reuse Strategy of Converting Waste Activated Sludge into Biochar for Contaminants Removal from Water: Modifications, Applications and Perspectives. *J. Hazard. Mater.* **2022**, *438*, 129437. [CrossRef] [PubMed]
20. Sun, S.; Liu, H.; Zhang, J.; Wang, W.; Pang, X.; Zhu, X.-G.; Wang, Y.; Wan, S. Application of a Novel Coagulant in Reservoir Water Treatment in Qingdao. *Desalination Water Treat.* **2023**, *284*, 49–60. [CrossRef]
21. Wang, Z.; Hu, L.; Zhao, M.; Dai, L.; Hrynsphan, D.; Savitskaya, T.; Chen, J. Bamboo Charcoal Fused with Polyurethane Foam for Efficiently Removing Organic Solvents from Wastewater: Experimental and Simulation. *Biochar* **2022**, *4*, 28. [CrossRef]
22. Thabede, P.M.; Shooto, N.D.; Xaba, T.; Naidoo, E.B. Magnetite Functionalized Nigella Sativa Seeds for the Uptake of Chromium(VI) and Lead(II) Ions from Synthetic Wastewater. *Adsorpt. Sci. Technol.* **2021**, *2021*, 6655227. [CrossRef]
23. Eze, S.I.; Abugu, H.O.; Odewole, O.A.; Ukwueze, N.N.; Alum, L.O. Thermal and Chemical Pretreatment of Terminalia Mantaly Seed Husk Biosorbent to Enhance the Adsorption Capacity for Pb^{2+}. *Sci. Afr.* **2022**, *15*, e01123. [CrossRef]
24. Castro, D.J.; Rosas-Laverde, N.M.; Aldás, M.B.; Almeida-Naranjo, C.E.; Pruna, A. Chemical Modification of Agro-Industrial Waste-Based Bioadsorbents for Enhanced Removal of Zn(II) Ions from Aqueous Solutions. *Materials* **2021**, *14*, 2134. [CrossRef]
25. Shah, G.M.; Imran, M.; Aiman, U.; Iqbal, M.; Akram, M.; Javeed, H.M.R.; Waqar, A.; Rabbani, F. Efficient Sequestration of Lead from Aqueous Systems by Peanut Shells and Compost: Evidence from Fixed Bed Column and Batch Scale Studies. *PeerJ Phys. Chem.* **2022**, *4*, 21. [CrossRef]
26. Azam, M.; Wabaidur, S.M.; Khan, M.R.; Al-Resayes, S.I.; Islam, M.S. Heavy Metal Ions Removal from Aqueous Solutions by Treated Ajwa Date Pits: Kinetic, Isotherm, and Thermodynamic Approach. *Polymers* **2022**, *14*, 914. [CrossRef] [PubMed]
27. Mariana, M.; Mulana, F.; Juniar, L.; Fathira, D.; Safitri, R.; Muchtar, S.; Bilad, M.R.; Shariff, A.H.M.; Huda, N. Development of Biosorbent Derived from the Endocarp Waste of Gayo Coffee for Lead Removal in Liquid Wastewater—Effects of Chemical Activators. *Sustainability* **2021**, *13*, 3050. [CrossRef]
28. Batool, F.; Akbar, J.; Iqbal, S.; Noreen, S.; Bukhari, S.N.A. Study of Isothermal, Kinetic, and Thermodynamic Parameters for Adsorption of Cadmium: An Overview of Linear and Nonlinear Approach and Error Analysis. *Bioinorg. Chem. Appl.* **2018**, *2018*, 3463724. [CrossRef]
29. Rahman, A.; Yoshida, K.; Islam, M.M.; Kobayashi, G. Investigation of Efficient Adsorption of Toxic Heavy Metals (Chromium, Lead, Cadmium) from Aquatic Environment Using Orange Peel Cellulose as Adsorbent. *Sustainability* **2023**, *15*, 4470. [CrossRef]
30. Suhaimi, N.; Kooh, M.R.R.; Lim, C.M.; Chao, C.-T.C.; Chau, Y.-F.C.; Mahadi, A.H.; Chiang, H.-P.; Hassan, N.H.H.; Thotagamuge, R. The Use of Gigantochloa Bamboo-Derived Biochar for the Removal of Methylene Blue from Aqueous Solution. *Adsorpt. Sci. Technol.* **2022**, *2022*, 8245797. [CrossRef]
31. Kooh, M.R.R.; Dahri, M.K.; Lim, L.B.L. Removal of Methyl Violet 2B Dye from Aqueous Solution Using Nepenthes Rafflesiana Pitcher and Leaves. *Appl. Water Sci.* **2017**, *7*, 3859–3868. [CrossRef]
32. Rahman, A.; Haque, M.A.; Ghosh, S.; Shinu, P.; Attimarad, M.; Kobayashi, G. Modified Shrimp-Based Chitosan as an Emerging Adsorbent Removing Heavy Metals (Chromium, Nickel, Arsenic, and Cobalt) from Polluted Water. *Sustainability* **2023**, *15*, 2431. [CrossRef]
33. Akpomie, K.G.; Conradie, J. Advances in Application of Cotton-Based Adsorbents for Heavy Metals Trapping, Surface Modifications and Future Perspectives. *Ecotoxicol. Environ. Saf.* **2020**, *201*, 110825. [CrossRef] [PubMed]
34. Güzel, F.; Sayğılı, H.; Sayğılı, G.A.; Koyuncu, F.; Yılmaz, C. Optimal Oxidation with Nitric Acid of Biochar Derived from Pyrolysis of Weeds and Its Application in Removal of Hazardous Dye Methylene Blue from Aqueous Solution. *J. Clean. Prod.* **2017**, *144*, 260–265. [CrossRef]
35. Wei, S.; Chen, T.; Hou, H.; Xu, Y. Recent Advances in Electrochemical Sterilization. *J. Electroanal. Chem.* **2023**, *937*, 117419. [CrossRef]
36. Yang, H.; Li, F.; Hu, N.; Fu, S.-Y. Frictional Characteristics of Graphene Oxide-Modified Continuous Glass Fiber Reinforced Epoxy Composite. *Compos. Sci. Technol.* **2022**, *223*, 109446. [CrossRef]
37. Chen, D.; Wang, Q.; Li, Y.; Li, Y.; Zhou, H.; Yue, F. A General Linear Free Energy Relationship for Predicting Partition Coefficients of Neutral Organic Compounds. *Chemosphere* **2020**, *247*, 125869. [CrossRef]
38. Sikam, K.G.; Dzouemo, L.C.; Happi, G.M.; Wansi, J.D. Recent advances on pharmacology and chemistry of *Pycnanthus angolensis* Over the last decade (2012–2021). *Nat. Resour. Hum. Health* **2022**, *2*, 322–326. [CrossRef] [PubMed]
39. Simic, A.; Kroepfl, D.; Simic, N.; Ogunwande, I.A. *Pycnanthus angolensis* (Welw) Excell: Volatile Oil Constituents and Antimicrobial Activity. *Nat. Prod. Commun.* **2006**, *1*, 651–654. [CrossRef]
40. Bello, O.A.; Akinyele, A.O. Qualitative morphology of *Pycnanthus angolensis* (Welw.) fruits and seeds. *Afr. J. Agric. Technol. Environ.* **2018**, *7*, 90–98.
41. Oladimeji, O.; Ahmadu, A.A. Antioxidant activity of compounds isolated from *Pycnanthus angolensis* (Welw.) Warb and *Byrophyllum pinnatum* (Lam.) Oken. *Eur. Chem. Bull.* **2019**, *8*, 96–100. [CrossRef]

42. Gustafson, K.; Wu, Q.; Asante-Dartey, J.; Simon, J.E. Pycnanthus angolensis: Bioactive Compounds and Medicinal Applications. In *African Natural Plant Products Volume II: Discoveries and Challenges in Chemistry, Health, and Nutrition*; ACS Symposium Series; American Chemical Society: Washington, DC, USA, 2013; pp. 63–78. [CrossRef]
43. Guo, B.; Wang, Y.; Yu, F.; Liang, C.; Li, T.; Yao, X.; Hu, F. The Effects of Environmental Tax Reform on Urban Air Pollution: A Quasi-Natural Experiment Based on the Environmental Protection Tax Law. *Front. Public Health* **2022**, *10*, 967524. [CrossRef] [PubMed]
44. Song, Z.; Han, D.; Yang, M.; Huang, J.; Shao, X.; Li, H. Formic Acid Formation via Direct Hydration Reaction (CO + $H_2O \rightarrow$ HCOOH) on Magnesia-Silver Composite. *Appl. Surf. Sci.* **2023**, *607*, 155067. [CrossRef]
45. Zhang, G.; Zhao, Z.; Yin, X.; Zhu, Y. Impacts of Biochars on Bacterial Community Shifts and Biodegradation of Antibiotics in an Agricultural Soil during Short-Term Incubation. *Sci. Total Environ.* **2021**, *771*, 144751. [CrossRef]
46. Imran, M.; Khan, Z.U.H.; Iqbal, M.; Iqbal, J.; Shah, N.S.; Munawar, S.; Ali, S.; Murtaza, B.; Naeem, M.A.; Rizwan, M. Effect of Biochar Modified with Magnetite Nanoparticles and HNO_3 for Efficient Removal of Cr(VI) from Contaminated Water: A Batch and Column Scale Study. *Environ. Pollut.* **2020**, *261*, 114231. [CrossRef]
47. Diraki, A.; Mackey, H.R.; McKay, G.; Abdala, A. Removal of Oil from Oil–Water Emulsions Using Thermally Reduced Graphene and Graphene Nanoplatelets. *Chem. Eng. Res. Des.* **2018**, *137*, 47–59. [CrossRef]
48. Tariq, M.; Nadeem, M.; Iqbal, M.; Imran, M.; Siddique, M.H.; Iqbal, Z.; Amjad, M.; Rizwan, M.; Ali, S. Effective Sequestration of Cr (VI) from Wastewater Using Nanocomposite of ZnO with Cotton Stalks Biochar: Modeling, Kinetics, and Reusability. *Environ. Sci. Pollut. Res.* **2020**, *27*, 33821–33834. [CrossRef]
49. Waqar, R.; Kaleem, M.; Iqbal, J.; Minhas, L.A.; Haris, M.S.; Chalgham, W.; Ahmad, A.; Mumtaz, A.S. Kinetic and Equilibrium Studies on the Adsorption of Lead and Cadmium from Aqueous Solution Using *Scenedesmus* sp. *Sustainability* **2023**, *15*, 6024. [CrossRef]
50. Dada, A.O.; Olalekan, A.P.; Olatunya, A.M.; Dada, O.J.I.J.C. Langmuir, Freundlich, Temkin and Dubinin-Radushkevich Isotherms Studies of Equilibrium Sorption of Zn^{2+} unto Phosphoric Acid Modified Rice Husk. *IOSR J. Appl. Chem.* **2012**, *3*, 38–45. [CrossRef]
51. Shahbeig, H.; Bagheri, N.; Ghorbanian, S.A.; Hallajisani, A.; Pourkarimi, S. A new adsorption isotherm model of aqueous solutions on granular activated carbon. *World J. Model. Simul.* **2013**, *9*, 243–254.
52. Theivarasu, C.; Mylsamy, S. Removal of Malachite Green from Aqueous Solution by Activated Carbon Developed from Cocoa (*Theobroma Cacao*) Shell—A Kinetic and Equilibrium Studies. *J. Chem.* **2011**, *8*, S363–S371. [CrossRef]
53. Israel, U.; Eduok, U. Biosorption of zinc from aqueous solution using coconut (*Cocos nucifera* L) coir dust. *Arch. Appl. Sci. Res.* **2012**, *4*, 809–819.
54. Amin, M.; Alazba, A.A.; Shafiq, M. Adsorptive Removal of Reactive Black 5 from Wastewater Using Bentonite Clay: Isotherms, Kinetics and Thermodynamics. *Sustainability* **2015**, *7*, 15302–15318. [CrossRef]
55. Akpomie, K.G.; Conradie, J. Ultrasonic Aided Sorption of Oil from Oil-in-Water Emulsion onto Oleophilic Natural Organic-Silver Nanocomposite. *Chem. Eng. Res. Des.* **2021**, *165*, 12–24. [CrossRef]
56. Ezekoye, O.M.; Akpomie, K.G.; Eze, S.I.; Chukwujindu, C.N.; Ani, J.U.; Ujam, O.T. Biosorptive Interaction of Alkaline Modified *Dialium Guineense* Seed Powders with Ciprofloxacin in Contaminated Solution: Central Composite, Kinetics, Isotherm, Thermodynamics, and Desorption. *Int. J. Phytoremediat.* **2020**, *22*, 1028–1037. [CrossRef]
57. Ayawei, N.; Ebelegi, A.N.; Wankasi, D. Modelling and Interpretation of Adsorption Isotherms. *J. Chem.* **2017**, *2017*, 3039817. [CrossRef]
58. Eze, S.I.; Abugu, H.O.; Ekowo, L.C. Thermal and Chemical Pretreatment of Cassia Sieberiana Seed as Biosorbent for Pb^{2+} Removal from Aqueous Solution. *Desalination Water Treat.* **2021**, *226*, 223–241. [CrossRef]
59. Beksissa, R.; Tekola, B.; Ayala, T.; Dame, B. Investigation of the Adsorption Performance of Acid Treated Lignite Coal for Cr (VI) Removal from Aqueous Solution. *Environ. Chall.* **2021**, *4*, 100091. [CrossRef]
60. Tan, C.-Y.; Gan, L.; Lu, X.; Chen, Z. Biosorption of Basic Orange Using Dried A. Filiculoides. *Ecol. Eng.* **2010**, *36*, 1333–1340. [CrossRef]
61. Fonseca, J.G.; Albis, A.; Montenegro, A.R. Evaluation of Zinc Adsorption Using Cassava Peels (Manihot Esculenta) Modified with Citric Acid. *Contemp. Eng. Sci.* **2018**, *11*, 3575–3585. [CrossRef]
62. Khan, M.A.; Otero, M.; Kazi, M.; Alqadami, A.A.; Wabaidur, S.M.; Siddiqui, M.R.; Alothman, Z.A.; Sumbul, S. Unary and Binary Adsorption Studies of Lead and Malachite Green onto a Nanomagnetic Copper Ferrite/Drumstick Pod Biomass Composite. *J. Hazard. Mater.* **2019**, *365*, 759–770. [CrossRef]
63. Li, X.; Tang, Y.; Cao, X.; Lu, D.; Luo, F.; Shao, W. Preparation and Evaluation of Orange Peel Cellulose Adsorbents for Effective Removal of Cadmium, Zinc, Cobalt and Nickel. *Colloids Surf. A Physicochem. Eng. Asp.* **2008**, *317*, 512–521. [CrossRef]
64. Ahmad, M.; Ahmad, M.; Usman, A.R.A.; Al-Faraj, A.S.; Abduljabbar, A.S.; Al-Wabel, M.I. Biochar Composites with Nano Zerovalent Iron and Eggshell Powder for Nitrate Removal from Aqueous Solution with Coexisting Chloride Ions. *Environ. Sci. Pollut. Res.* **2017**, *25*, 25757–25771. [CrossRef]
65. Shafiq, M.; Alazba, A.A.; Amin, M. Kinetic and Isotherm Studies of Ni^{2+} and Pb^{2+} Adsorption from Synthetic Wastewater Using Eucalyptus Camdulensis—Derived Biochar. *Sustainability* **2021**, *13*, 3785. [CrossRef]
66. Das, D.; Samal, D.P.; Bc, M. Preparation of Activated Carbon from Green Coconut Shell and Its Characterization. *J. Chem. Eng. Process Technol.* **2015**, *6*, 248. [CrossRef]

67. Agarry, S.E.; Ogunleye, O.O.; Ajani, O.A. Biosorptive Removal of Cadmium (II) Ions from Aqueous Solution by Chemically Modified Onion Skin: Batch Equilibrium, Kinetic and Thermodynamic Studies. *Chem. Eng. Commun.* **2015**, *202*, 655–673. [CrossRef]
68. Adetoye, M.D.; Adeojo, S.O.; Ajiboshin, B.F. Photocatalyst utilization of garlic waste as adsorbentfor heavy metal removal from aqueous solution. *J. Pure Appl. Sci.* **2018**, *6*, 6–13.
69. Zhou, P.; Regenstein, J.M. Effects of Alkaline and Acid Pretreatments on Alaska Pollock Skin Gelatin Extraction. *J. Food Sci.* **2006**, *70*, c392–c396. [CrossRef]
70. Yusuff, A.S. Adsorptive Removal of Lead and Cadmium Ions from Aqueous Solutions by Aluminium Oxide Modified Onion Skin Wastes: Adsorbent Characterization, Equilibrium Modelling and Kinetic Studies. *Energy Environ.* **2021**, *33*, 152–169. [CrossRef]
71. Huang, J.; Fang, Y.; Zeng, G.; Xue, L.; Gu, Y.; Shi, L.; Liu, W.; Shi, Y. Influence of PH on Heavy Metal Speciation and Removal from Wastewater Using Micellar-Enhanced Ultrafiltration. *Chemosphere* **2017**, *173*, 199–206. [CrossRef]
72. Tang, S.; Lin, L.-C.; Wang, X.; Yu, A.; Sun, X. Interfacial Interactions between Collected Nylon Microplastics and Three Divalent Metal Ions (Cu(II), Ni(II), Zn(II)) in Aqueous Solutions. *J. Hazard. Mater.* **2021**, *403*, 123548. [CrossRef] [PubMed]
73. Awual, M.R.; Hasan, M.M.; Eldesoky, G.E.; Khaleque, M.A.; Rahman, M.M.; Naushad, M. Facile Mercury Detection and Removal from Aqueous Media Involving Ligand Impregnated Conjugate Nanomaterials. *Chem. Eng. J.* **2016**, *290*, 243–251. [CrossRef]
74. Wang, R.-S.; Li, Y.; Shuai, X.; Liang, R.; Chen, J.; Liu, C. PEctin/Activated Carbon-Based Porous Microsphere for PB^{2+} Adsorption: Characterization and Adsorption Behaviour. *Polymers* **2021**, *13*, 2453. [CrossRef] [PubMed]
75. Radha, E.; Gomathi, T.; Sudha, P.N.; Latha, S.; Ghfar, A.A.; Hossain, N. Adsorption Studies on Removal of Pb(II) and Cd(II) Ions Using Chitosan Derived Copoymeric Blend. *Biomass Convers. Biorefin.* **2021**. [CrossRef]
76. Taşar, Ş.; Kaya, F.; Özer, A. Biosorption of Lead(II) Ions from Aqueous Solution by Peanut Shells: Equilibrium, Thermodynamic and Kinetic Studies. *J. Environ. Chem. Eng.* **2014**, *2*, 1018–1026. [CrossRef]
77. Thi, T.T.L.; Van, K. Adsorption behavior of Pb (II) in aqueous solution using coffee husk-based activated carbon modified by nitric acid. *Am. J. Eng. Res.* **2016**, *5*, 120–129.
78. Tangtubtim, S.; Saikrasun, S. Adsorption Behavior of Polyethyleneimine-Carbamate Linked Pineapple Leaf Fiber for Cr(VI) Removal. *Appl. Surf. Sci.* **2019**, *467–468*, 596–607. [CrossRef]
79. Ghanim, B.; Leahy, J.J.; O'Dwyer, T.F.; Pembroke, J.T.; Murnane, J.G. Removal of Hexavalent Chromium (Cr(VI)) from Aqueous Solution Using Acid-modified Poultry Litter derived Hydrochar: Adsorption, Regeneration and Reuse. *J. Chem. Technol. Biotechnol.* **2021**, *97*, 55–66. [CrossRef]
80. Chukwuemeka-Okorie, H.O.; Ekuma, F.K.; Akpomie, K.G.; Nnaji, J.C.; Okereafor, A.G. Adsorption of Tartrazine and Sunset Yellow Anionic Dyes onto Activated Carbon Derived from Cassava Sievate Biomass. *Appl. Water Sci.* **2021**, *11*, 27. [CrossRef]
81. Wei, W.; Li, A.; Ma, F.; Pi, S.; Yang, J.; Wang, Q.; Ni, B.-J. Simultaneous Sorption and Reduction of Cr(VI) in Aquatic System by Microbial Extracellular Polymeric Substances from Klebsiella Sp. J1. *J. Chem. Technol. Biotechnol.* **2018**, *93*, 3152–3159. [CrossRef]
82. Cui, D.; Tan, C.; Deng, H.; Gu, X.; Pi, S.; Chen, T.; Zhou, L.; Li, A. Biosorption Mechanism of Aqueous Pb^{2+}, Cd^{2+}, and Ni^{2+} Ions on Extracellular Polymeric Substances (EPS). *Archaea* **2020**, *2020*, 8891543. [CrossRef]
83. Hu, L.; Guang, C.; Liu, Y.; Su, Z.; Gong, S.; Yao, Y.; Wang, Y. Adsorption Behavior of Dyes from an Aqueous Solution onto Composite Magnetic Lignin Adsorbent. *Chemosphere* **2020**, *246*, 125757. [CrossRef]
84. Liu, X.; Xu, X.; Dong, X.; Park, J. Competitive Adsorption of Heavy Metal Ions from Aqueous Solutions onto Activated Carbon and Agricultural Waste Materials. *Pol. J. Environ. Stud.* **2019**, *29*, 749–761. [CrossRef]
85. Ghanim, B.; Murnane, J.G.; O'Donoghue, L.M.T.; Courtney, R.; Pembroke, J.T.; O'Dwyer, T.F. Removal of Vanadium from Aqueous Solution Using a Red Mud Modified Saw Dust Biochar. *J. Water Process Eng.* **2020**, *33*, 101076. [CrossRef]
86. Shi, Y.; Zhang, T.; Ren, H.; Kruse, A.; Cui, R. Polyethylene Imine Modified Hydrochar Adsorption for Chromium (VI) and Nickel (II) Removal from Aqueous Solution. *Bioresour. Technol.* **2018**, *247*, 370–379. [CrossRef] [PubMed]
87. Dehmani, Y.; Sellaoui, L.; Al-Ghamdi, Y.O.; Lainé, J.; Badawi, M.; Amhoud, A.; Bonilla-Petriciolet, A.; Lamhasni, T.; Abouarnadasse, S. Kinetic, Thermodynamic and Mechanism Study of the Adsorption of Phenol on Moroccan Clay. *J. Mol. Liq.* **2020**, *312*, 113383. [CrossRef]
88. Barros, L.; Carvalho, A.M.; Ferreira, I.C.F.R. The Nutritional Composition of Fennel (Foeniculum Vulgare): Shoots, Leaves, Stems and Inflorescences. *Lebensm.-Wiss. Technol.* **2010**, *43*, 814–818. [CrossRef]
89. Anwar, J.; Shafique, U.; Waheed-Uz-Zaman; Salman, M.; Dar, A.; Anwar, S. Removal of Pb(II) and Cd(II) from Water by Adsorption on Peels of Banana. *Bioresour. Technol.* **2010**, *101*, 1752–1755. [CrossRef]
90. Gupta, V.K.; Ali, I. Removal of Lead and Chromium from Wastewater Using Bagasse Fly Ash—A Sugar Industry Waste. *J. Colloid Interface Sci.* **2004**, *271*, 321–328. [CrossRef]
91. Peláez-Cid, A.-A.; Romero-Hernández, V.; Herrera-González, A.M.; Bautista-Hernández, A.; Coreño-Alonso, O. Synthesis of Activated Carbons from Black Sapote Seeds, Characterization and Application in the Elimination of Heavy Metals and Textile Dyes. *Chin. J. Chem. Eng.* **2020**, *28*, 613–623. [CrossRef]
92. Lalhruaitluanga, H.; Jayaram, K.M.; Prasad, M.N.V.; Kumar, K.K. Lead(II) Adsorption from Aqueous Solutions by Raw and Activated Charcoals of *Melocanna baccifera* Roxburgh (Bamboo)—A Comparative Study. *J. Hazard. Mater.* **2010**, *175*, 311–318. [CrossRef] [PubMed]
93. Shukla, S.R.; Pai, R.S. Removal of Pb(II) from Solution Using Cellulose-Containing Materials. *J. Chem. Technol. Biotechnol.* **2005**, *80*, 176–183. [CrossRef]

94. Gaur, N.; Kukreja, A.; Yadav, M.; Tiwari, A. Adsorptive Removal of Lead and Arsenic from Aqueous Solution Using Soya Bean as a Novel Biosorbent: Equilibrium Isotherm and Thermal Stability Studies. *Appl. Water Sci.* **2018**, *8*, 98. [CrossRef]
95. Abugu, H.O.; Okoye, P.A.C.; Ajiwe, V.I.E.; Omuku, P.E.; Umeobika, U.C. Preparation and Characterization of Activated Carbon Produced from Oil Bean (Ugba or Ukpaka) and Snail Shell. *J. Environ. Anal. Chem.* **2015**, *2*, 165. [CrossRef]
96. Chijioke, U.; Benjamin, D.; Anayochukwu, E.; Onyeka, A.; Ogechi, A.; Ifeanyi, S.; Abiola, O. Chromium Adsorption Using Modified Locust Bean and Maize Husk. *Pharma Chem.* **2020**, *12*, 7–14.
97. Abugu, H.O.; Okoye, P.A.C.; Ajiwe, V.I.E.; Ofordile, P.C. Preparation and Characterisation of Activated Carbon from Agrowastes Peanut Seed (African Canarium) and Palm Kernel Shell. *Int. J. Innov. Res. Dev.* **2014**, *3*, 418–441.

Disclaimer/Publisher's Note: The statements, opinions and data contained in all publications are solely those of the individual author(s) and contributor(s) and not of MDPI and/or the editor(s). MDPI and/or the editor(s) disclaim responsibility for any injury to people or property resulting from any ideas, methods, instructions or products referred to in the content.

Article

Effects of Domestic Sewage on the Photosynthesis and Chromium Migration of *Coix lacryma-jobi* L. in Chromium-Contaminated Constructed Wetlands

Yu Nong, Xinyi Liu, Zi Peng, Liangxiang Li, Xiran Cheng, Xueli Wang, Zhengwen Li, Zhigang Li * and Suli Li *

College of Agriculture, Guangxi University, Nanning 530004, China; nongyugxu@163.com (Y.N.); himiyosa@163.com (X.L.); pengyingzimm@163.com (Z.P.); 18589909986@163.com (L.L.); cxrgxu@163.com (X.C.)
* Correspondence: lizhigangnn@163.com (Z.L.); lisuli88@163.com (S.L.)

Abstract: To investigate the effects of domestic sewage on the photosynthesis and chromium migration of plants in chromium-contaminated constructed wetlands, small vertical flow constructed wetlands of *Coix lacryma-jobi* L. were set up. These wetlands were used to treat wastewater containing 0, 20, and 40 mg/L of hexavalent chromium (Cr (VI)), prepared with domestic sewage (DS), 1/2 Hoagland nutrient solution (NS), and 1/2 Hoagland nutrient solution prepared with domestic wastewater (DN), respectively. The aim was to investigate the effects of domestic sewage on indicators, such as plant growth and chromium accumulation. The results were as follows: (1) Plant heights were significantly inhibited under 20 mg/L and 40 mg/L Cr (VI) treatments, and stem diameters were not significantly affected. The use of domestic sewage in treatment alleviated the inhibition of Cr (VI) on the growth of *Coix lacryma-jobi* L. (2) Indicators such as root activity, photosynthetic gas exchange, and chlorophyll fluorescence properties significantly decreased with the increase in Cr (VI) concentration. The values of these photosynthetic gas exchange parameters under the DN treatment were the greatest, followed by NS and DS. On the 70th day of Cr (VI) treatment, the net photosynthetic rate (Pn) under the DN treatment was significantly higher than that under NS and DS treatments. (3) Glutathione (GSH) content in roots, stems, and leaves of *Coix lacryma-jobi* L. significantly increased with the increase in Cr concentration, and it increased more significantly under the DN and DS treatments than under the NS treatment. (4) With the same Cr treatment, the Cr content in roots, stems, and leaves of *Coix lacryma-jobi* L. under the NS treatment was the highest, followed by DS and DN. The total Cr content in the substrate under the DN treatment was the highest, followed by DS and NS. (5) The addition of domestic sewage reduced the Cr (VI) content in the water sample and increased the organic matter content. The Cr (VI) content in the water sample under the NS treatment was the highest, followed by DS and DN. The addition of domestic sewage increased the accumulation of chromium in the substrate, decreased the absorption of chromium by plants, increased GSH content in roots, stems, and leaves, alleviated the damage of Cr (VI) to plants, and thus benefited the growth of *Coix lacryma-jobi* L. in the constructed wetlands and ensured the sustainable and stable operation of the wetlands.

Keywords: hexavalent chromium; glutathione; domestic sewage; *Coix lacryma-jobi* L.; constructed wetland; photosynthetic parameter

Citation: Nong, Y.; Liu, X.; Peng, Z.; Li, L.; Cheng, X.; Wang, X.; Li, Z.; Li, Z.; Li, S. Effects of Domestic Sewage on the Photosynthesis and Chromium Migration of *Coix lacryma-jobi* L. in Chromium-Contaminated Constructed Wetlands. *Sustainability* **2023**, *15*, 10250. https://doi.org/10.3390/su151310250

Academic Editor: Said Muhammad

Received: 13 May 2023
Revised: 19 June 2023
Accepted: 26 June 2023
Published: 28 June 2023

Copyright: © 2023 by the authors. Licensee MDPI, Basel, Switzerland. This article is an open access article distributed under the terms and conditions of the Creative Commons Attribution (CC BY) license (https://creativecommons.org/licenses/by/4.0/).

1. Introduction

Chromium (Cr) exists mainly in the forms of Cr (III) and Cr (VI) in nature. Cr (III) is relatively stable in soil. A small amount of it in the human body can promote carbohydrate metabolism, but an excessively high level can be toxic to the human body [1]. Cr (VI) is highly mobile in soil and easily enriched in crops and may ultimately be absorbed by the human body. Cr (VI) is highly toxic to the human body, with strong mutagenic, carcinogenic, and teratogenic effects [2,3].

Cr is widely used in industries, such as electroplating, printing and dyeing, papermaking, and leathermaking. Both urban and rural industrialization accelerate amid social development, and improper discharge of Cr-containing wastewater can still be seen in China due to supervision, technology, and other reasons. For example, in January 2019, several business owners in Henan Province were reported by Dahebao to have discharged wastewater with an excessive level of Cr during sheepskin dyeing, causing serious environmental pollution [4]. In March 2022, the Wanxiu Court of Wuzhou City reported an incident of illegal discharge of pollutants containing elements, such as nickel and Cr from an electroplating workshop in Wuzhou City [5]. Improper discharge of industrial Cr not only threatens the safety of crop production, but also poses a risk to human health.

Taking technical measures to reduce the Cr content in wastewater is an important way to prevent and control Cr pollution. Constructed wetlands, with the advantages of low construction and operation costs, easy maintenance and management, and high ecological benefits [6], have been widely used in the purification of industrial wastewater and domestic sewage. However, in underdeveloped regions or countries, such as India, Brazil, and South Africa, it is common to find mixed discharge of industrial wastewater and domestic sewage [7–9]. Therefore, some researchers have noticed the influence of domestic sewage on the role of constructed wetlands in removing heavy metals from wastewater. For example, a study by Lou suggests that mixing domestic sewage can improve the ability of constructed wetlands to remove Fe, Zn, Cu, Cd, and Ni from acid mine wastewater [10]. A study by Chen et al. shows that adding domestic sewage or straw decomposition products can improve the ability of constructed wetlands to remove heavy metals from wastewater [11]. Wang et al. found that the coexistence of domestic sewage and acid mine wastewater can improve the ability of constructed wetlands to remove heavy metals from wastewater [12]. Constructed wetlands play a good role in purifying Cr in wastewater [13]. Li Kai found that a higher content of domestic sewage in mixed polluted wastewater can better benefit the purification of Cr (VI) and Ni (II) by Leersia hexandra Swartz constructed wetlands [14]. A study by Li Shuai shows that the use of domestic sewage to prepare 1/2 Hoagland nutrient solution can significantly improve the ability of *Coix lacryma-jobi* L. constructed wetlands to remove Cr from wastewater, which has a better treatment effect and results in better plant growth than irrigation with only domestic sewage or 1/2 Hoagland nutrient solution [15]. Organic matter and microorganisms in domestic sewage may facilitate the transformation of Cr (VI) to Cr (III) in the substrate, increase the adsorption of Cr by the substrate, and reduce the absorption of Cr by plants [16]. Therefore, it is believed that adding domestic sewage may reduce the absorption of Cr by plants, promote the accumulation of Cr in the substrate, and thus alleviate the harm of Cr on plants and benefit the sustainable and efficient operation of constructed wetlands. However, not many studies have been conducted in this field.

Constructed wetlands mainly rely on the absorption of plants and substrates and the effects of microorganisms to remove pollutants from wastewater. How the plants grow is crucial for the continuous and efficient treatment of wastewater by constructed wetlands [17]. Photosynthetic capacity directly reflects the growth of plants, and photosynthetic parameters are extremely sensitive to adversity and stress. Sun et al. [18] found that Cr (VI) stress significantly reduced stomatal conductance (Gs) and the activity of photosystem II (PSII) in wheat leaves, leading to a significant decrease in photosynthetic rate (Pn). Under Cr (VI) stress, plants produce excessive reactive oxygen species (ROS), which can cause serious damage to the photosynthetic system of plants. As a result, the photosynthetic gas exchange and chlorophyll fluorescence properties of plants are inhibited, and the degree of inhibition can reflect the degree of Cr (VI) stress on plants [19]. Glutathione (GSH) generated in plants is a low-molecular-weight antioxidant that can enhance important functions, such as reducing intracellular peroxides, scavenging free radicals, and chelating heavy metals. Fatma et al. [20] found that the toxicity of Cr to the photosynthesis of wheat can be mitigated by the exogenous addition of S and NO, essentially by increasing the level of GSH, and thus mitigating the toxic effects of Cr on photosynthesis and the activity of

Calvin cycle enzymes in leaves. Whether the effect of Cr (VI) on the photosynthesis of *Coix lacryma-jobi* L. is mitigated by regulating the GSH content in leaves under the condition of domestic sewage added needs to be studied in depth.

In this study, miniature *Coix lacryma-jobi* L. vertical flow constructed wetlands were set up, and wastewater containing different concentrations of Cr (VI) was prepared by adding domestic sewage, 1/2 Hoagland nutrient solution, and 1/2 Hoagland nutrient solution prepared with domestic wastewater. The effects of adding domestic sewage and nutrients on plant growth, photosynthetic system, and GSH content of the plants in chromium-contaminated constructed wetlands were studied to reveal the physiological mechanism of domestic wastewater mitigating the effect of Cr (VI) on plants. The results can provide a theoretical basis and new ideas for the efficient treatment of Cr-polluted wastewater by constructed wetlands.

2. Material and Methods

2.1. Material

Coix lacryma-jobi L., a wild wetland plant in Guangxi, is provided by the Crops Research Institute of Guangxi Academy of Agricultural Sciences.

2.2. Experiment Design

This study was conducted from March to September 2022 at the teaching and research base of the College of Agriculture, Guangxi University. Miniature vertical flow constructed wetlands were constructed with a study by Li et al. as a reference [21]. Large plastic barrels with a top diameter of 71 cm, a bottom diameter of 45 cm, and a height of 61 cm were used. The barrels were filled with 10 cm of large cobblestones (diameters ranging from 3 to 5 cm) and 40 cm of river sand (particle size ranging from 0.25 to 0.5 mm) from the bottom up, and a tap was installed 10 cm above the bottom for drainage. Twelve uniformly grown *Coix lacryma-jobi* L. seedlings were planted in each barrel. Within 1 month after planting, the seedlings were irrigated with 1/2 Hoagland nutrient solution. Cr (VI)-containing wastewater treatment was performed after the seedlings grew to 20 cm. Domestic wastewater (DS, main indicators, COD: 200~205 mg/L, TP: 0.78~1.18 mg/L, TN: 8.16~12.33 mg/L, NH_3-N: 15.98~22.56 mg/L. Wastewater was collected from sewage pipes on each water inflow day, with its main indicators measured. The wastewater was not pretreated). Thereafter, 1/2 Hoagland nutrient solution (NS) and 1/2 Hoagland nutrient solution prepared with domestic wastewater (DN) were used as mother liquor, respectively. To ensure the stability of Cr (VI) in the incoming water, $K_2Cr_2O_7$ solution was added to the mother liquor 2 h before each inflow to simulate wastewater containing 0, 20, and 40 mg/L of Cr (VI) (no significant change in Cr (VI) content was detected within 2 h). After entering the constructed wetlands, the wastewater containing Cr (VI) remained for 3 days after the water inflow, followed by a 4-day drying cycle, making a 7-day intermittent cycle. Each treatment was repeated three times, on five wetland pools per repetition. Each wetland tank was filled with 30 L of water by each water inflow.

2.3. Sample Collection

Coix lacryma-jobi L. plant samples and water samples were collected 10, 40, and 70 days after Cr (VI) treatment, and substrate samples were collected 48 days after the treatment. The roots, stems, and leaves of the samples were separated, rinsed with deionized water, and filled into sample bags after the surface was dried. They were kept fresh at 4 °C and brought to the laboratory. The samples were cut into small pieces and stored in sealed bags in a freezer with an ultra-low temperature of −80 °C for further measurement of the total chromium content in various organs of *Coix lacryma-jobi* L. The root tip was used for root activity measurement. The water samples were collected at the outlet and stored in PE plastic bottles with a capacity of 100 mL, and the bottles were stored in an insulation box with ice packs and sent to the laboratory. After being filtered through a 0.45 μm water filter membrane, the water samples were placed in centrifuge tubes with a capacity of 50 mL at

4 °C for the determination of total Cr content in the water discharged. The surface substrate was randomly sampled to obtain 200 g of fresh sample, which was placed in sealed bags and air-dried in a drying room. The substrate samples were used for the measurement of total Cr and total organic carbon content in the substrate.

Root, stem, and leaf samples were collected 20 and 50 days after the Cr (VI) treatment for GSH content measurement in different organs.

2.4. Experimental Methods

2.4.1. Measurement of Stem Diameters and Plant Heights

Six uniformly grown plants were selected from each repetition 10, 40, and 70 days after Cr (VI) treatment, and plant heights and stem diameters were measured. The plant height is the above-ground length, and the stem diameter is the diameter of the second above-ground node. According to the method developed by Long et al. [22], the inhibition rate of Cr (VI) treatment on plant height and stem diameter was calculated as follows:

$$\text{Inhibition rate of plant}(\%) = \frac{\text{Plant height of CK} - \text{Height of plants treated with Cr (VI)}}{\text{Plant height of CK}} \times 100\% \quad (1)$$

$$\text{Inhibition rate of stem diameter}(\%) = \frac{\text{Stem diameter of CK} - \text{Stem diameter treated with Cr (VI)}}{\text{Stem diameter of CK}} \times 100 \quad (2)$$

2.4.2. Measurement of Root Activity

The TTC (triphenyl tetrazolium chloride) method was used to measure root activity [23]. Briefly, 0.5 g of *Coix lacryma-jobi* L. root tip sample was cut into small pieces and added to 5 mL of 0.4% TTC and 5 mL of phosphate buffer (pH = 7.0). The sample was incubated at 37 °C for 3 h and extracted in ethyl acetate for 15 min. The characteristic absorption peak at 458 nm was measured using a spectrophotometer.

2.4.3. Measurement of Photosynthetic Parameters and Chlorophyll Fluorescence Parameters

Photosynthetic parameters were measured using an LI-6400XT portable photosynthesis meter (LI-COR, Lincoln, NE, USA) from 9 to 11 a.m. 10, 40, and 70 days after Cr (VI) treatment, respectively [24]. The chlorophyll fluorescence parameters were measured using an AMP-2100 portable chlorophyll fluorometer (Walz, Effeltrieh, Germany), and the initial fluorescence (Fo), maximum fluorescence (Fm), and maximum photochemical quantum yield of PS II (Fv/Fm) were measured. The same parts of nine effective leaves were selected for each treatment, and one leaf was recorded three times to obtain the mean value for analysis.

2.4.4. Measurement of GSH Content in Different Parts of *Coix lacryma-jobi* L.

GSH was measured using the colorimetric method developed by Xie et al. [25]. Briefly, 1 g of fresh leaf sample was placed in a mortar, and 3 mL of pre-cooled 5% TCA solution and a small amount of quartz sand were added. The sample was thoroughly ground in an ice bath and then centrifuged at a low temperature (4 °C) for 15 min. Then, 0.25 mL of the supernatant liquid was obtained and mixed with 0.5 mL of Tris-HCl buffer solution (0.25 mol·L^{-1}, pH = 8). Thereafter, 0.25 mL of 3% formaldehyde was added, and the solution was shaken well and stood at room temperature for 20 min. Moreover, 3 mL of the DTNB solution pre-warmed in a constant temperature water bath at 25 °C was added and mixed well. After 5 min of standing, the measurement was obtained at 412 nm using a spectrophotometer, and GSH content was calculated based on its standard curve.

2.4.5. Measurement of Cr Content in Different Parts of Coix lacryma-jobi L., Substrate, and the Water Discharged

Based on the method developed by Wang [26], 0.3 g of the sample was placed into a boiling tube and soaked with 6 mL of HNO_3 and 1.5 mL of $HClO_4$ overnight. Thereafter, the sample was boiled and neutralized in a graphite furnace. After cooling, the solution was diluted to 50 mL with 0.2% HNO_3 and filtered. The chromium content was measured using an inductively coupled plasma emission spectrometer (model: ICP-5000, manufacturer: Focused Photonics Inc., origin: Beijing, China).

Based on the method developed by Liu [27], the total Cr content in substrate was measured using an inductively coupled plasma emission spectrometer after the samples were boiled and filtered.

Based on the method developed by Ribas [28], the Cr (VI) content in water was measured using diphenylcarbazide spectrophotometry. The colorimetric tubes used were soaked in 10% dilute nitric acid for over 16 h to prevent Cr (VI) adsorption on the inner walls of the tubes. Based on the method developed by Yue [29], the spectrophotometric method was used to measure the total Cr content in the water discharged.

2.4.6. Measurement of Organic Matter Content in Substrate

Based on the method developed by Wang [30], the potassium dichromate oxidation-external heating method was used to measure the total organic carbon content in the soil samples.

2.5. Statistical Analysis

Origin 2021 was used for plotting, Excel 2019 was used to collate data, SPSS Statistics 25 was used for calculation and statistical analysis, and Duncan's test was used for multiple comparisons of significant differences ($p < 0.05$).

3. Results and Analysis

3.1. Plant Height and Stem Diameter

As shown in Table 1, the plant heights of CK *Coix lacryma-jobi* L. ranged from 22.16 to 140.03 cm, and the plant heights under 20 mg/L and 40 mg/L Cr (VI) treatments ranged from 18.67 to 129.4 cm and 16.23 to 109.23 cm, respectively, with all being significantly shorter than CK. The degree of inhibition was greater with a higher Cr (VI) concentration ($p < 0.05$). The differences between treatments with different Cr (VI) concentrations were significant. With the same Cr concentration, the plant heights of *Coix lacryma-jobi* L. were greatest under 1/2 Hoagland nutrient solution prepared with domestic wastewater (DN), followed by 1/2 Hoagland nutrient solution (NS) and domestic sewage (DS). The plants under the DN treatment were significantly taller than those under NS and DS treatments ($p < 0.05$). With 20 mg/L of Cr (VI), the inhibition rates of *Coix lacryma-jobi* L. plant heights under DN, NS, and DS were 5.33%~75.9%, 15.64%~16.7%, and 13.88%~15.31%, respectively. With 40 mg/L of Cr (VI), the inhibition rates under DN, NS, and, DS were 16.74%~24.45%, 22.81%~28.52%, and 22.39%~26.7%, respectively. The higher the Cr concentration, the more significantly the plant growth is inhibited. The plant heights of *Coix lacryma-jobi* L. irrigated with domestic sewage were less inhibited than those under NS.

The stem diameters of *Coix lacryma-jobi* L. ranged from 8.4 to 12.73 mm, and the diameters under DN were the largest, followed by NS and DS, with insignificant differences.

Table 1. Plant heights and stem diameters under different treatments.

Treatment Time (d)	Irrigation Moisture	Cr (VI) Concentration (mg/L)	Plant Height (cm)	Plant Height Inhibition Rate (%)	Stem Diameter (mm)	Stem Diameter Inhibition Rate (%)
10	DN	CK	22.5 ± 0.72 a	-	8.77 ± 0.21 a	-
		20	21.3 ± 0.72 a	5.33	8.7 ± 0.1 abc	0.76
		40	18.73 ± 0.57 b	16.74	8.47 ± 0.12 cd	3.42
	NS	CK	22.17 ± 0.49 a	-	8.73 ± 0.15 ab	-
		20	18.67 ± 0.68 b	15.79	8.5 ± 0.1 bcd	2.67
		40	16.23 ± 0.7 c	26.77	8.4 ± 0.1 d	3.82
	DS	CK	22.33 ± 1.07 a	-	8.67 ± 0.06 abc	-
		20	19.23 ± 0.45 b	13.88	8.47 ± 0.15 cd	2.31
		40	17.33 ± 0.55 c	22.39	8.4 ± 0.1 d	3.08
40	DN	CK	92.3 ± 1.74 a	-	11.43 ± 0.25 a	-
		20	83.23 ± 1.48 b	9.82	11.17 ± 0.06 abc	2.33
		40	69.73 ± 1.04 d	24.45	11.03 ± 0.15 bcd	3.5
	NS	CK	91.6 ± 1.73 a	-	11.33 ± 0.25 ab	-
		20	76.3 ± 0.8 c	16.70	10.93 ± 0.15 cd	3.53
		40	65.47 ± 1.4 e	28.53	10.83 ± 0.15 cde	4.41
	DS	CK	83.37 ± 0.9 b	-	10.87 ± 0.25 cde	-
		20	70.6 ± 1.47 d	15.31	10.53 ± 0.21 de	3.07
		40	64.53 ± 0.87 e	22.59	10.7 ± 0.2 e	3.68
70	DN	CK	140.03 ± 3.01 a	-	12.77 ± 0.06 a	-
		20	129.4 ± 1.05 b	7.59	12.3 ± 0.2 b	3.66
		40	109.23 ± 1.29 e	21.99	12.03 ± 0.06 bcd	5.74
	NS	CK	139.7 ± 1.41 a	-	12.73 ± 0.15 a	-
		20	118.43 ± 0.83 d	15.64	12.17 ± 0.31 bc	4.45
		40	102.4 ± 0.7 f	22.81	11.93 ± 0.06 cde	6.28
	DS	CK	126.57 ± 2.15 c	-	12.27 ± 0.25 b	-
		20	106.77 ± 1.19 e	15.22	11.73 ± 0.15 de	4.35
		40	98.2 ± 1.47 g	22.41	11.67 ± 0.06 e	4.89

Note: Inhibition rate (%) = $\frac{\text{Plant height of CK} - \text{Height of plants treated with Cr (VI)}}{\text{Plant height of CK}} \times 100\%$. Different lowercase letters indicate significant differences between treatments at the same time ($p < 0.05$).

3.2. Root Activity

The root activity of *Coix lacryma-jobi* L. ranged from 0.475 to 2.418 mg/gFW-h (Figure 1) and decreased with the increase in Cr (VI) concentration. The root activity of the plant showed a trend of increasing and then decreasing as the time of Cr (VI) treatment extended. In each period, the root activity under DN was the highest, followed by NS and DS, and the 40th day of treatment saw the highest root activity. On the 10th day of Cr treatment and with 20 and 40 mg/L of Cr (VI), the differences in root activity under different inflow conditions were not significant. On the 70th day, the root activity under DN was significantly greater than that under NS and DS ($p < 0.05$). Compared with CK, the root activity of *Coix lacryma-jobi* L. under DN, NS, and DS with 20 mg/L of Cr (VI) decreased by 18.53%~42.67%, 23.74%~53.62%, and 18.61%~41.08%, respectively, and that with 40 mg/L of Cr (VI) decreased by 39.07%~61.09%, 42.62%~72.27%, and 36.85%~65.76%, respectively. With the same Cr treatment, the root activity under NS dropped to a greater degree than under DN and DS.

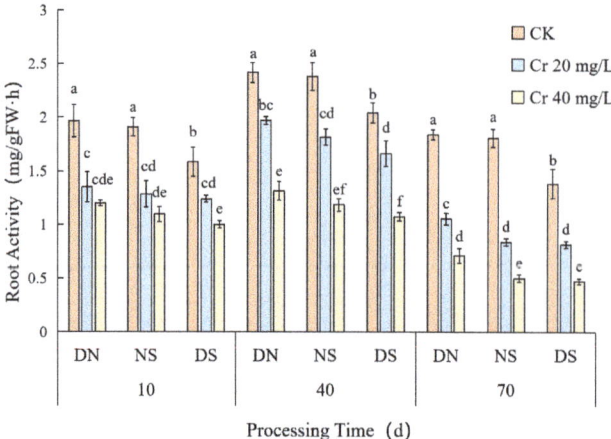

Figure 1. Root activity under different treatments. Note: Different lowercase letters indicate significant differences between treatments at the same time ($p < 0.05$).

3.3. Photosynthetic Gas Exchange Parameters of Coix lacryma-jobi L.

As shown in Table 2, the net photosynthetic rate (Pn) of *Coix lacryma-jobi* L. leaves ranged from 6.787 to 24.32 µmolCO$_2$/m^2·s, stomatal conductance (Gs) ranged from 0.043 to 0.183 mmol/m^{-2}·s^{-1}, intercellular CO$_2$ concentration (Ci) ranged from 115.188 to 273.836 µmol/mol, and transpiration rate (Tr) ranged from 1.32 to 3.69 µmol/m^2·s. All these parameters significantly dropped as the Cr (VI) concentration rose, with similar changing trends. The various photosynthetic parameters increased and then decreased as the Cr (VI) treatment time extended, reaching the highest value on the 40th day of Cr (VI) treatment. With the same Cr concentration, the values of these photosynthetic gas exchange parameters under the DN treatment were the greatest, followed by NS and DS. On the 70th day of Cr (VI) treatment, the net photosynthetic rate (Pn) under DN was significantly greater than that under NS and DS. The gas exchange level under the condition of adding domestic sewage increased more significantly than that under NS from 10 to 40 d of Cr (VI) treatment. From 40 to 70 d, the gas exchange level under DN and DS decreased more significantly than that under NS, indicating that the gas exchange of *Coix lacryma-jobi* L. leaves under the condition of adding domestic sewage maintained a relatively high increase when the plants grew fast, and the rate of gas exchange slightly decreased when the plants grew relatively slow. As the concentration of Cr (VI) increased, the degree to which gas exchange of *Coix lacryma-jobi* L. leaves irrigated by domestic sewage or the mixture containing domestic sewage was inhibited was smaller than that of leaves irrigated by the pure nutrient solution.

Table 2. Photosynthetic gas exchange parameters under different treatments.

Photosynthetic Gas Exchange Parameters	Treatment Time (d)	Water Management	CK	Cr 20 mg/L	Cr 40 mg/L
Net Photosynthetic Rate (µmol/m^2·s)	10	DN	19.2 ± 0.96 a	14.73 ± 0.51 b	11.56 ± 0.48 c
		NS	19 ± 0.98 a	12.12 ± 1.05 c	9.23 ± 0.73 d
		DS	13.96 ± 1.54 b	9.56 ± 1.52 d	6.79 ± 1.09 e
	40	DN	24.32 ± 1.12 a	19.52 ± 0.46 bc	16.13 ± 0.35 ef
		NS	24.22 ± 1.1 a	18.14 ± 1.25 cd	14.34 ± 1.23 fg
		DS	21.29 ± 1.31 b	17.53 ± 0.93 de	14.06 ± 1.43 g
	70	DN	18.6 ± 0.54 a	14.17 ± 0.9 b	11.77 ± 0.75 c
		NS	18.16 ± 1 a	10.17 ± 1.02 d	9.38 ± 0.92 d
		DS	13.94 ± 1.12 b	8.71 ± 1.07 d	6.9 ± 0.66 e

Table 2. Cont.

Photosynthetic Gas Exchange Parameters	Treatment Time (d)	Water Management	CK	Cr 20 mg/L	Cr 40 mg/L
Stomatal conductivity ($mmol \cdot m^{-2} \cdot s^{-1}$)	10	DN	0.177 ± 0.005 a	0.126 ± 0.011 b	0.116 ± 0.005 bc
		NS	0.169 ± 0.013 a	0.118 ± 0.014 bc	0.098 ± 0.016 cd
		DS	0.131 ± 0.017 b	0.109 ± 0.015 bc	0.085 ± 0.008 d
	40	DN	0.183 ± 0.016 a	0.143 ± 0.013 bc	0.124 ± 0.008 bcd
		NS	0.182 ± 0.016 a	0.138 ± 0.009 bc	0.119 ± 0.019 cd
		DS	0.15 ± 0.013 b	0.124 ± 0.012 bcd	0.105 ± 0.014 d
	70	DN	0.101 ± 0.007 a	0.073 ± 0.004 bc	0.064 ± 0.007 c
		NS	0.097 ± 0.007 a	0.064 ± 0.003 c	0.049 ± 0.004 d
		DS	0.081 ± 0.011 b	0.053 ± 0.004 d	0.043 ± 0.004 d
Intercellular CO_2 Concentration (μmol/mol)	10	DN	228.7 ± 15.97 a	200.16 ± 17.52 ab	163.49 ± 19.5 cde
		NS	225.49 ± 17.25 a	191.74 ± 16.27 bc	155.87 ± 17.39 de
		DS	182.08 ± 16.38 bcd	167.51 ± 22.77 bcde	147.29 ± 14.94 e
	40	DN	273.84 ± 24.18 a	238.79 ± 16.91 ab	213.15 ± 26.66 bc
		NS	272.93 ± 22.5 a	231.44 ± 18.14 bc	207.63 ± 24.05 bc
		DS	244 ± 22.26 ab	217.99 ± 17.75 bc	192.75 ± 20.76 c
	70	DN	176.43 ± 15.26 a	150.15 ± 19.67 abc	124.49 ± 12.42 cd
		NS	176.4 ± 16.15 a	147.33 ± 19.8 abc	120.74 ± 11.32 cd
		DS	163.66 ± 16.05 ab	136.98 ± 16.92 bcd	115.19 ± 14.79 d
Transpiration Rate (μmol/m²·s)	10	DN	2.64 ± 0.22 a	2.1 ± 0.15 b	1.8 ± 0.24 bcd
		NS	2.61 ± 0.21 a	1.97 ± 0.18 bc	1.62 ± 0.24 cd
		DS	2.13 ± 0.21 b	1.69 ± 0.23 cd	1.46 ± 0.19 d
	40	DN	3.69 ± 0.17 a	3.36 ± 0.16 ab	2.99 ± 0.19 bcd
		NS	3.66 ± 0.19 a	3.23 ± 0.19 b	2.82 ± 0.25 cd
		DS	3.13 ± 0.19 bc	2.98 ± 0.28 bcd	2.66 ± 0.23 d
	70	DN	2.42 ± 0.29 a	1.99 ± 0.25 b	1.66 ± 0.04 bcd
		NS	2.41 ± 0.3 a	1.83 ± 0.19 bc	1.52 ± 0.17 cd
		DS	1.96 ± 0.17 b	1.57 ± 0.21 cd	1.34 ± 0.11 d

Note: Different lowercase letters indicate significant differences between treatments at the same time ($p < 0.05$).

3.4. Chlorophyll Fluorescence Properties of Coix lacryma-jobi L.

Initial Fluorescence (Fo) and Maximum Fluorescence (Fm)

The initial fluorescence (*Fo*) of leaves reflects the fluorescence yield when the PSII reaction center is fully open, and the *Fo* of *Coix lacryma-jobi* L. leaves increased significantly ($p < 0.05$) with the increase in Cr (VI) concentration and decreased. Then, it increased as the time of Cr (VI) treatment extended (Table 3). The maximum fluorescence (*Fm*) can show the electron transfer of PSII, and the trend of *Fm* is opposite to that of *Fo*. *Fv/Fm* is the light energy conversion efficiency in the PSII reaction center, which is related to the degree of photoinhibition of *Coix lacryma-jobi* L. The *Fv/Fm* of *Coix lacryma-jobi* L. leaves ranged from 0.65 to 0.888, which decreased significantly ($p < 0.05$) with the increasing Cr (VI) concentration, indicating a decrease in photosynthetic capacity. Compared with the control, the maximum photochemical quantum yields under DN, NS, and DS under the 20 mg/L Cr (VI) treatment dropped by 4.14%~9.56%, 6.04%~12.6%, and 1.9%~5.2%, respectively, and by 7.21%~15.38%, 9.07%~18.53%, and 5.57%~13.44%, respectively under the 40 mg/L Cr (VI) treatment. With the same Cr concentration, the decrease under NS was the most significant, followed by DN and DS.

Table 3. Chlorophyll fluorescence parameters under different treatments.

Chlorophyll Fluorescence Parameters	Treatment Time (d)	Water Management	CK	Cr 20 mg/L	Cr 40 mg/L
F_o	10	DN	0.225 ± 0.016 d	0.249 ± 0.006 cd	0.276 ± 0.02 bc
		NS	0.226 ± 0.014 d	0.264 ± 0.011 bc	0.292 ± 0.019 ab
		DS	0.264 ± 0.013 bc	0.278 ± 0.017 bc	0.314 ± 0.021 a
	40	DN	0.189 ± 0.012 e	0.229 ± 0.01 d	0.258 ± 0.012 abc
		NS	0.198 ± 0.017 e	0.246 ± 0.013 bcd	0.271 ± 0.019 ab
		DS	0.241 ± 0.011 cd	0.25 ± 0.015 bcd	0.282 ± 0.012 a
	70	DN	0.25 ± 0.014 d	0.318 ± 0.01 c	0.333 ± 0.009 bc
		NS	0.251 ± 0.011 d	0.331 ± 0.029 bc	0.358 ± 0.012 ab
		DS	0.31 ± 0.016 c	0.328 ± 0.025 bc	0.37 ± 0.027 a
F_m	10	DN	1.444 ± 0.062 a	1.291 ± 0.061 b	1.19 ± 0.029 bc
		NS	1.431 ± 0.065 a	1.245 ± 0.074 bc	1.121 ± 0.075 cd
		DS	1.178 ± 0.078 bc	1.126 ± 0.068 cd	1.014 ± 0.074 d
	40	DN	1.779 ± 0.182 a	1.601 ± 0.042 b	1.511 ± 0.017 bc
		NS	1.773 ± 0.163 a	1.485 ± 0.059 bc	1.41 ± 0.053 c
		DS	1.476 ± 0.064 bc	1.395 ± 0.051 c	1.343 ± 0.057 c
	70	DN	1.362 ± 0.067 a	1.216 ± 0.055 bc	1.078 ± 0.056 cd
		NS	1.357 ± 0.072 a	1.16 ± 0.098 bcd	1.071 ± 0.095 d
		DS	1.245 ± 0.089 ab	1.148 ± 0.093 bcd	1.06 ± 0.04 d
F_v/F_m	10	DN	0.844 ± 0.017 a	0.807 ± 0.006 b	0.768 ± 0.018 cd
		NS	0.841 ± 0.016 a	0.788 ± 0.011 bc	0.738 ± 0.029 d
		DS	0.775 ± 0.026 bc	0.752 ± 0.03 cd	0.69 ± 0.003 e
	40	DN	0.894 ± 0.006 a	0.857 ± 0.007 b	0.829 ± 0.01 cd
		NS	0.888 ± 0.013 a	0.834 ± 0.015 bc	0.807 ± 0.021 de
		DS	0.837 ± 0.014 bc	0.821 ± 0.017 cd	0.79 ± 0.013 e
	70	DN	0.816 ± 0.018 a	0.738 ± 0.012 bc	0.69 ± 0.012 cde
		NS	0.815 ± 0.017 a	0.712 ± 0.043 bcd	0.664 ± 0.039 de
		DS	0.751 ± 0.017 b	0.712 ± 0.043 bcd	0.65 ± 0.039 e

Note: Different lowercase letters indicate significant differences between treatments at the same time ($p < 0.05$).

3.5. Glutathione (GSH) Content in Different Organs of Coix lacryma-jobi L.

According to Table 4, GSH content in the roots, stems, and leaves of *Coix lacryma-jobi* L. ranged from 36.56 to 103.14 μg/gFW, 18.64 to 53.13 μg/gFW, and 200.47 to 434.79 μg/gFW, respectively. With the same Cr concentration, the values of GSH content in roots, stems, and leaves of the plant were the highest under NS, followed by DS and DN (except for CK). Specifically, the GSH content in roots and leaves under NS was significantly higher than that under DN and DS ($p < 0.05$). GSH content in various parts of the plant increased to certain degrees as the time of treatment extended. With the same Cr concentration, the GSH content under NS increased from 5.57% to 18% over time, while that under DS and DN increased from 8.28% to 35.92% (except for CK). With domestic sewage added, the GSH content increased more significantly over time than under NS.

Compared with the control, values of GSH content in roots, stems, and leaves significantly increased after treatment with both 20 mg/L and 40 mg/L of Cr (VI). On the 20th and 50th days of Cr treatment with 40 mg/L of Cr (VI), the values of GSH content in roots, stems, and leaves under DN were 30.27% and 20.56%, 13.96% and 13.58%, and 30.28% and 13.08% higher than those under the treatment with 20 mg/L of Cr (VI); 9.98% and 7.13%, 11.46% and 11.14%, and 5.88% and 6.33% higher than the latter under NS; 29.8% and 27.67%, 19.06% and 15.26%, and 24.63% and 14.29% higher than the latter under DS. With domestic sewage added, the GSH content increased more significantly than under NS.

Table 4. The GSH content (μg/gFW) in different parts of *Coix lacryma-jobi* L. under different treatments.

Organs	Treatment Time (d)	Water Management	CK	Cr 20 mg/L	Cr 40 mg/L
Root	20	DN	36.915 ± 1.954 e	42.296 ± 2.197 de	55.099 ± 4.183 c
		NS	36.563 ± 2.489 e	74.323 ± 5.431 b	90.784 ± 5.431 a
		DS	40.959 ± 1.847 de	47.022 ± 2.409 d	61.037 ± 4.262 c
	50	DN	40.024 ± 6.117 f	57.49 ± 4.041 e	69.312 ± 2.781 cd
		NS	41 ± 4.818 f	86.561 ± 6.624 b	103.137 ± 7.757 a
		DS	41.532 ± 4.356 f	61.265 ± 6.836 de	78.218 ± 5.273 bc
Stem	20	DN	36.915 ± 1.954 e	42.296 ± 2.197 cd	55.099 ± 4.183 bc
		NS	36.563 ± 2.489 de	74.323 ± 5.431 ab	90.784 ± 5.431 a
		DS	40.959 ± 1.847 de	47.022 ± 2.409 bc	61.037 ± 4.262 a
	50	DN	40.024 ± 6.117 e	57.49 ± 4.041 d	69.312 ± 2.781 b
		NS	41 ± 4.818 e	86.561 ± 6.624 bc	103.137 ± 7.757 a
		DS	41.532 ± 4.356 e	61.265 ± 6.836 c	78.218 ± 5.273 a
Leaf	20	DN	36.915 ± 1.954 f	42.296 ± 2.197 e	55.099 ± 4.183 bc
		NS	36.563 ± 2.489 f	74.323 ± 5.431 c	90.784 ± 5.431 a
		DS	40.959 ± 1.847 f	47.022 ± 2.409 d	61.037 ± 4.262 b
	50	DN	40.024 ± 6.117 e	57.49 ± 4.041 d	69.312 ± 2.781 bc
		NS	41 ± 4.818 e	86.561 ± 6.624 b	103.137 ± 7.757 a
		DS	41.532 ± 4.356 e	61.265 ± 6.836 cd	78.218 ± 5.273 b

Note: Different lowercase letters indicate significant differences between treatments at the same time ($p < 0.05$).

3.6. Cr Content in Roots, Stems, and Leaves of Coix lacryma-jobi L.

The total Cr content in roots, stems, and leaves of *Coix lacryma-jobi* L. ranged from 9.16 to 236.16, 1.94 to 80.14, and 3.22 to 91.67 mg/kg, respectively, with the highest Cr content in roots, followed by that in leaves and stems. The Cr content in roots, stems, and leaves increased as the Cr (VI) concentration increased and the treatment time extended. In addition, there were significant differences between treatments with different Cr (VI) concentrations ($p < 0.05$) (Figure 2).

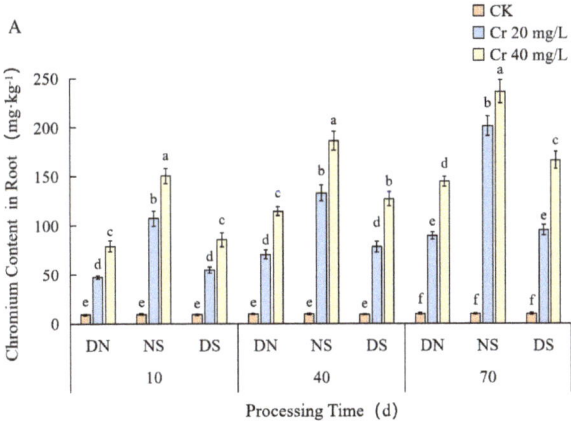

Figure 2. Cont.

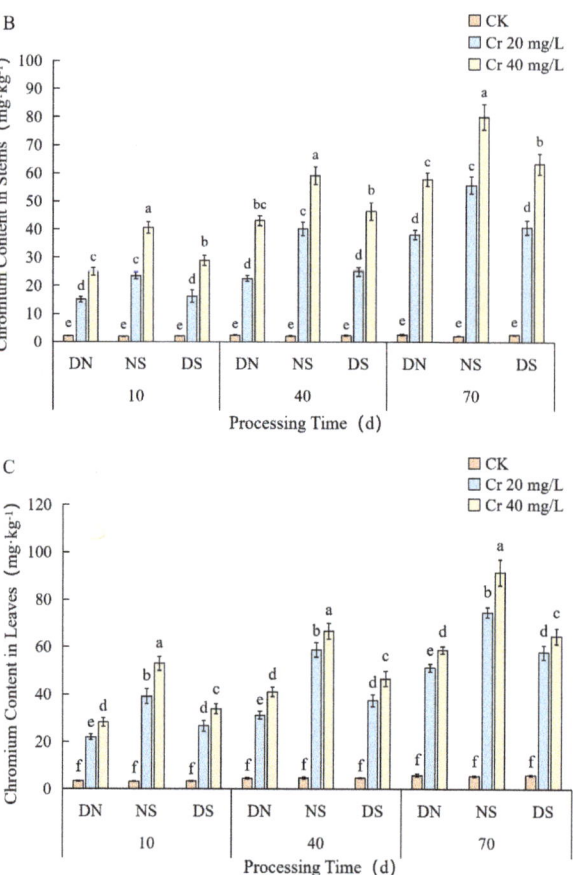

Figure 2. Total Cr content in roots, stems, and leaves of *Coix lacryma-jobi* L. under irrigation conditions with different Cr concentrations. Note: (**A**) shows the changes in the total Cr content in the roots of *Coix lacryma-jobi* L. with different treatments, (**B**) shows the changes in the stems of the plant, and (**C**) shows the changes in the leaves of the plant. Different lowercase letters indicate significant differences between treatments at the same time ($p < 0.05$).

Under DN, the Cr content in roots, stems, and leaves of *Coix lacryma-jobi* L. ranged from 9.16 to 144.7, 2.09 to 57.99, and 3.39 to 58.66 mg/kg, respectively. The figures under NS and DS were 9.48~236.16, 1.94~80.14, and 3.22~91.67 mg/kg, and 9.16~166.16, 2.21~63.45, and 3.34~64.7 mg/kg. With the same Cr (VI) concentration, the values of Cr content in roots, stems, and leaves of *Coix lacryma-jobi* L. under NS were higher than those under DN and DS.

3.7. Organic Matter Content in Substrate

As shown in Table 5, the organic matter content under the three treatments ranged from 18.17 to 34.12 mg/kg (DN), 11.27 to 27.11 mg/kg (NS), and 16.94 to 31.03 mg/kg (DS). Organic matter content decreased as the Cr (VI) concentration increased and increased as the treatment time extended. The organic matter content under both DN and DS was greater than that under NS with the same treatment time and Cr (VI) concentration. The organic matter content was significantly greater ($p < 0.05$) under DN and DS than under NS on the 10th and 40th days of the Cr (VI) treatment, but on the 70th day, the differences in organic matter content between different conditions were not significant.

Table 5. Organic matter content (mg/kg) under different treatments.

Treatment Time (d)	Water Management	CK	Cr 20 mg/L	Cr 40 mg/L
10	DN	26.32 ± 2.13 a	21.77 ± 2.37 bc	18.17 ± 1.87 cde
	NS	15.17 ± 1.23 ef	12.31 ± 2.76 f	11.27 ± 1.58 f
	DS	24.95 ± 2.49 ab	19.91 ± 2.58 cd	16.94 ± 2.23 de
40	DN	32.69 ± 1.74 a	25.98 ± 2.81 bc	23.24 ± 2.01 cd
	NS	20.15 ± 2.9 de	16.16 ± 2.57 ef	14.23 ± 3.32 f
	DS	30.33 ± 1.77 ab	24.06 ± 2.6 cd	21.66 ± 3.64 cd
70	DN	34.12 ± 3.62 a	27.31 ± 3.29 c	24.94 ± 3.58 bcd
	NS	27.11 ± 4.11 bc	21.07 ± 3.65 cd	18.78 ± 2.82 d
	DS	31.03 ± 3.14 ab	25.68 ± 4.09 bc	22.38 ± 3.62 cd

Note: Different lowercase letters indicate significant differences between treatments at the same time ($p < 0.05$).

3.8. Total Cr Content in Substrate and the Water Discharged

The total Cr content in substrate (Table 6) ranged from 6.74 to 183.44 mg/kg under DN, 3.8 to 174.67 mg/kg under NS, and 7.21 to 189.78 mg/kg under DS, and the total Cr content in substrate increased significantly ($p < 0.05$) as the concentration of Cr (VI) treatment increased and the treatment time extended. The total Cr content in the discharged water ranged from 0.06 to 0.72 mg/L under DN, 0.03~0.86 mg/L under NS, and 0.07~0.79 mg/L under DS, and the Cr (VI) content in the discharged water under the three treatments was 0.03~0.24 (DN), 0.03~0.38 (NS), and 0.03~0.25 mg/L (DS), respectively (Table 6). NS saw the highest total Cr content in substrate and Cr and Cr (VI) content in the discharged water, followed by DS and DN. Therefore, the DN treatment had the best effect in treating chromium-containing sewage.

Table 6. Total Cr content in substrate and total Cr content and Cr (VI) content in the water discharged under different treatments.

Treatment Time (d)	Water Management	Cr Content in the SUBSTRATE (mg/kg)			Cr content in the Water Discharged (mg/L)				Cr (VI) Content in the Water Discharged (mg/L)		
		CK	Cr 20 mg/L	Cr 40 mg/L	CK	Cr 20 mg/L	Cr 40 mg/L		CK	Cr 20 mg/L	Cr 40 mg/L
10	DN	6.74 ± 1.57 d	42.82 ± 3.16 bc	66.55 ± 5.23 a	0.06 ± 0.02 d	0.24 ± 0.03 c	0.43 ± 0.05 b		0.03 ± 0.01 d	0.08 ± 0.01 c	0.15 ± 0.02 b
	NS	3.8 ± 0.59 d	39.79 ± 5.46 c	62.78 ± 6.16 a	0.03 ± 0.01 d	0.3 ± 0.04 c	0.56 ± 0.03 a		0.03 ± 0.01 d	0.16 ± 0.01 b	0.19 ± 0.01 a
	DS	7.21 ± 1.4 d	47.89 ± 3.8 b	69.23 ± 5.86 a	0.07 ± 0.03 d	0.27 ± 0.06 c	0.5 ± 0.07 ab		0.03 ± 0.01 d	0.15 ± 0.02 b	0.15 ± 0.02 b
40	DN	8.69 ± 1.48 c	76 ± 4.37 b	141.87 ± 6.59 a	0.08 ± 0.03 e	0.36 ± 0.04 d	0.65 ± 0.05 b		0.04 ± 0.01 d	0.14 ± 0.02 c	0.19 ± 0.02 b
	NS	4.79 ± 0.59 c	72.88 ± 9.11 b	136.16 ± 7.28 a	0.04 ± 0.02 e	0.45 ± 0.04 c	0.77 ± 0.05 a		0.04 ± 0.02 d	0.22 ± 0.02 b	0.31 ± 0.02 a
	DS	9.95 ± 0.96 c	82.27 ± 6.09 b	147.45 ± 10.94 a	0.1 ± 0.03 e	0.41 ± 0.09 cd	0.69 ± 0.05 ab		0.05 ± 0.01 d	0.2 ± 0.02 b	0.2 ± 0.02 b
70	DN	9.7 ± 1.17 d	135.27 ± 5.44 c	183.44 ± 11.62 ab	0.11 ± 0.05 e	0.44 ± 0.04 d	0.72 ± 0.06 b		0.05 ± 0.02 e	0.19 ± 0.02 d	0.24 ± 0.02 c
	NS	4.95 ± 0.62 d	129.67 ± 8.4 c	174.67 ± 7.81 b	0.04 ± 0.01 e	0.57 ± 0.02 c	0.86 ± 0.06 a		0.04 ± 0.01 e	0.29 ± 0.02 b	0.38 ± 0.02 a
	DS	10.62 ± 0.83 d	140.47 ± 8.47 c	189.78 ± 10.48 a	0.12 ± 0.04 e	0.53 ± 0.06 c	0.79 ± 0.08 ab		0.05 ± 0.01 e	0.24 ± 0.02 c	0.25 ± 0.02 c

Note: Different lowercase letters indicate significant differences between treatments at the same time ($p < 0.05$).

4. Discussion

Cr is not an essential element for plants, and excessive accumulation can inhibit plant growth. Cr (VI) can greatly inhibit plant growth due to its strong oxidative properties [31]. A study by Adhikari et al. showed that Cr (VI) stress increases reactive oxygen species (ROS) in maize seedlings and significantly decreases biomass in sand culture [32]. Liu et al. found that the growth of *Coix lacryma-jobi* L. is significantly inhibited when the Cr (VI) content reaches 20 mg/L [27]. However, studies have also shown that adding domestic sewage to prepare 1/2 Hoagland nutrient solution to constructed wetlands can alleviate the inhibitory effect of Cr (VI) on plant growth [11]. The present study showed that the growth of *Coix lacryma-jobi* L. was significantly inhibited by treatment with both 20 mg/L and 40 mg/L of Cr (VI) and the degree of inhibition increased significantly with the increase in Cr (VI) concentration. Hussain et al. also found that the growth of rice was severely inhibited under Cr (VI) stress [33]. However, the degree of inhibition of the growth of *Coix lacryma-jobi* L. was lower under DN and DS conditions than under the NS conditions, indicating that the use of domestic sewage to prepare 1/2 Hoagland nutrient solution can effectively alleviate the inhibitory effect of Cr (VI) on the growth of the plant. Similarly, a study by Saleem et al. found that the use of press mud as an organic amendment can effectively reduce the impact of Cr (VI) on the growth of *Helianthus annuus* L. [34]. Photosynthesis is an important factor determining how plants grow, and the growth of *Coix lacryma-jobi* L. is inhibited under Cr stress, which is closely related to the inhibition of photosynthesis [35].

Photosynthesis is highly sensitive to heavy metal stress, and many studies have used changes in the parameters of photosynthetic gas exchange as important indicators of plant response to heavy metal stress [36,37]. When plants are subjected to heavy metal stress, the efficiency of photosynthetic gas exchange is significantly inhibited, leading to a decrease in their photosynthetic capacity [38]. A study by Vernay et al. showed that Cr (VI) stress leads to a significant decrease in photosynthetic gas exchange parameters in *Datura innoxia* [39]. The present study found that Pn, Gs, Ci, and Tr of *Coix lacryma-jobi* L. leaves were significantly inhibited under Cr (VI) treatment, which is consistent with the results of previous studies. Chlorophyll fluorescence parameters are important indicators for assessing the effects of adversity and stress on the photosynthetic system of plants. Elevated Fo means reversible inactivation or irreversible damage of PS II reaction centers [40]. Fv/Fm indicates the maximum photochemical efficiency of PS II when all reaction centers are open. When plants are subjected to heavy metal stress, Fv/Fm significantly decreases, and photoinhibition of PS II increases [41]. In this study, Fo of *Coix lacryma-jobi* L. increased under Cr (VI) stress, indicating that the PSII reaction center was damaged to some extent. Fv/Fm decreased with the increasing intensity of Cr (VI) stress, indicating that the plants were subjected to photoinhibition. The parameters of light energy conversion efficiency of *Coix lacryma-jobi* L. leaves decreased significantly as the Cr (VI) increased, indicating that Cr stress causes photoinhibition of plants, which is consistent with previous findings. This may be an important reason for the inhibition of photosynthesis in this study. The DN and DS treatments with domestic sewage can reduce the extent to which photosynthetic gas exchange and chlorophyll fluorescence parameters are inhibited under Cr (VI) stress, indicating that domestic sewage may maintain high photosynthetic capacity by alleviating the damage of Cr (VI) to the photosystem, which may be an important reason for normal plant growth. Domestic sewage contains a large amount of organic matter and microorganisms. Previous studies have shown that organic matter and microorganisms can promote the conversion of Cr (VI) to Cr (III). Cr (III) easily binds with anions or organic matter in substrate to generate an insoluble form, which reduces the absorption of Cr by plants [42]. This may be an important reason why plant growth is less inhibited with domestic sewage added.

The root system of plants is an important part that directly contacts and accumulates Cr, which directly affects the growth of the aboveground parts. Root activity reflects the ability of plants to absorb water and nutrients, and is usually inhibited under heavy metal stress [43]. Guo et al. found that under Cd and Pb stress, the root activity of *Matricaria chamomilla* L. significantly decreased and was positively correlated with the degree of stress [44]. The extent to which the growth of *Coix lacryma-jobi* L. is inhibited by Cr (VI) can be obtained by studying the effect of Cr (VI) on root activity [45]. In this study, the root activity of *Coix lacryma-jobi* L. decreased with the increase in Cr (VI) concentration. Compared with CK, the degree of root activity inhibition under DS and DN was lower than that under NS, indicating that the root activity of *Coix lacryma-jobi* L. was less inhibited with domestic sewage added. The root activity under DN was the highest, and that under NS was higher than under DS. Root activity is closely related to the amount of heavy metals absorbed by plants. A study by Liu et al. suggested that with the increase in Cd accumulation in rice roots, root activity significantly decreased [46]. This study found a similar trend. Root activity under DN and NS was greater than that under DS, indicating that good nutritional conditions facilitate root growth. Under the condition with domestic sewage added, the root activity of *Coix lacryma-jobi* L. was less inhibited, which may be due to the fact that the plant absorbed less Cr.

In this study, Cr content in different parts of *Coix lacryma-jobi* L. under DS and DN was lower than that under NS, while Cr content in substrate under DS and DN was higher than that under NS, indicating that adding domestic sewage to chromium-containing wastewater may promote the conversion of Cr into forms in substrate that are difficult for plants to use. Meanwhile, the absorption of Cr by the root system and transfer of Cr from roots to aboveground parts both decrease, which may be an important reason why the growth, photosynthesis, and root activity of *Coix lacryma-jobi* L. were less inhibited under the condition with domestic sewage added. A study by Yu showed that an increase in organic matter content could promote the chemical form transformation of cadmium and reduce the toxicity of the element in soil [47]. Cr (VI) content in constructed wetlands may be closely related to organic matter content. Xiao found that soluble organic matter is an excellent reducer of Cr (VI), which can promote the conversion of Cr (VI) to Cr (III) [48]. A study by Choppala et al. showed that cattle manure and S can promote the reduction in Cr (VI) and its transformation into forms that are difficult for plants to absorb [49]. Microorganisms play an important role in the chemical form transformation of Cr, which can promote the transformation of Cr (VI) into Cr (III) and reduce its mobility [50]. Domestic sewage contains abundant organic matter and microorganisms, which facilitate the chemical form transformation of Cr. In this study, the organic matter content in the substrate under the treatment with domestic sewage added was significantly higher than when the plants were irrigated with nutrient solution only, which may promote the conversion of Cr (VI) and reduce Cr absorption by plants in the constructed wetlands. Finally, the decrease in the amount of Cr absorbed by *Coix lacryma-jobi* L. may be an important reason why the plant was less inhibited under the condition with domestic sewage added.

Cr (VI) can produce reactive oxygen species in plants due to its strong oxidative properties, leading to peroxidation of cell membrane lipids and causing chain reactions that lead to cell damage [51]. GSH generated in plants can act as an antioxidant and complex with heavy metals, which is very pivotal for plants to alleviate toxicity [52]. Jan et al. found that Cr (VI) treatment could maintain a high level of GSH content, and GSH content in plants indicates the degree of Cr (VI) stress on plants [53]. The results of this study showed that GSH content in roots, stems, and leaves increased under Cr treatment. The GSH content in various organs of *Coix lacryma-jobi* L. under DN and DS was lower than that under NS, and the DN treatment saw the lowest values, indicating that plants under NS were subjected to the highest degree of Cr (VI) stress, which is similar to the previous findings. When plants are under adversity or stress, they adapt to or mitigate the damage caused by stress by regulating the generation of GSH [54]. A study by Qiu showed that exogenous GSH can effectively alleviate Cr toxicity in rice [55]. Jan's study showed that exogenous 24-Epibrassinolide can improve the antioxidative ability and increase the GSH content of plants under Cr stress [56]. The results of this study showed that domestic sewage improved the capability of *Coix lacryma-jobi* L. leaves to generate GSH, significantly increased GSH content, and improved the antioxidative ability of the plant, which is consistent with previous findings. Additionally, photosynthetic gas exchange parameters and chlorophyll fluorescence parameters of leaves under DN were higher than those under NS, indicating that domestic sewage may mitigate the effect of Cr (VI) on photosynthesis by increasing GSH content. This may be related to the complex composition of domestic sewage. However, further research is needed to investigate how domestic sewage increases GSH content.

5. Conclusions

Adding domestic sewage to chromium-containing wastewater increased the organic matter content in substrate, promoted Cr accumulation in substrate, reduced Cr uptake by plants, increased GSH content in roots, stems, and leaves of *Coix lacryma-jobi* L., improved indicators, such as photosynthetic gas exchange parameters of leaves, chlorophyll fluorescence parameters, and root activity, and alleviated the inhibition of plant growth by Cr (VI) stress. Under the condition with 1/2 Hoagland nutrient solution, indicators, such as plant growth, photosynthetic parameters, and fluorescence parameters of leaves were better in the constructed wetlands with domestic sewage added than under treatment with the nutrient solution or domestic sewage alone. The capability to remove Cr from wastewater under DN was better than that under NS and DS, indicating that under good nutritional conditions, adding domestic wastewater from external sources can improve the capacity of constructed wetlands to remove Cr (VI) from wastewater and contribute to the sustainable and efficient operation of the wetlands. However, the mechanism of promoting the chemical form transformation of Cr to mitigate the damage of Cr (VI) to plants and improve the effect of Cr (VI) treatment under a condition with domestic wastewater added needs further research.

Author Contributions: Validation, X.L.; Formal analysis, L.L.; Resources, X.C., X.W. and Z.L. (Zhengwen Li); Writing—original draft, Y.N.; Writing—review & editing, Y.N., Z.L. (Zhigang Li) and S.L.; Supervision, Z.L. (Zhigang Li); Funding acquisition, Z.P. All authors have read and agreed to the published version of the manuscript.

Funding: This work was supported by the National Natural Science Foundation of China (41867023, 21167002) and the Natural Science Foundation of Guangxi Province (2018GXNSFAA281214).

Data Availability Statement: The data that support the findings of this study are available from the corresponding author, [Li, Z.], upon reasonable request.

Acknowledgments: The authors thank the College of Agriculture of Guangxi University for providing the research platform.

Conflicts of Interest: The authors declare that they have no known competing financial interests or personal relationships that could have appeared to influence the work reported in this paper.

References

1. Ray, A.; Jankar, J.S. A Comparative Study of Chromium: Therapeutic Uses and Toxicological Effects on Human Health. *J. Pharmacol. Pharmacother.* **2022**, *13*, 239–245. [CrossRef]
2. Shahid, M.; Shamshad, S.; Rafiq, M.; Khalid, S.; Bibi, I.; Niazi, N.K.; Dumat, C.; Rashid, M.I. Chromium speciation, bioavailability, uptake, toxicity and detoxification in soil-plant system: A review. *Chemosphere* **2017**, *178*, 513–533. [CrossRef] [PubMed]
3. Zhou, X.; Hu, H.; Ying, C.; Zheng, J.; Zhou, F.; Jiang, H.; Ma, Y. Study on Chromium Uptake and Transfer of Different Maize Varieties in Chromium-Polluted Farmland. *Sustainability* **2022**, *14*, 14311. [CrossRef]
4. Duan, W. In the "Fur Capital of China" Excessive Chromium Discharge of Sewage? Sentenced and given an industry restraining order! *Dahe News*, 4 June 2020.
5. Mei, Y. Sentenced for Environmental Pollution by Excessive Discharge of Electroplating Wastewater. 12 March 2022. Available online: https://www.163.com/dy/article/H27UCC7A0514JN6C.html (accessed on 19 June 2023).
6. Li, Z.; Li, S.; Mei, L.; Wan, X.; Liang, H.; Chen, W.; Chen, H.; Zhou, Z. Purification effect of canna (*Canna indica* Linn.) and reed (*Phragmites australis* L.) constructed wetlands on chromium-containing domestic sewage and the physiological and ecological changes of plants. *J. Agro-Environ. Sci.* **2011**, *30*, 358–365.
7. Reis, M.M.; Tuffi Santos, L.D.; da Silva, A.J.; de Pinho, G.P.; Montes, W.G. Metal Contamination of Water and Sediments of the Vieira River, Montes Claros, Brazil. *Arch. Environ. Contam. Toxicol.* **2019**, *77*, 527–536. [CrossRef]

8. Addo-Bediako, A.; Rasifudi, L. Spatial distribution of heavy metals in the Ga-Selati River of the Olifants River System, South Africa. *Chem. Ecol.* **2021**, *37*, 450–463. [CrossRef]
9. Mohanty, B.; Anirban, D. Heavy metals in agricultural cultivated products irrigated with wastewater in India: A review. *AQUA-Water Infrastruct. Ecosyst. Soc.* **2023**, *1*, 1–12. [CrossRef]
10. Lou, Y. Research on Pollutant Removal Effect of Joint Treatment of Domestic Sewage and Acidic Mine Wastewater. Master's Thesis, Guizhou University, Guiyang, China, 16 February 2017.
11. Chen, J.; Deng, S.; Jia, W.; Li, X.; Chang, J. Removal of multiple heavy metals from mining-impacted water by biochar-filled constructed wetlands: Adsorption and biotic removal routes. *Bioresour. Technol.* **2021**, *331*, 125061. [CrossRef]
12. Wang, H.; Zhang, M.; Lv, Q.; Xue, J.; Yang, J.; Han, X. Effective co-treatment of synthetic acid mine drainage and domestic sewage using multi-unit passive treatment system supplemented with silage fermentation broth as carbon source. *J. Environ. Manag.* **2022**, *310*, 114803. [CrossRef]
13. Peng, Z.; Li, Z.; He, B.; Li, S.; Yang, P.; Li, Z.; Liang, H. Root decomposition and chromium release and chemical form changes of barley constructed wetland. *J. Environ. Sci.* **2015**, *35*, 238–244.
14. Li, K.; Gu, C.; Liu, J.; Huang, H.; Gao, Y. Experiment of Li Shihe purifying domestic sewage containing heavy metals. *Environ. Sci. Technol.* **2014**, *37*, 151–155.
15. Li, S. Study on the Purification Mechanism of Chromium (VI) Containing Wastewater by Barley Constructed Wetland. Master's Thesis, Guangxi University, Nanning, China, 16 January 2016.
16. Xu, R.; Wang, Y.-N.; Sun, Y.; Wang, H.; Gao, Y.; Li, S.; Guo, L.; Gao, L. External sodium acetate improved Cr (VI) stabilization in a Cr-spiked soil during chemical-microbial reduction processes: Insights into Cr (VI) reduction performance, microbial community and metabolic functions. *Ecotoxicol. Environ. Saf.* **2023**, *251*, 114566. [CrossRef]
17. Li, Y.; Cheng, C.; Li, X. Research progress on water purification efficiency of multiplant combination in constructed wetland. In Proceedings of the IOP Conference Series, Earth and Environmental Science, Philadelphia, PA, USA, 11 October 2020.
18. Sun, C.; Tan, Q.; Liu, X.; Zhang, Z.; Sun, J. Effects of exogenous melatonin on photosynthetic characteristics and nutrient uptake of wheat seedlings under chromium (Cr~(6+)) stress. *J. Wheat Crops* **2022**, *42*, 1535–1542.
19. Wu, M.; Jia, Y.; Li, H.; Yang, L.; Wang, G. Effects of chromium stress on chlorophyll fluorescence characteristics and active oxygen metabolism system of tobacco leaves. *Jiangsu Agric. Sci.* **2014**, *42*, 92–95.
20. Fatma, M.; Sehar, Z.; Iqbal, N.; Alvi, A.F.; Abdi, G.; Proestos, C.; Khan, N.A. Sulfur supplementation enhances nitric oxide efficacy in reversal of chromium-inhibited Calvin cycle enzymes, photosynthetic activity, and carbohydrate metabolism in wheat. *Sci. Rep.* **2023**, *13*, 6858. [CrossRef] [PubMed]
21. Li, Y.; Jiang, Y.; Li, S.; Huang, H.; Chen, W.; Chen, H. Effects of constructed wetland wastewater treatment on photosynthesis and chlorophyll fluorescence characteristics of three plants. *Ecol. Environ.* **2008**, *17*, 2187–2191.
22. Long, H.; Zhang, D. Preliminary results of drought resistance identification of 22 Stylophyllum materials at seedling stage. *Trop. Agric. Sci.* **2015**, *35*, 26–30.
23. Castañares, J.L.; Bouzo, C.A. Effect of exogenous melatonin on seed germination and seedling growth in melon (*Cucumis melo* L.) under salt stress. *Hortic. Plant J.* **2019**, *5*, 79–87. [CrossRef]
24. Bu, Y. Physiological Mechanism of Barley in Response to Different Concentrations of Chromium (VI). Master's Thesis, Guangxi University, Nanning, China, 16 June 2011.
25. Xie, B.; Jin, T.; Liu, P.; Jin, H.; Kong, L. Changes of phytochelatin and metallothionein in soybean roots and leaves under aluminum stress. *Chin. J. Oil Crops* **2008**, *116*, 191–197.
26. Wang, A.; Huang, S.; Zhong, G.; Xu, G.; Liu, Z.; Shen, X. Effects of chromium stress on growth and chromium accumulation of three herbaceous plants. *Environ. Sci.* **2012**, *33*, 2028–2037.
27. Liu, X.; Nong, Y.; Huang, J.; Li, S.; Li, L.; Cheng, X.; Wang, X.; Li, Z.; Li, Z. Effects of Cr^{6+} on Photosynthetic Characteristics and Trace Element Absorption of *Coix* in Constructed Wetlands. *Guangxi Plant* **2022**, *42*, 1959–1970.
28. Ribas, T.C.; Mesquita, R.B.; Machado, A.; Miranda, J.L.; Marshall, G.; Bordalo, A.; Rangel, A.O. A Robust Flow-Based System for the Spectrophotometric Determination of Cr (VI) in Recreational Waters. *Molecules* **2022**, *27*, 2073. [CrossRef]
29. Le, H.; Luo, L.; Liu, S. Condition optimization for the simultaneous determination of Cr(VI), Cr(III) and total chromium in water by spectrophotometry. *Ind. Water Treat.* **2007**, *195*, 73–75.
30. Wang, M. Discussion on the determination of organic carbon content in soil by potassium dichromate oxidation-external heating method. *Xinjiang Nonferrous Met.* **2019**, *42*, 98–99.
31. Stambulska, U.Y.; Bayliak, M.M.; Lushchak, V.I. Chromium (VI) Toxicity in Legume Plants: Modulation Effects of Rhizobial Symbiosis. *BioMed Res. Int.* **2018**, *2018*, 8031213. [CrossRef]
32. Adhikari, A.; Adhikari, S.; Ghosh, S.; Azahar, I.; Shaw, A.; Roy, D.; Roy, S.; Saha, S.; Hossain, Z. Imbalance of redox homeostasis and antioxidant defense status in maize under chromium (VI) stress. *Environ. Exp. Bot.* **2020**, *169*, 103873. [CrossRef]
33. Hussain, A.; Ali, S.; Rizwan, M.; Zia ur Rehman, M.; Hameed, A.; Hafeez, F.; Alamri, S.A.; Alyemeni, M.N.; Wijaya, L. Role of zinc-lysine on growth and chromium uptake in rice plants under Cr stress. *J. Plant Growth Regul.* **2018**, *37*, 1413–1422. [CrossRef]
34. Saleem, M.; Asghar, H.N.; Khan, M.Y.; Zahir, Z.A. Gibberellic acid in combination with pressmud enhances the growth of sunflower and stabilizes chromium (VI)-contaminated soil. *Environ. Sci. Pollut. Res.* **2015**, *14*, 10610–10617. [CrossRef]

35. Sharma, A.; Kumar, V.; Shahzad, B.; Ramakrishnan, M.; Singh Sidhu, G.P.; Bali, A.S.; Handa, N.; Kapoor, D.; Yadav, P.; Khanna, K.; et al. Photosynthetic Response of Plants Under Different Abiotic Stresses: A Review. *J. Plant Growth Regul.* **2020**, *39*, 509–531. [CrossRef]
36. Zhang, Y. Research progress on the effect of heavy metal chromium on plant photosynthetic system. *Mod. Hortic.* **2016**, *313*, 25. [CrossRef]
37. Ali, S.; Rizwan, M.; Bano, R.; Bharwana, S.A.; Rehman, M.Z.U.; Hussain, M.B.; Al-Wabel, M.I. Effects of biochar on growth, photosynthesis, and chromium (Cr) uptake in *Brassica rapa* L. under Cr stress. *Arab. J. Geosci.* **2018**, *11*, 507. [CrossRef]
38. Emamverdian, A.; Ding, Y.; Mokhberdoran, F.; Xie, Y. Growth Responses and Photosynthetic Indices of Bamboo Plant (*Indocalamus latifolius*) under Heavy Metal Stress. *Sci. World J.* **2018**, *2018*, 121936. [CrossRef] [PubMed]
39. Vernay, P.; Gauthier-Moussard, C.; Jean, L.; Bordas, F.; Faure, O.; Ledoigt, G.; Hitmi, A. Effect of chromium species on phytochemical and physiological parameters in *Datura innoxia*. *Chemosphere* **2008**, *72*, 763–771. [CrossRef] [PubMed]
40. Yu, A.; Yang, T.; Gao, X.; Han, W. Effects of high temperature stress on fluorescence parameters of silver leaf tree. *Hunan For. Sci. Technol.* **2022**, *49*, 25–30.
41. Luo, J.; Li, Y.; Li, Y.; Zhao, W.; Xu, Y.; Zhao, S.; Zhang, Z.; Gao, H. Effects of six plant growth regulators on photoinhibition of photosystem II and photosystem I in isolated cucumber leaves under light and temperature stress. *Acta Plant Physiol.* **2021**, *57*, 178–186.
42. Wang, X.; Lei, L.; Yan, X.; Meng, X.; Chen, Y. Processes of chromium (VI) migration and transformation in chromate production site: A case study from the middle of China. *Chemosphere* **2020**, *257*, 127282. [CrossRef]
43. López-Bucio, J.S.; Ravelo-Ortega, G.; López-Bucio, J. Chromium in plant growth and development: Toxicity, tolerance and hormesis. *Environ. Pollut.* **2022**, *312*, 120084. [CrossRef]
44. Guo, R.; Fan, M.; Tao, Y.; Wu, H.; Jiang, C. Root activity and heavy metal migration of chamomile chamomile under lead and cadmium stress. *Heilongjiang Agric. Sci.* **2020**, *311*, 42–46.
45. Yu, X.-Z.; Lin, Y.-J.; Fan, W.-J.; Lu, M.-R. The role of exogenous proline in amelioration of lipid peroxidation in rice seedlings exposed to Cr (VI). *Int. Biodeterior. Biodegrad.* **2017**, *123*, 106–112. [CrossRef]
46. Liu, C.; Liu, Y.; Luo, S.; Cui, L. Effects of different cadmium concentrations on rice root activity and grain quality in cold regions. *Anhui Agric. Sci.* **2013**, *41*, 5758–5760.
47. Yu, S.; Gao, S.; Qu, Y.; Chen, Y.; Wang, G. Toxic effects of cadmium on tomato roots under different soil conditions and its toxic critical value. *J. Agric. Environ. Sci.* **2014**, *33*, 640–646.
48. Xiao, W. Chromium Migration and Transformation Rules and Pollution Diagnostic Indicators in Typical Soils. Ph.D. Thesis, Zhejiang University, Hangzhou, China, 16 November 2014.
49. Choppala, G.; Kunhikrishnan, A.; Seshadri, B.; Park, J.H.; Bush, R.; Bolan, N. Comparative sorption of chromium species as influenced by pH, surface charge and organic matter content in contaminated soils. *J. Geochem. Explor.* **2018**, *184*, 255–260. [CrossRef]
50. Xiao, W.; Ye, X.; Ye, Z.; Zhang, Q.; Zhao, S.; Chen, D.; Gao, N.; Huang, M. Responses of microbial community composition and function to biochar and irrigation management and the linkage to Cr transformation in paddy soil. *Environ. Pollut.* **2022**, *304*, 119232. [CrossRef] [PubMed]
51. Zeng, F. Physiological and Molecular Mechanisms of Rice Chromium Toxicity and Tolerance. Ph.D. Thesis, Zhejiang University, Hangzhou, China, 16 July 2010.
52. Wang, X. Study on the Mechanism of Reduced Glutathione (GSH) Alleviating Cadmium Stress in Wheat Seedlings. Master's Thesis, Northwest A & F University, Xianyang, China, 15 January 2021.
53. Gupta, P.; Seth, C.S. 24-Epibrassinolide Regulates Functional Components of Nitric Oxide Signalling and Antioxidant Defense Pathways to Alleviate Salinity Stress in *Brassica juncea* L. cv. Varuna. *J. Plant Growth Regul.* **2022**, *1*, 1–16. [CrossRef]
54. Hasanuzzaman, M.; Bhuyan, M.H.M.B.; Anee, T.I.; Parvin, K.; Nahar, K.; Mahmud, J.A.; Fujita, M. Regulation of ascorbate-glutathione pathway in mitigating oxidative damage in plants under abiotic stress. *Antioxidants* **2019**, *8*, 384. [CrossRef] [PubMed]
55. Qiu, B. Genetic Analysis of Rice Stress Tolerance to Chromium Stress and Mechanism of Reduced Glutathione Alleviating Chromium Toxicity. Ph.D. Thesis, Zhejiang University, Hangzhou, China, 15 August 2012.
56. Jan, S.; Noman, A.; Kaya, C.; Ashraf, M.; Alyemeni, M.N.; Ahmad, P. 24-Epibrassinolide alleviates the injurious effects of Cr (VI) toxicity in tomato plants: Insights into growth, physio-biochemical attributes, antioxidant activity and regulation of Ascorbate—Glutathione and Glyoxalase cycles. *J. Plant Growth Regul.* **2020**, *39*, 1587–1604. [CrossRef]

Disclaimer/Publisher's Note: The statements, opinions and data contained in all publications are solely those of the individual author(s) and contributor(s) and not of MDPI and/or the editor(s). MDPI and/or the editor(s) disclaim responsibility for any injury to people or property resulting from any ideas, methods, instructions or products referred to in the content.

Article

Appraisal of Heavy Metals Accumulation, Physiological Response, and Human Health Risks of Five Crop Species Grown at Various Distances from Traffic Highway

Shakeel Ahmad [1], Fazal Hadi [1], Amin Ullah Jan [2], Raza Ullah [1,3], Bedur Faleh A. Albalawi [4] and Allah Ditta [5,6,*]

[1] Laboratory of Molecular Stress Physiology and Phytotechnology, Department of Biotechnology, Faculty of Biological Science, University of Malakand, Chakdara 18800, Pakistan
[2] Department of Biotechnology, Faculty of Science, Shaheed Benazir Bhutto University, Sheringal, Dir (upper) 18000, Pakistan
[3] Laboratory of Plant Molecular Biology and Biotechnology, Department of Biology, School of Arts and Science, University of North Carolina at Greensboro, Greensboro, NC 27412, USA
[4] Department of Biology, University of Tabuk, Tabuk 47512, Saudi Arabia
[5] Department of Environmental Sciences, Shaheed Benazir Bhutto University, Sheringal, Dir (upper) 18000, Pakistan
[6] School of Biological Sciences, The University of Western Australia, 35 Stirling Highway, Perth, WA 6009, Australia
* Correspondence: allah.ditta@sbbu.edu.pk or allah.ditta@uwa.edu.au

Abstract: Road surfaces and vehicular traffic contribute to heavy metals (HM) contamination of soil and plants, which poses various health risks to humans by entering the food chain. It is imperative to evaluate the status of contamination with HM and associated health risks in soils and plants, especially food crops. In this regard, five crop species, i.e., strawberry (*Fragaria ananassa*), wheat (*Triticum aestivum*), tomato (*Lycopersicon esculentum*), sugar cane (*Saccharum officinarum*), and tobacco (*Nicotiana tabacum*), were evaluated at 0–10, 10–50, and 50–100 m distance from the highway near the urban area (Takht Bhai) of Mardan, Khyber Pakhtunkhwa, Pakistan. Lead (Pb) and cadmium (Cd) accumulation, phenolics, carotenoids, chlorophyll, and proline contents in plant parts were assessed. Pb and Cd in plants decreased with an increase in distance. Pb was above the critical limit in all plants except wheat, Cd exceeded the permissible level of the World Health Organization in all plants except wheat and tomato. Pb and Cd were higher in strawberries. Tomato and strawberry fruits, tobacco leaves, and sugarcane stems showed higher Pb contents at a 0–10 m distance. Phenolic contents in leaves were higher than in roots. The target hazard quotient (THQ) in edible parts of most crops has been greater than one, which presents a threat to human health upon consumption. To the best of our knowledge, this study presents the first holistic approach to assess metal contamination in the selected area, its accumulation in field-grown edible crops, and associated health risk.

Keywords: lead; cadmium; crops; soil contamination; health risk index; phenolics; proline; urbanization

1. Introduction

Food security and safety have been a special concern worldwide due to the rise of natural lands contaminated with heavy metals (HM) and other classes of emerging contaminants, which are inextricably linked with human health [1–3]. The root causes of this issue are widely linked to the rapid pace of urbanization and land use for industrialization and roads, particularly in developing countries with high population growth [4]. Among various anthropogenic sources, road surfaces and vehicular traffic add HM and ultimately contaminate the soil [5,6]. Heavy metals such as lead (Pb), cadmium (Cd), copper (Cu), and Zinc (Zn) are released into the environment during various operations of road transport [7,8]. Road transport mainly deposits Pb and Cd from fuel burning, wear out of tires, and leakage of oils, which contaminate soil and edible crops [9,10]. The bioaccumulation of these toxic metals in crop plants has received global attention because of their negative effects on human health [11–13] and phytotoxicity [14]. Pb in humans causes neurological disorders, anemia, hypertension, and impaired renal function [15]. Cd is mutagenic and carcinogenic and its elevated levels in human causes damage to the kidneys, liver, and bones [16]. In plants, HM induce certain effects such as growth reduction, nitrogen assimilation, the inhibition of chlorophyll, and enzyme activities [17–19]. HM (Pb and Cd) initiate oxidative stress and some plants combat the oxidative stress through anti-oxidative defense systems by producing phenolics, carotenoids [20–22], and free proline [23,24].

The edible crops grown in the agricultural fields near the roadside tend to accumulate HM and are considered a serious threat to human health [10]. Therefore, it is essential to assess the existing levels of accumulated metals in edible parts and their comparison to the safe levels/thresholds recommended by international organizations such as the World Health Organization (WHO). The permissible levels for Pb and Cd in edible plants are 0.3 mg kg^{-1} and 0.1 mg kg^{-1}, respectively [25,26]. Numerous studies have been conducted on the perspective of metal contamination in edible plants and its adverse health effects on humans if consumed [27,28]; however, there is still a lack of insight to investigate the toxic metals in food commodities and their risk assessment in many developing countries [29]. In the same domain, qualitative and quantitative analysis of HM has not been focused and no efforts have been made to fully assess the associated human health risk [30]. As a proactive step to formulating suitable prevention strategies for soil contamination by HM, accurate mapping of pollutants in a given area is needed [8,10].

To the best of our knowledge, it is the first reported study to assess the concentration of toxic metal in roadside farmlands and edible crops and ascertain risk assessment in the selected study area, and this research is significant to lay the foundation for further studies regarding metals accumulation in edible plants growing in metal-contaminated sites near the traffic highways and other polluted areas. For this purpose, five important crops species, i.e., strawberry (*Fragaria ananassa*), tomato (*Lycopersicon esculentum* L.), wheat (*Triticum aestivum*), sugar cane (*Saccharum officinarum* L.) and tobacco (*Nicotiana tabacum*) growing under natural environmental conditions were selected and these crops are popularly cultivated in the study area under consideration because farmers prefer to grow them for their high income. Pb and Cd accumulation from HM-contaminated soils has been reported in strawberries [31], tomatoes [32], wheat [33], sugarcane [34], and tobacco [35]. The current study was conducted with an aim to (1) evaluate Pb and Cd concentrations in soil samples and in various parts of selected crop plants, (2) to investigate the antioxidants and biochemical parameters (phenolic, carotenoids, proline, and chlorophyll contents) in leaves and roots and to examine their correlation with metals concentration in plants (3) to compare Pb and Cd concentration in plants with WHO threshold level. (iv) to evaluate the risk assessment based on HRI and bio-concentration factor (BCF).

2. Materials and Methods

2.1. Site Description and Samples Collection

The selection of the study site was based on the distance from the highway in an urban area (Takht Bhai) of Mardan, Khyber Pakhtunkhwa, Pakistan (Figure 1). Mardan is located at 34.1989° N, 72.0231° E. The road is surfaced with tar coal and is experiencing a huge traffic density including trucks, passenger buses, private cars, and others. Three plots on the bases of distances (0–10, 10–50, and 100 m) from the main road were selected.

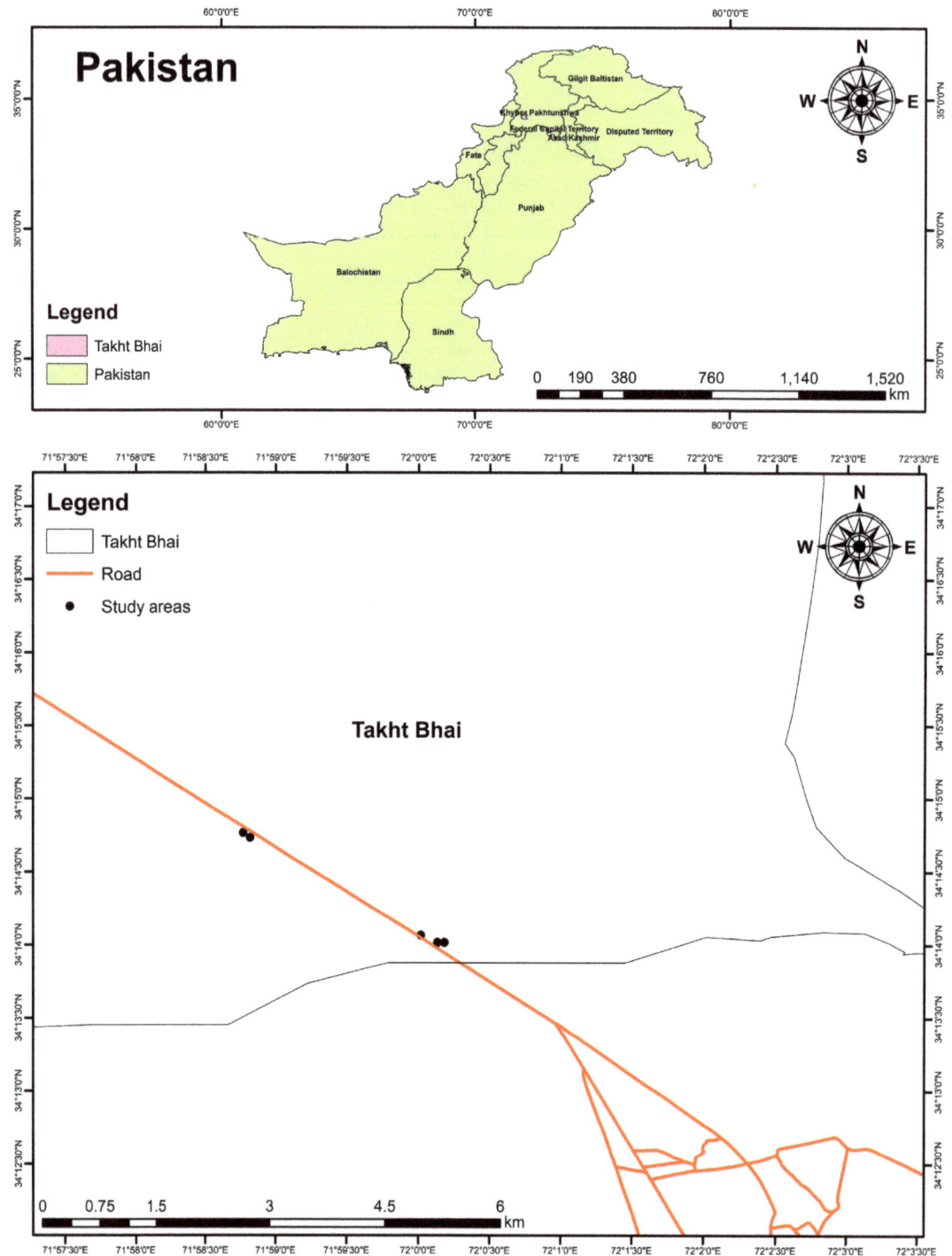

Figure 1. Study area map of sampling sites of the plants.

Five different plants species were selected for the current study, i.e., *Fragaria ananassa* (Strawberry), *Lycopersicon esculentum* (Tomato), *Triticum aestivum* (Wheat), *Saccharum officinarum* (Sugar cane), and *Nicotiana tabacum* (Tobacco) (Figure S1). These crops are popularly grown in this area as these are basic and high-income commodities. The plants were cultivated in the growing season under natural field conditions. The selected plants and associated soil samples were collected from each plot. Fifteen replicates for each plant species were collected randomly from each of the selected plots and were transported immediately to the laboratory for further analysis. Plants used per species (3 plots × 15 replicates = 45) and total plants (3 plots × 15 replicates × 5 species = 225). Plants were rinsed with dH_2O to remove the adhered soil and contaminants. Then plants were rinsed with 5 mM Tris HCl pH 6.0 and 5 mM EDTA solution to remove surface-bound metals [36]. All plant samples were arranged into fruits, leaves, stems, and roots and kept in labeled paper bags for further analysis.

2.2. Soil and Plant Analysis

The soil samples collected from all sites were kept in the lab. To dry completely. These samples were crushed in pestle and mortar and sieved through 2 mm mesh to remove coarse particles. The fine soil samples obtained were kept in labeled zipper bags for further analysis. The physicochemical parameters, i.e., pH, electrical conductivity (EC), and texture of the samples were evaluated using standard methods of APHA [37]. The dried soil sample was taken in a flask and thoroughly shaken to dissolve in 30 mL of distilled water. Then, the mixture was filtered via Whatman filter paper to collect the filtrate for analyzing EC and pH using an EC meter (Jenco 3173, Chatsworth, CA, USA) and pH meter (Jenco 6175, Chatsworth, CA, USA), respectively. The soil texture was determined by using the hydrometer method. The plant parts were enclosed in labeled envelopes and then kept in an oven at 80 °C for complete drying. Once dried, the samples were blended into powder using a commercial blender and kept in labeled zipper bags in desiccators until further analysis. Powdered samples (0.25 g) of fruit, leaf, stem, and root were taken in conical flasks having a volume of 50 mL. Soil and plant tissues were acid-digested [38]. Briefly, 10 mL H_2SO_4 was added into the flasks containing plant or soil sample (0.1 g) and kept overnight in the fume hood. The next day, the flasks were heated on a hot plate set at a high temperature (250 °C) inside the fume hood until black color appeared. Then, H_2O_2 was added to the flasks frequently till a clear solution appeared. The digested samples were cooled down at room temperature and the volume was raised to 50 mL using distilled water. It was followed by filtration through Whatman filter paper into labeled bottles. Heavy metal concentrations in all the samples were analyzed using Atomic Absorption Spectrophotometer (SP-IAA320, Jinan, China). Metals translocation and bioconcentration factors were measured to assess the transfer potential of selected HM from soil to the plant [39]. Bio-concentration factor is the ratio of a particular heavy metal in the plant as compared to that in the soil [40]. The translocation factor is the ratio of the concentration of heavy metal in the stem versus the concentration in the root [41].

2.3. Physiological and Biochemical Analyses of Plants Samples

For estimation of chlorophyll contents, 200 mg fresh leaves were ground in 2 mL of 80% acetone solution and centrifuged for 5 min at 1000 rpm. The supernatant was pipetted and transferred to a test tube. Further, acetone was added to the test tubes to get a final volume of 6 mL. The obtained extracts were analyzed through a UV-visible spectrophotometer. Aqueous acetone (80%) was run as blank. The absorbance was measured at 663 nm and 645 nm for chlorophyll a and b, respectively [42].

To evaluate carotenoid contents, extraction from leaves was performed with 90% acetone. The absorbance was measured at 480 nm with a spectrophotometer. Acetone (90%) was used as a blank [43]. The carotenoid contents of all the samples were assessed using three biological replicates.

Proline concentrations were assessed according to Bates et al. [44]. Firstly, a 100 mg sample of fresh roots and leaves was weighed in 2 mL tubes and then homogenized with 1.5 mL sulfosalicylic acid (3%). The homogenate was centrifuged at 13,000 rpm for 5 min. Then, a 300 µL aliquot of the clear supernatant was taken and mixed with 2 mL of an equal volume of acetic acid and acid ninhydrin. The mixture was incubated in a boiling water bath for one hour. The reaction was stopped by transferring the tubes from the water bath to the ice bucket. It was followed by adding 1 mL of toluene and vigorous shaking. In the aqueous phase, the colored layer having toluene was taken through a micropipette into a tube, which was kept at room temperature for warming. For proline estimation, the absorbance of the mixture was determined at 520 nm while keeping toluene as a blank/reference. This reaction was run with three replicates for all the samples.

To estimate total phenolics contents, Folin-Ciocalteau (FC) reagent method was used with little modifications [45]. Each sample of 200 mg from overnight air-dried tissues was blended, mixed with 10 mL of ethanol (80%), and stirred vigorously in covered flasks for 30 min. 2 mL aliquot of the extract was centrifuged at 13,000 rpm for 3–5 min. Tenfold diluted FC reagent (250 µL) was added post-centrifugation to 100 µL methanolic extract and kept in dark for 3–5 min at room temperature. It was followed by adding 500 µL of 7% sodium carbonate solution and then DI water was added to get the final volume of 5 mL. Again, the samples were kept in dark for 2 h before measuring the absorbance at 760 nm. 80% methanol was used as a reference solution. Analysis for all samples was performed with three replicates.

2.4. Health Risk Assessment

2.4.1. Estimated Daily Intake (EDI)

The estimated daily intake of the metals was determined based on their mean concentration in each plant sample and the estimated daily consumption of the vegetables in grams. The EDI value of each metal of interest was determined by the formula used by Chen et al. [46] with slight modification as presented in the following equation:

$$EDI = \frac{E_f \times E_d \times F_{IR} \times C_m \times C_f}{B_w \times T_A} \times 0.001 \quad (1)$$

where E_f is exposure frequency (365 days/year), E_D is the exposure duration (65 years), equivalent to an average lifetime [47], and F_{IR} is the average food (vegetable) consumption (240 g/person/day) which were obtained from the World Health Report (WHO, 2002) for low vegetable intake; C_M is the metal concentration (mg/kg dry weight), C_f is the concentration conversion factor for fresh vegetable weight to dry weight, i.e., 0.085 [48], B_W is reference body weight for an adult, which is 70 kg [47], T_A is the average exposure time (65 years × 365 days) and 0.001 is the unit conversion factor.

2.4.2. Target Hazard Quotient (THQ)

The target hazard quotient (THQ) values were estimated to assess non-carcinogenic human health risks from the consumption of vegetables contaminated by heavy metals. The THQ values were calculated using the following equation as described by Chen et al. [46].

$$THQ = \frac{EDI}{RfD} \quad (2)$$

where EDI is the estimated daily metal intake of the population in mg/day/kg body weight and RfD is the oral reference dose (mg/kg/day) values which were 0.0035 for Pb and 0.001 for Cd (US-EPA). If the value of THQ is <1, it is generally presumed to be safe for the risk of non-carcinogenic effects and if it is >1, it is supposed that there is a chance of non-carcinogenic effects with an increasing probability as the value upsurges [46,49].

2.4.3. Hazard Index (HI)

It has been documented that the individual health risks of the analyzed heavy metals in the same vegetable are accumulative and that is expressed as a hazard index [46,49]. Accordingly, the HI of target metals considered in this study was calculated using the following equation proposed by Antoine et al. [49]:

$$HI = \sum_{n=1}^{i} THQ_n; i = 1, 2, 3, \ldots, n \qquad (3)$$

where HI is the sum of various metals hazards. If the HI value became <1.0, there is no apparent health impact due to the metals considered. However, an HI value of >1.0 indicates potential health impact implications. A serious chronic health impact has been suggested for HI > 10.0 [49].

2.4.4. Target Cancer Risk (TCR)

The cancer risk posed to human health due to the ingestion of individual possibly carcinogenic metals was estimated using the following equation as described by Sharma et al. [50]. Then, the target cancer risk (TCR) resulting from heavy metals (Pb and Cd) ingestion, which may promote carcinogenic effects depending on the exposure dose, was calculated using the following equation as described by Kamunda et al. [51].

$$CR = EDI \times CPS_o \qquad (4)$$

$$CR = \sum_{n=1}^{i} CR; i = 1, 2, 3, \ldots, n \qquad (5)$$

where CR represents cancer risk over a lifetime by individual heavy metal ingestion, EDI is the estimated daily metal intake of the population in mg/day/kg body weight, CPS_o is the oral cancer slope factor in (mg/kg/day)-1 and n is the number of heavy metals considered for cancer risk calculation. The CPS_o values used for Pb and Cd were 0.0035 and 0.001, respectively [51]. It has been pointed out that the slope factor converts the estimated daily intake of the metal averaged over a lifetime of exposure directly to the incremental risk of an individual developing cancer [52].

BCF is the ratio of the concentration of a particular heavy metal in the plant to its concentration in the soil [39].

2.5. Statistical Analysis

In this study, Microsoft Excel was used to find the means of the replicates and calculate the standard deviations. While ANOVA was done through GraphPad Prism, version 5. Significant difference among different values was obtained through the least significant difference (LSD) test.

3. Results and Discussion

3.1. Physicochemical Properties of Soil

The soils collected from different spots (plots) based on distance from the main highway and their physicochemical properties are given in Table 1. A slight difference in the pH of soils from the three plots was observed. Sharma and Prasad [53] reported pH values for soil samples from the roadside field almost in the same range. The electrical conductivity was higher (618 μS) in samples collected closest to the road and reduced (591 μS) with increasing distance from the road, which confirms the previous outcomes of the study by Sharma and Prasad [53]. Our current study demonstrates that the concentration of HM in roadside fields has an inverse correlation with distance from the road. Metals concentration (mg kg^{-1}) at sites nearest to the road was high and decreased with an increase in distance from the main road. The concentrations of Pb in soil were below 0.3 mg kg^{-1}, which is the permissible limit set by the World health organization (WHO), while Cd concentration was higher than the threshold value of 3 mg kg^{-1}, which is a standard set by WHO [54] (Table 1).

Table 1. Physicochemical properties and heavy metals concentrations in soils sampled at different distances from the main highway.

Distance from Highway	pH	EC (µS cm^{-1})	Soil Texture	Lead (mg kg^{-1})	Cadmium (mg kg^{-1})
100 m distance	6.89 ± 0.17	591 ± 1.21	Loamy	2.67 ± 1.93	0.47 ± 0.03
10–50 m distance	7.01 ± 0.91	612 ± 1.91	Loamy	18.84 ± 13.11	1.17 ± 0.15
0–10 m distance	6.94 ± 0.88	618 ± 0.84	Loamy	32.02 ± 17.37	4.51 ± 0.77

3.2. Lead (Pb) Concentration in Plant Tissues

The concentration of Pb in various parts of the experimental plants (strawberry, tomato, wheat, tobacco, and sugarcane) is presented in Figure 2A–E. Pb uptake was higher in plants nearest to the main highway. Pb concentration was highest in strawberries compared to other studied plants. The higher Pb concentration in parts observed in order, i.e., roots > leaves > stem > fruits in strawberry, tomato, and tobacco. At the nearest to the road, the strawberry and tomato fruits, leaves of tobacco and stem from sugarcane showed the highest Pb concentrations, which were above the WHO permissible limits (i.e., 0.3 mg kg^{-1}). Thus, our results depict that Pb concentration exceeded the threshold level [26] in various parts of the plants including edible fruit sections of strawberries and tomatoes. Khan et al. [55] found a higher concentration of Pb in leaves of certain crop plants such as cauliflower, spinach, tomatoes, and carrot that was above the permissible limit. These results support the findings that plants can accumulate high concentrations of lead in their tissues. Ahsan et al. [56] worked on some edible plants and observed that Pb and Cd contents were above the recommended thresholds in edible portions, while Arora et al. [57] had similar findings in edible portions of vegetables, which is of great concern because of potential health hazards to human beings. In tomato seedlings, Pb uptake, distribution, and accumulation were highest in the root, followed by the leaf, shoot, and fruits. The uptake of HM by crop plants depends upon the type of species as well as the physicochemical characteristics of the soils. Through different mechanisms, Pb accumulation in many plants exceeds a hundred times the maximum threshold levels according to WHO [26], thus posing a threat to public health.

3.3. Pb Translocation and Accumulation in Plants

The translocation, accumulation, and bioconcentration of Pb (µg dry biomass^{-1}) in various parts of selected plant species are given in Table 2. It shows that Pb accumulation decreases when the distance from the road increases. The highest accumulation of Pb (i.e., 121.52 µg dry biomass^{-1}) in the whole plant was observed in strawberries at a distance of 0–10 m, followed by sugarcane, tomato, tobacco, and wheat, respectively. Pb accumulation among reference plants (from 100 m distance from the road) was recorded at more than 1 µg except for wheat and tobacco. In all plants, roots were observed to have the highest Pb accumulation followed by aerial parts. The translocation of Pb from root to above-ground parts of the plant was <1 in all studied plants except sugarcane. The highest translocation of Pb from root to leaf was observed in strawberry, tomato, and tobacco plants, while higher Pb translocation from root to stem was recorded in wheat and sugarcane. The bioconcentration for Pb was also found to be less than one in all the plants. In tomatoes, the highest bioconcentration of Pb (0.70 µg dry biomass^{-1}) was found at 0–10 m from the road. The translocation and bio-concentration of Pb in all plants decreased as the distance from the road increased.

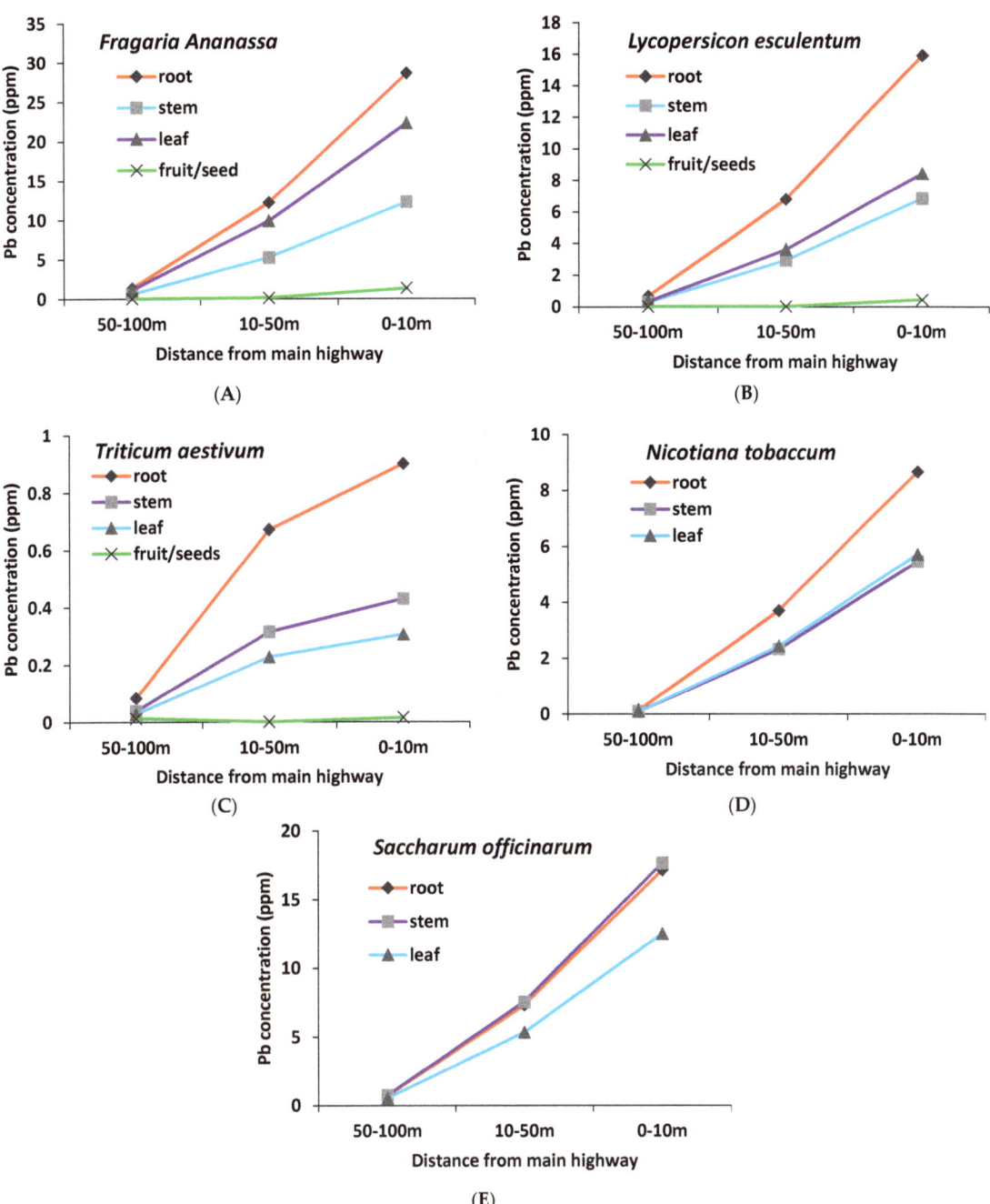

Figure 2. (**A–E**) Concentration of Pb (mg kg^{-1}) in roots, stems, leaves, and fruits of *Fragaria ananassa* (**A**), *Lycopersicon esculentum* (**B**), *Triticum aestivum* (**C**), *Nicotiana tabacum* (**D**), and *Saccharum officinarum* (**E**) plants. The permissible level for Pb in edible plants is 0.3 mg kg^{-1}.

Table 2. Pb accumulation (μg dry biomass^{-1}), translocation, and bioconcentration in plants (*Fragaria ananassa, Lycopersicon esculentum, Triticum aestivum, Nicotiana tabacum,* and *Saccharum officinarum*. Pb accumulation in roots, stems, leaves, and entire plants (EP). Pb translocation from roots to stems (R–S) and from roots to leaves (R–L).

Plant	Distance from the Road (m)	Lead (Pb) Accumulation (μg Dry Biomass^{-1})				Pb Translocation		Pb Bioconcentration
		Root	Stem	Leaves	EP	R–S	R–L	
Fragaria ananassa	50–100 m	1.57 ± 0.22	1.29 ± 0.18	2.45 ± 0.34	5.31 ± 0.74	0.43 ± 0.05	0.83 ± 0.04	0.24 ± 0.003
	10–50 m	9.53 ± 2.45	5.91 ± 1.52	16.14 ± 4.15	31.58 ± 8.13	0.48 ± 0.03	0.81 ± 0.07	0.30 ± 0.002
	0–10 m	50.80 ± 16.9	22.22 ± 7.42	48.50 ± 16.2	121.52 ± 40.5	0.38 ± 0.08	0.78 ± 0.06	0.34 ± 0.010
Lycopersicon esculentum	50–100 m	0.78 ± 0.11	0.65 ± 0.09	0.78 ± 0.11	2.21 ± 0.31	0.45 ± 0.06	0.59 ± 0.04	0.56 ± 0.011
	10–50 m	5.29 ± 1.36	3.28 ± 0.84	5.87 ± 1.51	14.45 ± 3.72	0.53 ± 0.03	0.53 ± 0.05	0.64 ± 0.010
	0–10 m	28.22 ± 9.42	12.34 ± 4.12	18.31 ± 6.11	58.87 ± 19.66	0.42 ± 0.07	0.45 ± 0.05	0.70 ± 0.008
Triticum aestivum	50–100 m	0.10 ± 0.01	0.09 ± 0.01	0.06 ± 0.01	0.25 ± 0.03	0.47 ± 0.04	0.31 ± 0.05	0.11 ± 0.003
	10–50 m	0.52 ± 0.13	0.35 ± 0.09	0.37 ± 0.10	1.25 ± 0.32	0.41 ± 0.08	0.36 ± 0.06	0.12 ± 0.005
	0–10 m	1.60 ± 0.53	0.76 ± 0.26	0.66 ± 0.22	3.03 ± 1.01	0.43 ± 0.06	0.39 ± 0.03	0.08 ± 0.000
Nicotiana tabacum	50–100 m	0.17 ± 0.02	0.21 ± 0.03	0.21 ± 0.03	0.59 ± 0.08	0.63 ± 0.02	0.66 ± 0.08	0.12 ± 0.008
	10–50 m	2.88 ± 0.74	2.62 ± 0.67	3.98 ± 1.02	9.47 ± 2.44	0.71 ± 0.05	0.60 ± 0.05	0.14 ± 0.003
	0–10 m	15.36 ± 5.13	9.84 ± 3.29	12.41 ± 4.14	37.60 ± 12.56	0.66 ± 0.02	0.71 ± 0.04	0.16 ± 0.001
Saccharum officinarum	50–100 m	0.84 ± 0.12	1.66 ± 0.23	1.16 ± 0.16	3.67 ± 0.51	1.01 ± 0.08	0.73 ± 0.07	0.44 ± 0.000
	10–50 m	5.70 ± 1.47	8.46 ± 2.18	8.70 ± 2.24	22.86 ± 5.88	1.12 ± 0.10	0.79 ± 0.09	0.34 ± 0.001
	0–10 m	30.38 ± 10.1	31.83 ± 10.6	27.15 ± 9.07	89.35 ± 29.84	1.03 ± 0.09	0.71 ± 0.03	0.43 ± 0.000

The values following the mean values as ± represent the standard deviation where $n = 3$.

3.4. Cadmium (Cd) Concentration in Plant Tissues

Cd concentrations in different parts of the five studied plants are shown in Figure 3A–E. The concentration of Cd in plants reduced with increasing distance from the road, i.e., 0–10 m > 10–50 m > 100 m. In strawberry plants, the concentration of Cd was highly significant. Like Pb, Cd concentration was highest in roots, followed by leaves in strawberries, tomato, and tobacco; however, in the case of wheat and sugarcane, it was higher in stems. Strawberry fruits and tobacco leaves have Cd concentrations above permissible levels [26]. The diminution in Cd concentrations with increasing distance from the road exhibited that emissions from automobiles contribute a substantial amount of metals in the roadside farmlands [58]. Ahsan et al. [56] observed Cd contents in edible parts of plants that were not according to the threshold limits of the WHO. Similarly, Liu et al. [59] reported higher concentrations of toxic HM in vegetables. Therefore, Cd build-up in edible portions of crops and vegetables is of increasing concern.

3.5. Cd Translocation and Bioconcentration in Plant Parts

The translocation, accumulation, and bioconcentration (μg dry biomass^{-1}) of various parts of the studied plants are shown in Table 3. From the table, it can be illustrated that the highest accumulation of Cd (20.85 μg dry biomass^{-1}) was found in strawberries, while the lowest accumulation of Cd was recorded in wheat at the site nearest to the road.

Cd translocation in different parts was found < 1 in all plants except sugarcane. Translocation of Cd from roots to stem at nearest soil to road was higher in sugarcane (i.e., 1.13 μg dry biomass^{-1}), while translocation from root to leaf was highest in strawberry, which is 0.78 μg dry biomass^{-1}. In sugar cane, the translocation factor- for Cd from root to stem (1.02 μg dry biomass^{-1}), and from root to leaves (0.71 μg dry biomass^{-1}) was the highest at a distance of 100 m from the road. Cd bioconcentration was found < 1 in selected plants in all the sites except for tobacco at 0-10 m distance, which was 1.08 μg dry biomass^{-1}. At all the selected distances from the road, Cd accumulation, translocation as well as bioconcentration decreased with increasing distance from the road. The BCF for Cd was higher compared to Pb in selected crop plants, with the highest value recorded in tobacco and sugarcane followed by strawberry, while the lowest concentrations have been noticed in wheat (Tables 2 and 3).

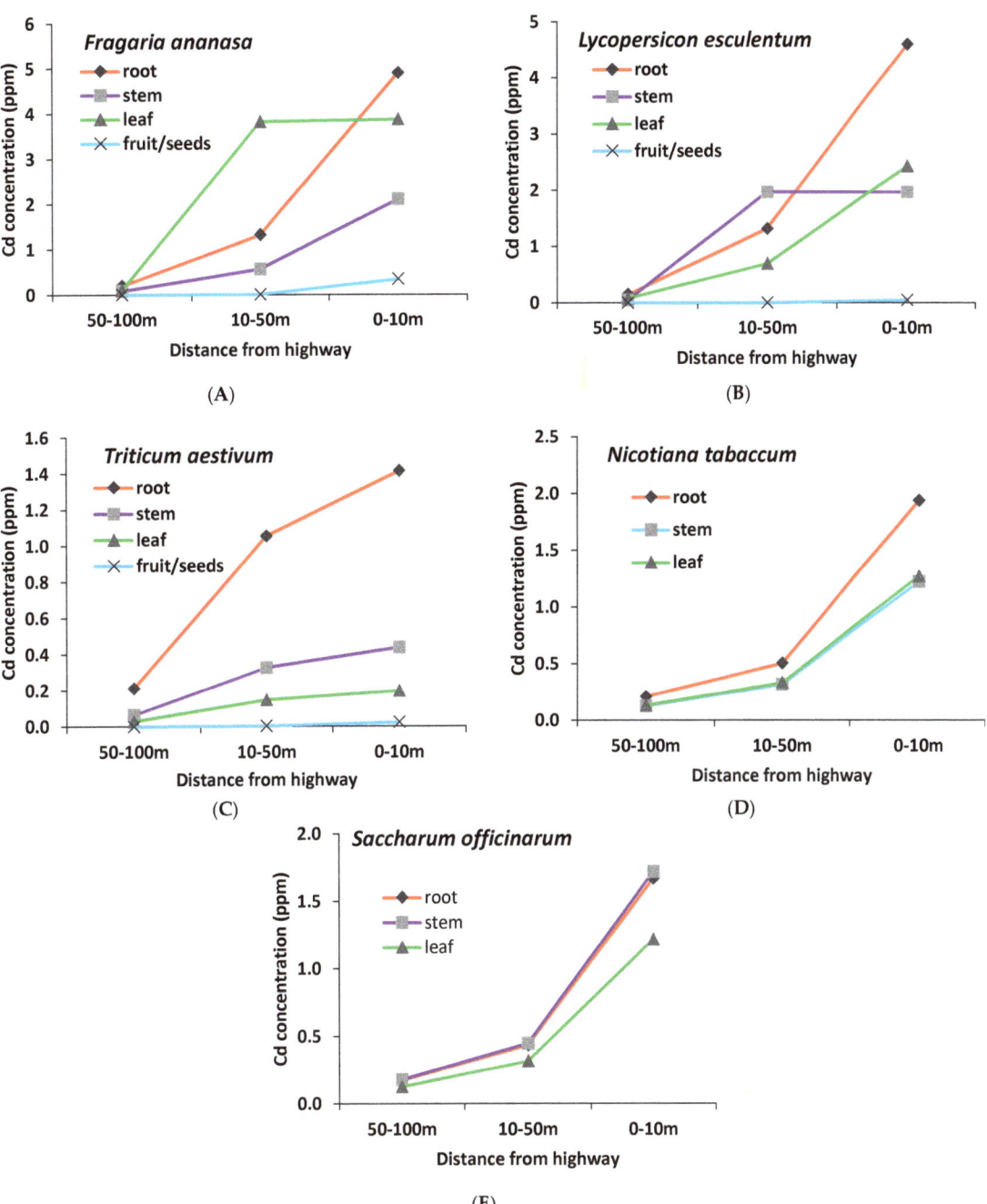

Figure 3. (**A**–**E**) Concentration of Cd (mg kg^{-1}) in root, stem, leaves, and fruits of *Fragaria ananassa* (**A**), *Lycopersicon esculentum* (**B**), *Triticum aestivum* (**C**), *Nicotiana tabacum* (**D**), and *Saccharum officinarum* (**E**). The permissible level for Cd in edible plants is 0.1 mg kg^{-1}.

Table 3. Cd translocation and accumulation (μg dry biomass^{-1}) in plants (*Fragaria ananassa, Lycopersicon esculentum, Triticum aestivum, Nicotiana tabacum,* and *Saccharum officinarum*). Accumulation of Cd in roots, stems, leaves, and entire plants (EP). Translocation of Cd from root to stem (R–S) and from root to leaves (R–L).

Plant	Distance from Road	Cd Accumulation (μg Dry Biomass^{-1})				Cd Translocation		Cd Bioconcentration
		Root	Stem	Leaves	EP	R–S	R–L	
Fragaria ananassa	50–100 m	0.23 ± 0.03	0.19 ± 0.03	0.23 ± 0.03	0.66 ± 0.09	0.43 ± 0.02	0.53 ± 0.05	0.80 ± 0.04
	10–50 m	1.03 ± 0.27	0.64 ± 0.16	1.75 ± 0.45	3.42 ± 0.88	0.55 ± 0.08	0.81 ± 0.03	0.83 ± 0.05
	0–10 m	8.71 ± 2.91	8.32 ± 1.27	3.81 ± 2.78	20.85 ± 6.96	0.40 ± 0.04	0.78 ± 0.08	0.85 ± 0.07
Lycopersicon esculentum	50–100 m	0.17 ± 0.02	0.11 ± 0.02	0.17 ± 0.02	0.46 ± 0.06	0.33 ± 0.03	0.51 ± 0.03	0.58 ± 0.03
	10–50 m	1.02 ± 0.26	0.63 ± 0.16	1.13 ± 0.29	2.79 ± 0.72	0.43 ± 0.06	0.58 ± 0.05	0.65 ± 0.08
	0–10 m	8.16 ± 2.72	5.29 ± 1.19	3.14 ± 1.77	17.01 ± 5.68	0.45 ± 0.08	0.53 ± 0.02	0.68 ± 0.05
Triticum aestivum	50–100 m	0.24 ± 0.03	0.14 ± 0.02	0.06 ± 0.01	0.45 ± 0.06	0.31 ± 0.06	0.14 ± 0.01	0.17 ± 0.07
	10–50 m	0.82 ± 0.21	0.37 ± 0.09	0.24 ± 0.06	1.43 ± 0.37	0.39 ± 0.03	0.19 ± 0.03	0.14 ± 0.07
	0–10 m	2.51 ± 0.84	0.79 ± 0.26	0.43 ± 0.14	3.73 ± 1.25	0.33 ± 0.07	0.21 ± 0.02	0.35 ± 0.04
Nicotiana tabacum	50–100 m	0.23 ± 0.03	0.28 ± 0.04	0.29 ± 0.04	0.80 ± 0.11	0.65 ± 0.08	0.66 ± 0.03	0.88 ± 0.08
	10–50 m	0.40 ± 0.22	0.37 ± 0.20	0.56 ± 0.30	1.33 ± 0.72	0.71 ± 0.05	0.69 ± 0.06	0.94 ± 0.07
	0–10 m	3.44 ± 1.15	2.20 ± 0.74	2.78 ± 0.93	8.42 ± 2.81	0.74 ± 0.03	0.63 ± 0.09	1.08 ± 0.09
Saccharum officinarum	50–100 m	0.20 ± 0.03	0.40 ± 0.06	0.27 ± 0.04	0.87 ± 0.12	1.02 ± 0.09	0.71 ± 0.05	0.85 ± 0.03
	10–50 m	0.34 ± 0.12	0.51 ± 0.18	0.52 ± 0.18	1.37 ± 0.48	1.08 ± 0.10	0.83 ± 0.07	0.95 ± 0.05
	0–10 m	3.10 ± 0.99	2.96 ± 1.04	2.64 ± 0.88	8.70 ± 2.91	1.13 ± 0.08	0.74 ± 0.05	0.91 ± 0.08

The values following the mean values as ± represent the standard deviation where $n = 3$.

3.6. Proline Contents in Plants at Various Distances from the Main Highway

To cope with the stress of the HM, plants accumulate free proline to prevent oxidative stress. It is regarded to have strong antioxidative potential due to which it prevents plant cell death [60]. Our present study demonstrates that both the studied HM triggered proline synthesis in the roots and leaves of plants collected at all three sites at varying distances from the road, as presented in Table 4. The results revealed that proline accumulation was higher in root tissues among all experimental plants compared to their leaves. In the nearest range, a very high concentration of proline was detected; however, its levels started to decrease as the distance from the road increased. A strawberry from site 0–10 m distance exhibited proline accumulation of 3.33 ppm, which is highly significant, whereas wheat shows a lower proline content of 0.90 ppm. In the case of leaves, the pattern of proline contents accumulation was similar, i.e., highest in strawberry (2.33 ppm) and lowest in wheat (0.72 ppm).

Table 4. Proline, carotenoids, phenolic, and chlorophyll contents (ppm) in various plants parts.

Plants	Distance from Road	Proline (ppm)		Phenolics (ppm)		Carotenoids (ppm)	Chlorophylls (ppm)		
		Root	Leaves	Root	Leaves		a	b	a + b
Fragaria ananassa	50–100 m	1.33 ± 0.09 c	0.93 ± 0.07 c	0.25 ± 0.02 c	0.45 ± 0.03 c	0.61 ± 0.02 a	11.11 ± 0.78 a	5.22 ± 0.37 a	16.33 ± 1.14 a
	10–50 m	2.00 ± 0.26 b	1.40 ± 0.18 b	0.30 ± 0.04 b	0.60 ± 0.08 b	0.45 ± 0.06 b	10.00 ± 1.30 b	3.65 ± 0.47 b	13.65 ± 1.77 b
	0–10 m	3.33 ± 0.57 a	2.33 ± 0.40 a	0.51 ± 0.09 a	1.34 ± 0.23 a	0.32 ± 0.10 c	8.00 ± 1.36 c	2.61 ± 0.44 c	10.61 ± 1.80 c
Lycopersicon esculentum	50–100 m	0.97 ± 0.07 c	0.78 ± 0.05 c	0.07 ± 0.00 c	0.12 ± 0.01 c	0.52 ± 0.03 a	10.23 ± 0.72 a	4.99 ± 0.35 a	15.22 ± 1.07 a
	10–50 m	1.56 ± 0.20 b	1.25 ± 0.16 b	0.08 ± 0.01 b	0.16 ± 0.02 b	0.40 ± 0.05 b	9.20 ± 1.20 b	3.75 ± 0.49 b	12.95 ± 1.68 b
	0–10 m	1.95 ± 0.33 a	1.56 ± 0.27 a	0.14 ± 0.02 a	0.36 ± 0.06 a	0.39 ± 0.09 c	7.36 ± 1.25 c	2.50 ± 0.42 c	9.86 ± 1.68 c
Triticum aestivum	50–100 m	0.45 ± 0.03 c	0.36 ± 0.03 c	0.17 ± 0.01 c	0.30 ± 0.02 c	0.48 ± 0.02 a	8.18 ± 0.57 a	4.11 ± 0.29 a	12.29 ± 0.86 a
	10–50 m	0.72 ± 0.09 b	0.58 ± 0.07 b	0.20 ± 0.03 b	0.40 ± 0.05 b	0.40 ± 0.06 b	7.36 ± 0.96 b	3.20 ± 0.42 b	10.57 ± 1.37 b
	0–10 m	0.90 ± 0.15 a	0.72 ± 0.12 a	0.34 ± 0.06 a	0.89 ± 0.15 a	0.33 ± 0.07 c	5.89 ± 1.00 c	2.05 ± 0.35 c	7.94 ± 1.35 c
Nicotiana tabacum	50–100 m	1.17 ± 0.08 c	0.93 ± 0.07 c	0.09 ± 0.01 c	0.16 ± 0.01 c	0.48 ± 0.02 a	5.79 ± 0.41 a	3.39 ± 0.24 a	9.18 ± 0.64 a
	10–50 m	1.87 ± 0.24 b	1.49 ± 0.19 b	0.11 ± 0.01 b	0.22 ± 0.03 b	0.39 ± 0.06 b	5.61 ± 0.73 b	3.32 ± 0.43 b	8.93 ± 1.16 b
	0–10 m	2.33 ± 0.40 a	1.87 ± 0.32 a	0.19 ± 0.03 a	0.49 ± 0.08 a	0.25 ± 0.07 c	5.22 ± 0.89 c	2.71 ± 0.46 c	7.93 ± 1.35 c
Saccharum officinarum	50–100 m	1.38 ± 0.10 c	1.10 ± 0.08 c	0.16 ± 0.01 c	0.28 ± 0.02 c	0.33 ± 0.01 a	7.34 ± 0.51 a	3.13 ± 0.22 a	10.47 ± 0.73 a
	10–50 m	2.21 ± 0.29 b	1.77 ± 0.23 b	0.19 ± 0.02 b	0.37 ± 0.05 b	0.30 ± 0.04 b	6.61 ± 0.86 b	2.50 ± 0.33 b	9.11 ± 1.18 b
	0–10 m	2.76 ± 0.47 a	2.21 ± 0.38 a	0.31 ± 0.05 a	0.83 ± 0.14 a	0.16 ± 0.05 c	5.28 ± 0.90 c	1.56 ± 0.27 c	6.85 ± 1.16 c

The values following the mean values as ± represent the standard deviation where $n = 3$. Mean values sharing the same letter (s) in a column are statistically non-significant with each other at $p \leq 0.05$

Similar findings have been previously documented for *Triticum aestivum* [61]. Bhattacharjee and Mukherjee [62] reported a higher accumulation of proline contents in roots of *Vigna unguiculata* compared to leaves under Cd and Pb stress. Proline accumulation in response to toxic HM has also been reported in tomatoes [63] and wheat plants [64], and certain weed plants [65]. From the present study, it is evident that Pb and Cd-induced stress in our experimental plants resulted in enhanced proline accumulation in their roots and leaves to survive under metal stress. We also found a strong positive correlation between proline accumulation in plants with both Pb and Cd concentrations. Proline prevents the inactivation of key enzymes by toxic metal ions [66,67]. In addition, proline acts as an osmoregulator and scavenger of free radicals thus protecting plants from oxidative injuries [68]. In this study, high production of proline confirmed that detoxification of ROS enables the plants to tolerate HM stress.

3.7. Phenolic Contents in Plants

The plants' exposure to HM stress may induce the production of a high level of phenolics [69]. Phenolic compounds act as potent antioxidants due to their ability to chelate HM and act as ROS quenchers and membrane stabilizers [70]. In our study, a noticeable rise in the production of phenolic content was observed in leaves compared to roots (Table 4). Phenolic content accumulation was negatively correlated to the distance from the road. At the nearest site, the phenolic contents in strawberry leaves were significantly high, i.e., 1.34 ppm, while tomato leaves showed the lowest accumulation of 0.36 ppm phenolic contents. Similarly, strawberry roots accumulated the highest phenolic contents (0.51 ppm) in an order of strawberry > wheat > sugarcane > tobacco > tomato. Rastgoo and Alemzadeh [71] identified that Gouan (*Aeluropus littoralis*) produced the highest amount of phenolic compounds in response to Pb and Cd, as compared to control plants. *Cannabis sativa* and *Ricinus communis* accumulated increased phenolics under Cd stress [72,73]. The highest accumulation of phenolic compounds was observed in the plant on exposure to HM and these compounds are known for their antioxidant activity. The ability of redox reactions enables them to play significant functions as hydrogen donors, reducing agents, reactive oxygen species (ROS) quenchers, and metal ions chelators [71].

3.8. Carotenoid and Chlorophyll Contents in Plants

Heavy metals are known to negatively affect the chlorophyll and carotenoid contents of plants, and adversely affect photosynthesis. Chl a, Chl b, total chlorophyll, and carotenoid contents in selected plants at various distances from the main road were assessed in this study. The current study shows that these contents significantly decreased with increasing metal (Pb and Cd) concentrations in plant tissues as shown in Table 4. The chlorophyll and carotenoid contents in plants exhibited the same trend of under Pb and Cd exposure as demonstrated by Emanuil et al. [74] and Rahman et al. [75]. We observed a decrease in carotenoid contents when the Pb and Cd concentrations were higher. A similar trend has been observed for chlorophyll contents in *Aeluropus littoralis* [71]. Öncel et al. [76] reported that both chlorophyll 'a' and 'b' of two varieties decreased significantly under Pb and Cd treatments. Similar findings have also been reported by Ullah et al. [65], John et al. [77], and Mobin and Khan. [78]. Vijayarengan [43] found that increased zinc level in the soil results in decreasing the chlorophyll and carotenoid contents in the leaves of radish plant. A similar change in chlorophyll contents was recorded with Pb and Cd treatments [74,75]. The carotenoid contents decreased with increasing Pb, Cd concentrations demonstrated in various treatments.

3.9. Correlations among Different Parameters

The proline and phenolics content of the leaves and roots showed a significantly positive correlation while photosynthetic pigments (chlorophyll and carotenoids) revealed a negative correlation with Pb and Cd concentration in all experimental plants' tissues (Table 5). The findings of the current research revealed that the study sites have high concentrations of Pb; however, the Cd levels are above the WHO-recommended level in these soils. All the plants considered in our study displayed decreasing trend of metal accumulation in their tissues with an increasing distance from the main road. Pb and Cd levels in edible parts of all the studied plants were above the permissible limits set by WHO. The only exceptions were *Triticum aestivum* (for Pb and Cd) and *Lycopersicon esculentum* (for Cd). Earlier, Mabood et al. [8] recorded a similar trend in different physiological and biochemical parameters in different crop species with heavy metals

Table 5. Pearson correlation among various physiological and biochemical parameters in different crop species with heavy metals (Cd and Pb).

Plant Sample	Heavy Metal	Chlorophyll (Leaf)	Carotenoid (Leaf)	Proline (Root)	Proline (Leaf)	Phenolics (Root)	Phenolics (Leaf)
Fragaria annanasa	Lead	−0.9627 *	−0.8672	+0.8679	+0.9973	+0.7975	+0.9802
	Cadmium	−0.9973	−0.8750	+0.8881	+0.9894	+0.9654	+0.9224
Lycopersicon esculentum	Lead	−0.9776	−0.8540	+0.8429	+0.887	+0.9674	+0.956
	Cadmium	−0.9863	−0.8516	+0.7293	+0.9077	+0.8304	+0.949
Triticum aestivum	Lead	−0.981	−0.888	+0.881	+0.9837	+0.7326	+0.8698
	Cadmium	−0.9811	−0.8789	+0.7841	+0.9756	+0.8809	+0.889
Nicotiana tabacum	Lead	−0.9852	−0.9688	+0.7775	+0.9004	+0.9281	+0.9699
	Cadmium	−0.9917	−0.9824	+0.6725	+0.7487	+0.9045	+0.9912
Saccharum officinarum	Lead	−0.9913	−0.9767	+0.7801	+0.8736	+0.9548	+0.9774
	Cadmium	−0.9211	−0.9192	+0.6791	+0.7234	+0.9456	+0.9911

* All the values either positive or negative were statically significant at $\alpha = 0.05$.

3.10. Risk Assessment

3.10.1. Estimated Daily Intake of Metals and Health Risk Assessment

It is crucial to assess human health risks, especially in developing countries such as Pakistan, where mostly wastewater is used for irrigating fields or HM polluted farmlands are used for agricultural activities. Numerous sources have been contributing to the contamination, which ultimately results in metal toxicity in human beings because, in our country, most people consume wheat, fruits, and vegetables in their diet. In the present study, the EDI of Pb and Cd through edible parts of crop plants and their associated health risks have been assessed and presented in Table 6.

Table 6. Estimated daily intake (EDI) of metals (mg person^{-1} day^{-1}) through edible parts of crop plants and their associated health risks.

Plant Species	Distance from Road	EDI Pb	EDI Cd	THQ Pb	THQ Cd	HI	CR Pb	CR Cd	TCR
Strawberry	50–100 m	0.0011	0.0001	0.311	0.122	0.434	3.815×10^{-6}	1.224×10^{-7}	3.937×10^{-6}
	10–50 m	0.0064	0.0007	1.824	0.697	2.521	2.235×10^{-5}	6.965×10^{-7}	2.304×10^{-5}
	0–10 m	0.0206	0.0035	5.889	3.535	9.424	7.213×10^{-5}	3.535×10^{-6}	7.567×10^{-5}
Tomato	50–100 m	0.0004	0.0001	0.119	0.082	0.201	1.459×10^{-6}	8.160×10^{-8}	1.540×10^{-6}
	10–50 m	0.0027	0.0005	0.762	0.513	1.275	9.333×10^{-6}	5.129×10^{-7}	9.846×10^{-6}
	0–10 m	0.0089	0.0025	2.552	2.457	5.009	3.126×10^{-5}	2.457×10^{-6}	3.372×10^{-5}
Wheat	50–100 m	0.0000	0.0001	0.012	0.058	0.071	1.530×10^{-7}	5.829×10^{-8}	2.113×10^{-7}
	10–50 m	0.0002	0.0002	0.060	0.178	0.238	7.344×10^{-7}	1.778×10^{-7}	9.122×10^{-7}
	0–10 m	0.0004	0.0004	0.118	0.356	0.474	1.448×10^{-6}	3.555×10^{-7}	1.804×10^{-6}
Tobacco	50–100 m	0.0001	0.0002	0.035	0.166	0.201	4×10^{-7}	1.661×10^{-7}	5.945×10^{-7}
	10–50 m	0.0019	0.0003	0.550	0.271	0.821	6.732×10^{-6}	2.710×10^{-7}	7.003×10^{-6}
	0–10 m	0.0065	0.0015	1.853	1.451	3.304	2.270×10^{-5}	1.451×10^{-6}	2.415×10^{-5}
Sugarcane	50–100 m	0.0008	0.0002	0.235	0.195	0.430	2.876×10^{-6}	1.953×10^{-7}	3.072×10^{-6}
	10–50 m	0.0050	0.0003	1.429	0.300	1.729	1.750×10^{-5}	3.002×10^{-7}	1.780×10^{-5}
	0–10 m	0.0172	0.0016	4.911	1.632	6.543	6.016×10^{-5}	1.632×10^{-6}	6.179×10^{-5}

Where THQ = target hazard quotient, HI = hazard index, CR = cancer risk, and TCR = target cancer risk.

The EDI for Pb and Cd varied from crop to crop. The THQ for Pb and Cd were greater than 1 in all selected crops except wheat plants at all distances. THQ values were higher in crops grown in the fields nearest to the roadsides and the maximum THQ value for Pb and Cd were 5.889 and 3.535, respectively in strawberries. Our results are comparable to previous studies performed in various regions of Pakistan [28,29]. In the case of target cancer risk (TCR), the maximum value was observed in the case of strawberry plants grown at a distance of 0–10 m from the road. Earlier, the values of cancer risk and TCR were decreased in plants with increasing distances [8].

3.10.2. Dry Biomass

The dry biomass decreased with a decrease in distance from the highway and the minimum value was found at 0–10 m from the highway (Figure 4A–E). Earlier it has been found that the biomass of the plants decreased under the increasing concentration of heavy metals [74,75]. This premise is supported by the concentration of Pb and Cd in different plant species which were increasing with decreasing distance from the road (Tables 2 and 3). In the plants studied, the level of metal contamination exceeds the safe limit, which presents a serious threat to human health. Thus, it is recommended that such contaminated fields should not be used for agricultural purposes or at least for the production of edible crops. Moreover, the concerned authorities are advised to monitor the concerned areas regularly by testing soil and crops for metal contamination. However, public awareness is also an important initiative to be established. These measures will not only ensure food safety and security but also ascertains good public health. Therefore, the study establishes to assess metal contamination levels and associated risks to human health both locally and globally.

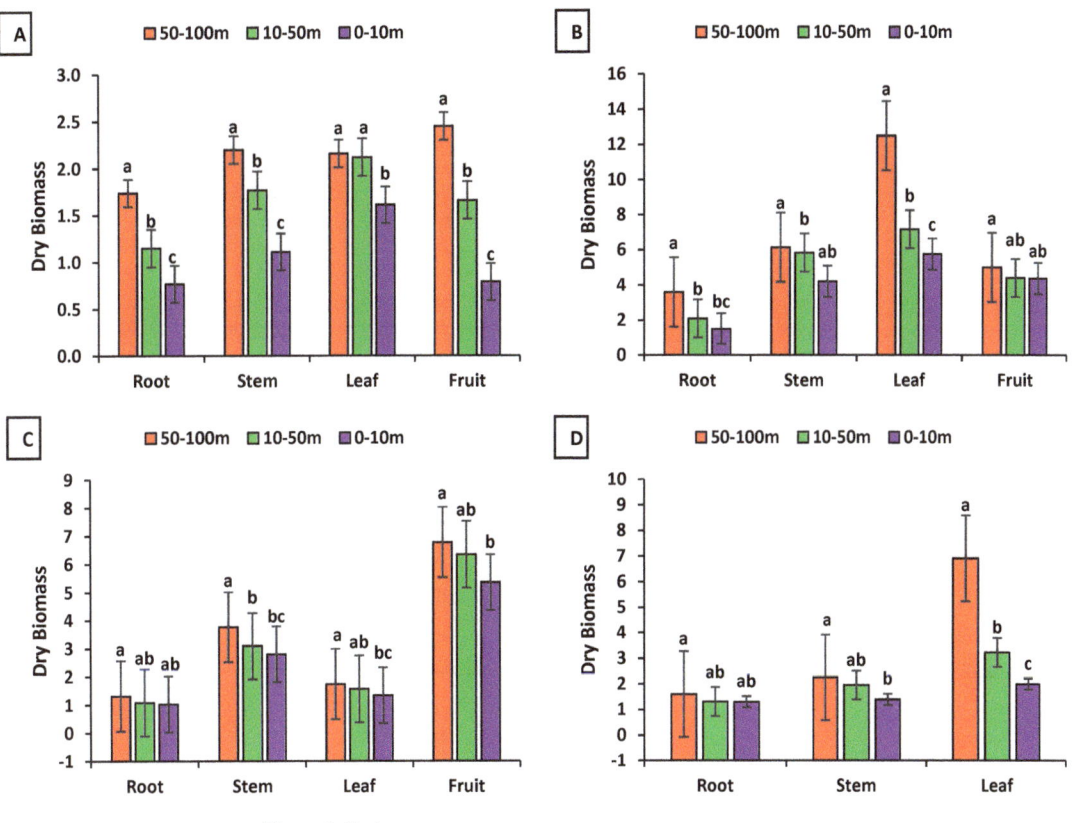

Figure 4. *Cont.*

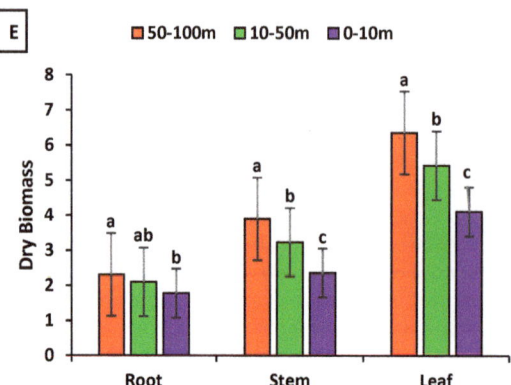

Figure 4. (**A–E**) Comparison of dry biomasses of selected plant species (**A**) = Strawberries, (**B**) = Tomatoes, (**C**) = Wheat, (**D**) = Tobacco, and (**E**) = Sugarcane at different distances from the road. The bars with different letters are significantly different from each other at $\alpha = 0.05$.

4. Conclusions

The present study showed that the metal concentrations tend to decrease in all selected plants when the distance from the road increased, i.e., the plants from the site nearest to the road had higher Pb and Cd concentrations and vice versa. Comparatively, Pb concentrations were higher in these plants than in Cd. In our experimental edible crops, Pb and Cd levels were exceeding the threshold values for these metals in edible portions while wheat grains and tomato fruit were exceptions for Pb + Cd and Cd, respectively. Moreover, proline and phenolic contents accumulation in roots and leaves were higher when the metal concentration was higher. The photosynthetic pigments (chlorophyll and carotenoids) in all the plants decreased, as the sampling site was getting closer to the road. The health risk indices such as target hazard quotient (THQ) and hazard index (HI) for edible parts of most plants (strawberry, tomato, tobacco, and sugarcane) are greater than the maximum permissible limits, i.e., >1. Future studies should be conducted to evaluate more toxic HM and other emerging contaminants in agricultural soils and edible crops for risk characterization and assessment. These will be crucial in understanding the metal contamination level and devising certain risk reduction and mitigation strategies.

Supplementary Materials: The following supporting information can be downloaded at: https://www.mdpi.com/article/10.3390/su142316263/s1, Figure S1: (A) *Fragaria ananassa* (B) *Lycopersicon esculentum* (C) *Nicotiana tabacum* (D) *Triticum aestivum* and (E) *Saccharum officinarum* plants.

Author Contributions: Conceptualization, S.A., F.H. and A.D.; Data curation, S.A., A.U.J. and R.U.; Formal analysis, S.A., F.H., A.U.J., R.U., B.F.A.A. and A.D.; Funding acquisition, B.F.A.A. and A.D.; Investigation, S.A.; Methodology, F.H.; Project administration, F.H., B.F.A.A. and A.D.; Resources, S.A., F.H., A.U.J., R.U., B.F.A.A. and A.D.; Software, F.H., R.U., B.F.A.A. and A.D.; Supervision, F.H. and A.U.J.; Validation, S.A., F.H., A.U.J., R.U., B.F.A.A. and A.D.; Visualization, S.A., F.H., A.U.J., R.U., B.F.A.A. and A.D.; Writing—original draft, S.A.; and Writing—review and editing, F.H., A.U.J., R.U., B.F.A.A. and A.D. All authors have read and agreed to the published version of the manuscript.

Funding: The study was financially supported by the Higher Education Commission (HEC) of Pakistan.

Institutional Review Board Statement: Not applicable.

Informed Consent Statement: Not applicable.

Data Availability Statement: Not applicable.

Acknowledgments: The Central Resources Laboratory (CRL) at the University of Peshawar and its staff are acknowledged for their support and the facilities provided.

Conflicts of Interest: The authors declare no conflict of interest.

References

1. Christou, A.; Karaolia, P.; Hapeshi, E.; Michael, C.; Fatta-Kassinos, D. Long-term wastewater irrigation of vegetables in real agricultural systems: Concentration of pharmaceuticals in soil, uptake and bioaccumulation in tomato fruits and human health risk assessment. *Water Res.* **2017**, *109*, 24–34. [CrossRef] [PubMed]
2. Muhammad, J.; Khan, S.; Su, J.Q.; Hesham, A.E.-L.; Ditta, A.; Nawab, J.; Ali, A. Antibiotics in poultry manure and their associated health issues: A systematic review. *J. Soils Sediments* **2020**, *20*, 486–497. [CrossRef]
3. Ullah, R.; Tsui, M.T.-K.; Chen, H.; Chow, A.; Williams, C.; Ligaba-Osena, A. Microplastics interaction with terrestrial plants and their impacts on agriculture. *J. Environ. Qual.* **2021**, *50*, 1024–1041. [CrossRef] [PubMed]
4. Montgomery, M. United Nations Human Settlements Programme: The State of the World's Cities, 2004/2005: Globalization and Urban Culture. *Popul. Dev. Rev.* **2005**, *31*, 592–594.
5. Bai, J.; Cui, B.; Wang, Q.; Gao, H.; Ding, Q. Assessment of heavy metal contamination of roadside soils in Southwest China. *Stoch. Environ. Res. Risk Assess.* **2009**, *23*, 341–347. [CrossRef]
6. Suzuki, K.; Yabuki, T.; Ono, Y. Roadside *Rhododendron pulchrum* leaves as bioindicators of heavy metal pollution in traffic areas of Okayama, Japan. *Environ. Monit. Assess.* **2008**, *149*, 133–141. [CrossRef] [PubMed]
7. Möller, A.; Müller, H.W.; Abdullah, A.; Abdelgawad, G.; Utermann, J. Urban soil pollution in Damascus, Syria: Concentrations and patterns of heavy metals in the soils of the Damascus Ghouta. *Geoderma* **2005**, *124*, 63–71. [CrossRef]
8. Mabood, F.; Hadi, F.; Jan, A.U.; Ditta, A.; Islam, Z.; Siddiqui, M.H.; Ali, H.M.; Sabagh, A.E.L. Assessment of Pb and Ni and potential health risks associated with the consumption of vegetables grown on the roadside soils in District Swat, Khyber Pakhtunkhwa, Pakistan. *Environ. Monit. Assess.* **2022**, *194*, 906. [CrossRef]
9. Loganathan, P.; Vigneswaran, S.; Kandasamy, J. Road-Deposited Sediment Pollutants: A Critical Review of their Characteristics, Source Apportionment, and Management. *Crit. Rev. Environ. Sci. Technol.* **2013**, *43*, 1315–1348. [CrossRef]
10. Ullah, I.; Ditta, A.; Imtiaz, M.; Mehmood, S.; Rizwan, M.; Rizwan, M.S.; Jan, A.U.; Ahmad, I. Assessment of health and ecological risks of heavy metal contamination: A case study of agricultural soils in Thall, Dir-Kohistan. *Environ. Monit. Assess.* **2020**, *192*, 786. [CrossRef]
11. Wang, Z.-X.; Chen, J.-Q.; Chai, L.-Y.; Yang, Z.-H.; Huang, S.-H.; Zheng, Y. Environmental impact and site-specific human health risks of chromium in the vicinity of a ferro-alloy manufactory, China. *J. Hazard. Mater.* **2011**, *190*, 980–985. [CrossRef] [PubMed]
12. Alengebawy, A.; Abdelkhalek, S.; Qureshi, S.; Wang, M.-Q. Heavy Metals and Pesticides Toxicity in Agricultural Soil and Plants: Ecological Risks and Human Health Implications. *Toxics* **2021**, *9*, 42. [CrossRef] [PubMed]
13. Naveed, M.; Tanvir, B.; Xiukang, W.; Brtnicky, M.; Ditta, A.; Kucerik, J.; Subhani, Z.; Nazir, M.Z.; Radziemska, M.; Saeed, Q.; et al. Co-composted biochar enhances growth, physiological, and phytostabilization efficiency of *Brassica napus* and reduces associated health risks under Chromium stress. *Front. Plant Sci.* **2021**, *12*, 775785. [CrossRef]
14. Amari, T.; Ghnaya, T.; Abdelly, C. Nickel, cadmium and lead phytotoxicity and potential of halophytic plants in heavy metal extraction. *S. Afr. J. Bot.* **2017**, *111*, 99–110. [CrossRef]
15. Khan, R.U.; Durrani, F.R.; Chand, N.; Anwar, H.; Naz, S.; Farooqi, F.A.; Manzoor, M.N. Effect of *Cannabis sativa* fortified feed on muscle growth and visceral organs in broiler chicks. *Int. J. Biol. Biotechnol.* **2009**, *6*, 179–182.
16. Raikwar, M.; Kumar, P.; Singh, M.; Singh, A. Toxic effect of heavy metals in livestock health. *Vet. World* **2008**, *1*, 28–30. [CrossRef]
17. Mehmood, S.; Ahmed, W.; Alatalo, J.M.; Mahmood, M.; Imtiaz, M.; Ditta, A.; Ali, E.F.; Abdelrahman, H.; Slaný, M.; Antoniadis, V.; et al. Herbal plants- and rice straw-derived biochars reduced metal mobilization in fishpond sediments and improved their potential as fertilizers. *Sci. Total. Environ.* **2022**, *826*, 154043. [CrossRef]
18. Majeed, A.; Muhmood, A.; Niaz, A.; Ditta, A.; Rajpar, M.N. Comparative efficacy of different biochars and traditional manures in the attenuation of cadmium toxicity in rice (*Oryza sativa* L.). *Arab. J. Geosci.* **2022**, *15*, 209. [CrossRef]
19. Mroczek-Zdyrska, M.; Wójcik, M. The Influence of Selenium on Root Growth and Oxidative Stress Induced by Lead in *Vicia faba* L. minor Plants. *Biol. Trace Element Res.* **2012**, *147*, 320–328. [CrossRef]
20. Hadi, F.; Ali, N.; Fuller, M.P. Molybdenum (Mo) increases endogenous phenolics, proline and photosynthetic pigments and the phytoremediation potential of the industrially important plant *Ricinus communis* L. for removal of cadmium from contaminated soil. *Environ. Sci. Pollut. Res.* **2016**, *23*, 20408–20430. [CrossRef]
21. Michalak, A. Phenolic compounds and their antioxidant activity in plants growing under heavy metal stress. *Pol. J. Environ. Stud.* **2006**, *15*, 523–530.
22. Ahmad, I.; Tahir, M.; Daraz, U.; Ditta, A.; Hussain, M.B.; Khan, Z.U.H. Responses and Tolerance of Cereal Crops to Metal and Metalloid Toxicity. In *Agronomic Crops*; Hassanuzzaman, M., Ed.; Springer: Singapore, 2020; pp. 235–264. [CrossRef]
23. Handique, G.K.; Handique, A.K. Proline accumulation in lemongrass (*Cymbopogon flexuosus* Stapf.) due to heavy metal stress. *J. Environ. Biol.* **2009**, *30*, 299–302. [PubMed]
24. Ahmad, R.; Hadi, F.; Jan, A.U.; Ditta, A. Straw incorporation in contaminated soil enhances drought tolerance but simultaneously increases the accumulation of heavy metals in rice. *Sustainability* **2022**, *14*, 10578. [CrossRef]
25. Joint FAO/WHO. *Expert Standards Program Codex Alimentation Commission*; WHO: Geneva, Switzerland, 2007.
26. *ALINORM 01/12A*; FAO/WHO Codex Alimentarius Commission. Food Additives and Contaminants. Joint FAO/WHO Food Standards Programme: Geneva, Switzerland, 2001; pp. 1–289.

27. Chaoua, S.; Boussaa, S.; El Gharmali, A.; Boumezzough, A. Impact of irrigation with wastewater on accumulation of heavy metals in soil and crops in the region of Marrakech in Morocco. *J. Saudi Soc. Agric. Sci.* **2019**, *18*, 429–436. [CrossRef]
28. Mahmood, A.; Malik, R.N. Human health risk assessment of heavy metals via consumption of contaminated vegetables collected from different irrigation sources in Lahore, Pakistan. *Arab. J. Chem.* **2014**, *7*, 91–99. [CrossRef]
29. Jan, F.A.; Ishaq, M.; Ihsanullah, I.; Asim, S. Multivariate statistical analysis of heavy metals pollution in industrial area and its comparison with relatively less polluted area: A case study from the City of Peshawar and district Dir Lower. *J. Hazard. Mater.* **2010**, *176*, 609–616. [CrossRef]
30. Han, L.; Chen, Y.; Chen, M.; Wu, Y.; Su, R.; Du, L.; Liu, Z. Mushroom residue modification enhances phytoremediation potential of *Paulownia fortunei* to lead-zinc slag. *Chemosphere* **2020**, *253*, 126774. [CrossRef]
31. Szteke, B.; Jedrzejczak, R.; Nieplocha, J.; Tych, W. Influence of the environmental factors on cadmium content in strawberry fruit. *Fruit Sci. Rep.* **1989**, *16*, 1–6.
32. Gebologlu, N.; Cetin, S.C.; Ece, A.; Yilmaz, E.; Flmastas, M. Assessment of lead and cadmium contents of tomatoes and beans grown in the vicinity of highway of Tokat, Turkey. *Asian J. Chem.* **2005**, *17*, 730–736.
33. Ul Abideen, S.N.; Abideen, S.A. Protein level and heavy metals (Pb, Cr, and Cd) concentrations in wheat (*Triticum aestivum*) and in oat (*Avena sativa*) plants. *Int. J. Innov. Appl. Stud.* **2013**, *3*, 284–289.
34. Udosen, E.D.; Ukpong, M.E.; Etim, E.E. Concentrations of heavy metals in soil samples within Mkpanak in Ibeno coastal area of Akwa Ibom State, Nigeria. *Int. J. Mod. Chem.* **2012**, *3*, 74–81.
35. Krystofova, O.; Zitka, O.; Krizkova, S.; Hynek, D.; Shestivska, V.; Adam, V.; Hubalek, J.; Mackova, M.; Macek, T.; Zehnalek, J.; et al. Accumulation of cadmium by transgenic tobacco plants (*Nicotiana tabacum* L.) carrying yeast metallothionein gene revealed by electrochemistry. *Int. J. Electrochem. Sci.* **2012**, *7*, 886–907.
36. Elless, M.P.; Blaylock, M.J. Amendment Optimization to Enhance Lead Extractability from Contaminated Soils for Phytoremediation. *Int. J. Phytoremediation* **2000**, *2*, 75–89. [CrossRef]
37. APHA. *Standard Methods for the Examination of Water and Wastewater*, 21st ed.; American Public Health Association: Washington, DC, USA; New York, NY, USA, 2005.
38. Allen, S.E.; Grimshaw, H.M.; Parkinson, J.A.; Quarmby, C. *Chemical Analysis of Ecological Materials*; Blackwell Scientific Publication: Oxford, UK, 1974.
39. Cui, Y.-J.; Zhu, Y.-G.; Zhai, R.-H.; Chen, D.-Y.; Huang, Y.-Z.; Qiu, Y.; Liang, J.-Z. Transfer of metals from soil to vegetables in an area near a smelter in Nanning, China. *Environ. Int.* **2004**, *30*, 785–791. [CrossRef] [PubMed]
40. Yoon, J.; Cao, X.; Zhou, Q.; Ma, L.Q. Accumulation of Pb, Cu, and Zn in native plants growing on a contaminated Florida site. *Sci. Total. Environ.* **2006**, *368*, 456–464. [CrossRef] [PubMed]
41. Su, R.; Ou, Q.; Wang, H.; Luo, Y.; Dai, X.; Wang, Y.; Chen, Y.; Shi, L. Comparison of Phytoremediation Potential of *Nerium indicum* with Inorganic Modifier Calcium Carbonate and Organic Modifier Mushroom Residue to Lead–Zinc Tailings. *Int. J. Environ. Res. Public Health* **2022**, *19*, 10353. [CrossRef] [PubMed]
42. Arnon, D.I. Copper enzyme in isolated chloroplasts, polyphenol oxidase in *Beta vulgaris*. *Plant Physiol.* **1949**, *130*, 267–272.
43. Vijayarengan, P. Growth and biochemical variations in radish under zinc applications. *Int. Res. J. Plant Sci.* **2012**, *2*, 43–49.
44. Bates, L.S.; Waldren, R.P.; Teare, I.D. Rapid determination of free proline for water-stress studies. *Plant Soil* **1973**, *39*, 205–207. [CrossRef]
45. Singleton, V.L.; Rossi, J.A. Colorimetry of total phenolics with phosphomolybdic-phosphotungstic acid reagents. *Am. J. Enol. Vitic.* **1965**, *16*, 144–158.
46. Chen, C.; Qian, Y.; Chen, Q.; Li, C. Assessment of Daily Intake of Toxic Elements Due to Consumption of Vegetables, Fruits, Meat, and Seafood by Inhabitants of Xiamen, China. *J. Food Sci.* **2011**, *76*, T181–T188. [CrossRef]
47. Sabir, M.; Baltrėnaitė-Gedienė, E.; Ditta, A.; Ullah, H.; Kanwal, A.; Ullah, S.; Faraj, T.K. Bioaccumulation of heavy metals in a soil–plant system from an open dumpsite and the associated health risks through multiple routes. *Sustainability* **2022**, *14*, 13223. [CrossRef]
48. Harmanescu, M.; Alda, L.; Bordean, D.; Gogoasa, I.; Gergen, I. Heavy metals health risk assessment for population via consumption of vegetables grown in old mining area; a case study: Banat County, Romania. *Chem. Cent. J.* **2011**, *5*, 64. [CrossRef] [PubMed]
49. Antoine, J.M.R.; Fung, L.A.H.; Grant, C.N. Assessment of the potential health risks associated with the aluminium, arsenic, cadmium and lead content in selected fruits and vegetables grown in Jamaica. *Toxicol. Rep.* **2017**, *4*, 181–187. [CrossRef] [PubMed]
50. Sharma, S.; Nagpal, A.K.; Kaur, I. Heavy metal contamination in soil, food crops and associated health risks for residents of Ropar wetland, Punjab, India and its environs. *Food Chem.* **2018**, *255*, 15–22. [CrossRef]
51. Fernández-Landero, S.; Giráldez, I.; Fernández-Caliani, J.C. Predicting the relative oral bioavailability of naturally occurring As, Cd and Pb from in vitro bioaccessibility measurement: Implications for human soil ingestion exposure assessment. *Environ. Geochem. Health* **2021**, *43*, 4251–4264. [CrossRef] [PubMed]
52. Kamunda, C.; Mathuthu, M.; Madhuku, M. Health Risk Assessment of Heavy Metals in Soils from Witwatersrand Gold Mining Basin, South Africa. *Int. J. Environ. Res. Public Health* **2016**, *13*, 663. [CrossRef] [PubMed]
53. Sharma, S.; Prasad, F.M. Accumulation of Lead and Cadmium in Soil and Vegetable Crops along Major Highways in Agra (India). *E-J. Chem.* **2010**, *7*, 1174–1183. [CrossRef]

54. WHO. *WHO Guidelines for Assessing Quality of Herbal Medicines with Reference to Contaminants and Residues*; World Health Organization: Geneva, Switzerland, 2007.
55. Khan, K.; Madhavan, T.P.V.; Kshirsagar, R.; Boosi, K.N.; Sadhale, P.; Muniyappa, K. N-terminal disordered domain of *Saccharomyces cerevisiae* Hop1 protein is dispensable for DNA binding, bridging, and synapsis of double-stranded DNA molecules but is necessary for spore formation. *Biochemistry* **2013**, *52*, 5265–5279. [CrossRef]
56. Perveen, S.; Shah, Z.; Nazif, W.; Shah, S.S.; Ihsanullah, I.; Shah, H.U. Study on accumulation of heavy metals in vegetables receiving sewage water. *J. Chem. Soc. Pak.* **2011**, *33*, 220–227.
57. Arora, M.; Kiran, B.; Rani, S.; Rani, A.; Kaur, B.; Mittal, N. Heavy metal accumulation in vegetables irrigated with water from different sources. *Food Chem.* **2008**, *111*, 811–815. [CrossRef]
58. Bakirdere, S.; Yaman, M. Determination of lead, cadmium and copper in roadside soil and plants in Elazig, Turkey. *Environ. Monit. Assess.* **2007**, *136*, 401–410. [CrossRef] [PubMed]
59. Liu, C.-P.; Luo, C.-L.; Gao, Y.; Li, F.-B.; Lin, L.-W.; Wu, C.-A.; Li, X. Arsenic contamination and potential health risk implications at an abandoned tungsten mine, southern China. *Environ. Pollut.* **2010**, *158*, 820–826. [CrossRef]
60. Chen, C.; Dickman, M.B. Proline suppresses apoptosis in the fungal pathogen *Colletotrichum trifolii*. *Proc. Natl. Acad. Sci. USA* **2005**, *102*, 3459–3464. [CrossRef]
61. Amani, A.L. Cadmium induced changes in pigment content, ion uptake, proline content and phosphoenolpyruvate carboxylase activity in *Triticum aestivum* Seedlings. *Aust. J. Basic Appl. Sci.* **2008**, *2*, 57–62.
62. He, L.; Su, R.; Chen, Y.; Zeng, P.; Du, L.; Cai, B.; Zhang, A.; Zhu, H. Integration of manganese accumulation, subcellular distribution, chemical forms, and physiological responses to understand manganese tolerance in *Macleaya cordata*. *Environ. Sci. Pollut. Res.* **2022**, *29*, 39017–39026. [CrossRef]
63. De, B.; Mukherjee, A.K. Mercury-induced Metabolic changes in Seedlings and Cultured cells of Tomato. *Geobios-Jodhpur* **1996**, *23*, 83–87.
64. Lalk, I.; Dorffling, K. Hardening, abscisic acid, proline and freezing resistance in two winter wheat varieties. *Physiol. Plant.* **1985**, *63*, 287–292. [CrossRef]
65. Ullah, R.; Hadi, F.; Ahmad, S.; Jan, A.U.; Rongliang, Q. Phytoremediation of lead and chromium contaminated soil improves with the endogenous phenolics and proline production in *Parthenium*, *Cannabis*, *Euphorbia*, and *Rumex* species. *Water Air Soil Pollut.* **2019**, *230*, 40. [CrossRef]
66. Awad, M.; El-Sayed, M.M.; Li, X.; Liu, Z.; Mustafa, S.K.; Ditta, A.; Hessini, K. Diminishing heavy metal hazards of contaminated soil via biochar supplementation. *Sustainability* **2021**, *13*, 12742. [CrossRef]
67. Mehmood, S.; Ahmed, W.; Rizwan, M.; Ditta, A.; Irshad, S.; Chen, D.-Y.; Bashir, S.; Mahmood, M.; Li, W.; Imtiaz, M. Biochar, slag and ferrous manganese ore affect lead, cadmium and antioxidant enzymes in water spinach (*Ipomoea aquatica*) grown in multi-metal contaminated soil. *Crop Pasture Sci.* **2022**. [CrossRef]
68. Hare, P.; Cress, W. Metabolic implications of stress-induced proline accumulation in plants. *Plant Growth Regul.* **1997**, *21*, 79–102. [CrossRef]
69. Winkel-Shirley, B. Biosynthesis of flavonoids and effects of stress. *Curr. Opin. Plant Biol.* **2002**, *5*, 218–223. [CrossRef] [PubMed]
70. Jung, C.; Maeder, V.; Funk, F.; Frey, B.; Sticher, H.; Frossard, E. Release of phenols from Lupinus albus L. roots exposed to Cu and their possible role in Cu detoxification. *Plant Soil* **2003**, *252*, 301–312. [CrossRef]
71. Rastgoo, L.; Alemzadeh, A. Biochemical responses of Gouan ('*Aeluropus littoralis*') to heavy metals stress. *Aust. J. Crop Sci.* **2011**, *5*, 375–383.
72. Ahmad, A.; Hadi, F.; Ali, N. Effective phytoextraction of cadmium (Cd) with increasing concentration of total phenolics and free proline in *Cannabis sativa* (L) plant under various treatments of fertilizers, plant growth regulators and sodium salt. *Int. J. Phytoremediation* **2015**, *17*, 56–65. [CrossRef]
73. Hadi, F.; Ahmad, A.; Ullah, R. Cadmium Phytoextraction Potential of *Ricinus communis* significantly Increased with Exogenous Application of Growth Regulators and Macronutrients. *Soil Sediment Contam. Int. J.* **2021**, *30*, 663–685. [CrossRef]
74. Emanuil, N.; Akram, M.S.; Ali, S.; Majrashi, A.; Iqbal, M.; El-Esawi, M.A.; Ditta, A.; Alharby, H.F. Exogenous caffeine (1,3,7-Trimethylxanthine) application diminishes cadmium toxicity by modulating physio-biochemical attributes and improving the growth of spinach (*Spinacia oleracea* L.). *Sustainability* **2022**, *14*, 2806. [CrossRef]
75. Rahman, S.U.; Xuebin, Q.; Riaz, L.; Yasin, G.; Shah, A.N.; Shahzad, U.; Jahan, M.S.; Ditta, A.; Bashir, M.A.; Rehim, A.; et al. The interactive effect of pH variation and cadmium stress on wheat (*Triticum aestivum* L.) growth, physiological and biochemical parameters. *PLoS ONE* **2021**, *16*, e0253798. [CrossRef]
76. Öncel, I.; Keleş, Y.; Üstün, A. Interactive effects of temperature and heavy metal stress on the growth and some biochemical compounds in wheat seedlings. *Environ. Pollut.* **2000**, *107*, 315–320. [CrossRef]
77. John, R.; Ahmad, P.; Gadgil, K.; Sharma, S. Heavy metal toxicity: Effect on plant growth, biochemical parameters and metal accumulation by *Brassica juncea* L. *Int. J. Plant Prod.* **2009**, *3*, 65–76.
78. Mobin, M.; Khan, N.A. Photosynthetic activity, pigment composition and antioxidative response of two mustard (*Brassica juncea*) cultivars differing in photosynthetic capacity subjected to cadmium stress. *J. Plant Physiol.* **2007**, *164*, 601–610. [CrossRef] [PubMed]

Article

Effects of Heavy Metal-Tolerant Microorganisms on the Growth of "Narra" Seedlings

Erny Yuniarti [1], Ida F. Dalmacio [2], Virginia C. Cuevas [2], Asuncion K. Raymundo [2], Erlinda S. Paterno [3], Nina M. Cadiz [2], Dwi N. Susilowati [1], Karden Mulya [1], Surono [4], Ikhwani [5], Heni Purwaningsih [6], Arif Anshori [5], Kristamtini [5,*] and Nani Radiastuti [7]

1. Research Center for Horticultural and Estate Crops, Research Organization for Agriculture and Food, Cibinong Science Center, National Research and Innovation Agency (BRIN), Bogor 16911, Indonesia
2. Institue of Biological Science, University of the Philippines Los Banos, Los Banos 4031, Philippines
3. Departement of Soil Science, Agricultural System Institute, College of Agriculture and Food Science, University of the Philippnes Los Banos, Los Banos 4031, Philippines
4. Research Center for Applied Microbiology, Research Organization for Life Sciences and Environment, Cibinong Science Center, National Research and Innovation Agency (BRIN), Bogor 16911, Indonesia
5. Research Center for Food Crops, Research Organization for Agriculture and Food, Cibinong Science Center, National Research and Innovation Agency (BRIN), Bogor 16911, Indonesia
6. Research Center for Food Technology and Processing, National Research and Innovation Agency (BRIN), Yogyakarta 55861, Indonesia
7. Faculty of Science and Technology, UIN Syarif Hidayatullah, Banten 15412, Indonesia
* Correspondence: kris035@brin.go.id

Abstract: The effectiveness of heavy metal-tolerant microorganisms for supporting plant growth needs to be understood before it can be used as a soil bioremediation agent. The purpose of this study was to determine the effect of heavy metal tolerant microorganisms on the growth of "Narra" seedling (*Pterocarpus indicus* Wild). Three heavy metals-resistant (Pb, Cd, and Cu) rhizobacteria from a copper (Cu) mined-out site in Marinduque, Philippines showed plant growth promotion in vitro. A treatment combination of formula inoculant A (CuNFbM 4.1, MGR 333), B (CuNFbM 4.1, MGR 333, PbSM 2.1), and O (Uninoculated); compost (0%, 4%); and lime + inorganic fertilizer {without or with lime and inorganic fertilizer (LF0; LF1)} were applied to Narra seedlings planted on 445 mg/kg Cu-contaminated soil. Lime (2 mg/ha) and the recommended dose of soybean inorganic fertilizer were used as positive controls to evaluate the ability of inoculations and composts to promote the growth and used as positive controls to evaluate the ability of inoculants and composts to promote the growth and copper accumulation of narra in greenhouse experiments. All treatment combinations resulted in significant differences in plant height, leaf number, stem diameter, shoot and root dry weight, as well as, shoot, root Cu content, and plant Cu uptake of 13-week-old "Narra". Inoculated "Narra" could thrive better in mine-degraded soil containing 445 ppm Cu with 4% compost. Inoculant B demonstrated the best plant performance while *Pseudomonas synxantha* (PbSM 2.1) probably increases the plant's growth due to 1-aminocyclopropane-1-carboxylate (ACC) deaminase it produces. Accumulation of Cu was higher in the root compared other plant parts. More research is necessary to elucidate the mechanism of plant growth promotion and heavy metal re mediation by *P. synxantha*.

Keywords: heavy metal; tolerance; microorganism; growth; "Narra" seedling

Citation: Yuniarti, E.; Dalmacio, I.F.; Cuevas, V.C.; Raymundo, A.K.; Paterno, E.S.; Cadiz, N.M.; Susilowati, D.N.; Mulya, K.; Surono; Ikhwani; et al. Effects of Heavy Metal-Tolerant Microorganisms on the Growth of "Narra" Seedlings. *Sustainability* 2022, 14, 9665. https://doi.org/10.3390/su14159665

Academic Editor: Said Muhammad

Received: 21 June 2022
Accepted: 1 August 2022
Published: 5 August 2022

Publisher's Note: MDPI stays neutral with regard to jurisdictional claims in published maps and institutional affiliations.

Copyright: © 2022 by the authors. Licensee MDPI, Basel, Switzerland. This article is an open access article distributed under the terms and conditions of the Creative Commons Attribution (CC BY) license (https://creativecommons.org/licenses/by/4.0/).

1. Introduction

Mining activities in many countries have brought about environmental issues such as physical damage to soil, and pollution of terrestrial and aquatic environments that threaten human, animal, and plant life. Heavy metal (HM) pollution and land marginalization due to removal of the original soil along with its organic matter (OM) and associated nutrients, as well as the unstable condition of the land are also problems posed by mining. These problems have raised great interest in coming up with environment-friendly and sustainable agriculture, since these are linked with food security. Hence, the restoration of these soils is essential to restore biodiversity and ecosystem integrity [1]. The use of plant growth-promoting rhizobacteria holds promise as a means of bringing back the ability of the soil to sustain plant growth.

Bioremediation is the process of removing contaminants from the environment, such as soil, or converting them into nontoxic forms using bacteria, fungi, and plants, in order to restore the soil to its pristine, pollution-free and productive environment. Increased beneficial soil microbes are important in nutrient recycling, degradation of pollutants and organic materials and maintenance of soil health. Improvement of soil health means a good amount of organic matter that help increase cation exchange capacity (CEC), good soil pore formation and water holding capacity (WHC) [2]. Phytoremediation efforts have been tried to remedy the situation of HM-contaminated soil [3–7]. Phytoremediation refers to the use of plants to absorb, accumulate, stabilize or volatilize HM contaminants from soil [3,8]. In phytoremediation, HM-resistant bacteria with growth-promoting properties contribute to plant fitness and its survival in harsh soil environments, such as drought and HM pollution, through nutrient and phytohormone supply, as well as biofilm formation and ACC deamination enzymes produce [9–11]. The synergic approach among microorganism-OM-plant in soil bioremediation is an effective strategy to rehabilitate mine-degraded soil and to improve the soil quality in terms of physical, chemical, and biologic characteristics [2,8].

Degraded soils from mines have been shown to exhibit high level of heavy metals, low pH, reduced supply of nutrients (i.e., nitrogen, phosphorus, potassium), in addition to lack of organic matter, malformed soil structure, and truncation of microbial communities. Under these conditions, plants used for phytoremediation often do not grow well and hardly survive [9].

Several studies reported that HM-resistant bacteria function as growth promoter and enhancer for a plant to accumulate HM [12,13], increasing the length of root and shoots, as well as increasing the fresh or dry weight plants [9,14]. Some copper-tolerant rhizosphere bacteria have been isolated from talahib (*Saccharum spontaneum*) growing in the rhizosphere of the Marinduque copper mine in the Philippines. These rhizobacteria displayed plant growth-promoting properties in vitro and were identified based on 16S rDNA sequence analysis as *Fulvimonas yonginensis* (CuNFb M4.1), *Rhizobium* sp. (MGR 333), and *Pseudomonas synxantha* (PbSM 2.1) [15,16].

"Narra" (Philippines) or Angsana (Indonesia) (*Pterocarpus indicus* Will) belongs to the family *Fabaceae* (Papilionoideae), is an endangered species native to Asian regions [16]. The natural distribution of this tree began from Burma to Southeast Asia up to the Philippines and the Pacific islands and it is widely cultivated in the tropics. "Narra", which is recommended as one of the trees in agroforestry provides shade to coffee and other crops and is adaptable in drought land areas [17]. This plant has ecological, industrial, and economic importance in Southeast countries due to its use in greening program, in medicine, in agroforestry, and as a source for natural dye, and furniture staff [18–27]. This plant has potential for phytoremediation of heavy metal compounds such as hexavalent chromium [28] or Cadmium [29].

This study was carried out because of the need to specifically test the synergistic effect of HM-resistant plant-promoting rhizosphere rhizobacteria, compost amendments on the growth of "Narra" trees in phytoremediation of copper-contaminated soils. Specifically, this research aimed to determine the effect of inoculation of the selected isolates on the growth of "Narra" by measuring different growth parameters such as plant height, midline length of stem, leaf number, as well as the dry weight of root and shoot part. In addition, the ability of narra to adsorb Cu on artificial copper-contaminated soil was tested by analyzing the presence of HM in the shoots, roots and soil before and after the pot experiment. The effect of inoculation on the pH and cation exchange capacity (CEC) of the soil before and after the experiment were also ascertained.

2. Materials and Methods

The research was conducted in a greenhouse at the Indonesian Agricultural Biotechnology and Genetic Resources Center in Bogor. The soil was collected from Kentrong, Banten Province, Indonesia. The soil composite at a depth of 0–20 cm was taken from ten points on an area of 1500 m^2 using scoops, labelled, and packed in airtight polypropylene woven sack bag and then transferred to the Soil Biology Laboratory of Indonesian Soil Research Institute, and stored at sample room (temperature at 21 °C, humidity at 50%, with 12 h:12 h (L:D)) until used for the next experiment. Prior to soil analysis and pot experiment, the soils were air-dried and passed through a 2 mm sieve before mixing. The homogenized soils were placed in airtight polypropylene woven sack bags stored in sample room.

The chemical characteristics of the soil before experiment were as follows: pH (H$_2$O) = 4.56; silty clay loam = 16% sand, 45% silt, 40% clay; total P$_2$O$_5$ = 14 mg·100 g^{-1}; available P = 2 ppm; soil organic carbon (SOC) = 2.13%; K$_2$O = 5 mg·100 g^{-1}; -1 CEC = 12.21 cmol$_{(+)}$·kg^{-1}; Cu = 14 ppm; Cd = 1.25 ppm; and Pb = 3.35 ppm. The low pH and nutrient content reflected the situation of the mine-degraded soil from where the bacteria were taken. Kentrong soil was used as artificial soil in this study because in preliminary work, we found that the soil from Marinduque copper mined-outside area showed Cu concentrations which varied from 167 ppm to 776 ppm, due to homogenized soil samples containing copper ores in the form of fine granules. In order to provide measured and uniform soil-contaminating Cu, artificial soil was employed.

Three rhizobacteria used in this experiment, i.e., CuNFbM 4.1, MGR 333, and PbSM 2.1 are HM resistant and isolated from the Cu mining site in Marinduque. They demonstrated in vitro plant growth promoting properties such as biofilm formation (PbSM 2.1 and CuNFbM 4.1), N fixation, phosphate (P) solubilization and IAA production (all isolates), stress controller, ACC deaminase production (PbSM 2.1) [15,16]. The experiment used twelve treatments in a Randomized Complete Block Design with three replications and two subsamples. The treatments consisted of: (1). No need for lime inorganic fertilizer (LIF), compost and inoculation; (2). Without LIF and compost, use of inoculant consortium A (IA); (3). Without LIF and compost, with inoculant consortium B (IB); (4). Without LIF, with compost, and without inoculation (I0); (5). Without LIF, with compost and inoculant consortium A (IA); (6). Without LIF, with compost and inoculant consortium B (IB); (7) LIF, without compost and inoculation; (8). LIF, without compost and with inoculant consortium A (IA); (9). LIF, without compost and with inoculant consortium B (IB); (10). with LIF, compost, and without inoculation (I0); (11). With LIF, compost, and inoculant consortium A (IA); (12). With LIF, compost, and inoculant consortium B (IB).

The pot experiment consisted of 12 treatment combinations, 3 replicates, and 2 seedling sub samples, forming 72 experimental units. The experiment was carried out in a pot (3 × 3 × 13 cm) filled with 1 kg of air-dried soil. Soil preparation: the CuCl$_2$ solution was added to each kg of soil until the desired concentration was attained. Lime was applied two weeks before planting which coincided with the addition of the HM solution. The copper-treated soils were incubated for two weeks in order for the lime and copper to come to equilibrium. Fertilizers and compost were incorporated into the soil in the afternoon, a day before planting.

Seedling preparation: the seeds were soaked in water overnight and were uncoated. The uncoated seeds were placed in a container containing a mixture of sterile soil and compost (1:1) for germination. When the seedlings reach a height of about 6–10 cm, which as around 2–3 weeks, they were transplanted to the pots. Inoculant preparation and inoculation. The rhizobacteria were grown in nutrient broth (NB) up to 12 h or the log phase stage (10^9 cells/mL). Inoculant A consisted of MGR 333 and CuNFbM4.1, sterile dH$_2$O (1:1:1), whereas Inoculant B was a consortium of MGR 333, CuNFbM 4.1, and PbSM 2.1 (1:1:1). The total volume of both inoculants was 3 mL. Prior to mixing three bacterial cultures for each inoculant consortium, each broth culture was centrifuged (6000 rpm, 10 min) and the pellet was washed once with sterile dH$_2$O. To achieve the original volume of the broth, sterile dH$_2$O was added to the pellet and mixed with vortex.

The roots of the seedlings were soaked in the corresponding inoculants for 10 min, and then transplanted into soil holes in experimental pots. However, before adding soil to the pot, 10 mL of the inoculant was poured into the roots of the seedlings to further soak them. Afterwards, the roots of the seedling in the hole were covered with soil.

Planting and plant maintenance. Each inoculated seedling was transplanted immediately to a pot containing 1 kg of soil, one "Narra" seedling per pot. The potted experimental setups were placed in the greenhouse of ICABIOGRAD, Ministry of Agriculture, Indonesia. The plants were maintained for 13 weeks and were watered once every two days during the first four weeks and once a day during the last period by using tap water. About 100 mL of water was added, and no dripping was observed.

Growth parameters: the study assessed plant growth by measuring plant height, stem diameter, and counting the number of leaves, a day before the experiment was termited. After 13 weeks, the shoots were cut, and the plants were fully uprooted. Any remaining soil that adhered to the roots was removed and soil-detached the roots were cleaned with tap water. The shoot and root parts were separated and were placed individually in paper bags. All plant material were then placed in an oven at 70 °C for 3–5 days until a constant weight (dry weight) was attained.

Soil and plant tissue analysis. The soil, shoot, and root samples were analyzed for total Cu, soil pH, and CEC in the Soil Analytical Chemistry Laboratory of the Indonesian Soil Research Institute (ISRI). Soil parameters assay used in this study have been previously described in [30].

Statistical analysis. Data were analyzed by analyses of variance (ANOVA) using SPSS ver. 25 (www.ibm.com/legal/copytrade.shtml (accessed on 5 January 2022)). Differences among all parameters were tested using Duncan's Multiple Range Test (DMRT) if the ANOVA F showed a significant difference. The correlation between soil chemical characteristics and plant growth parameters, as well as the correlation test between Cu content in the soil and plant Cu concentration were evaluated.

3. Results

3.1. Soil Characteristics Used in the Greenhouse Experiment

The soil in the Cu mined-out site in Mogpog, Marinduque is characterized by low pH, low nutrient (i.e., N and K) contents, low OM, and high Cu contamination. As a consequence, there is a loss of biodiversity and biological productivity since most of the area is barren. Kentrong soil which was used in the greenhouse experiment (Table 1) reflected Marinduque soil from where the isolates were taken. In this experiment, the Kentrong soil was artificially contaminated with 445 ppm of Cu to reflect the situation of the Cu-contaminated soil in Marinduque.

Physico-chemical analyses of the Kentrong soil show that it has silty clay loam texture, acidic with pH 4.6, and high exchangeable aluminum content (4.01 cmol$_{(+)}$ kg^{-1}) which led to the high capability of soil in phosphorus fixation resulting inefficient phosphorus fertilization in the soil. Also, the high Al content level observed could be toxic to plant and microorganism growth. In acidic soil, such as Kentrong's soil (pH 4.6), ion H+ as exchangeable H$^+$ but could also attack the structure of mineral soil, releasing Al^{3+} in the soil which toxic to many organisms and has potency to fix P in soil [31].

Low CEC (12.21 cmol$_{(+)}$ kg^{-1}) and base saturation (31%), as well as high Al saturation value, indicated the inability of the soil to hold essential nutrients as shown by low to very low soil nutrient content, namely (N = 1.8%; P$_2$O$_5$ = 14 mg·100 g^{-1}; K$_2$O = 5 mg·100 g^{-1}; P available = 2 ppm). Liming and compost were needed to improve the fertility of this soil. Liming treatment was intended to neutralize exchangeable H$^+$ and Al^{3+}, and replacing the exchange sites with Ca, as well as toxicity Al, Mn, and Cu. Treatment of 4% compost was given as a pH-dependent CEC which would improve the ability of the soil in holding essential nutrients needed for plant growth, also as C for the growth and activity of the microorganisms. Fertilizer was included as a soil treatment to assess the efficacy of inoculation and compost application in improving soil productivity and plant growth. The effects of treatment to soil chemical characteristics are presented in Table 2.

Table 1. Chemical and Physical Characteristic of Ultisol Soil Sample of Kentrong, Banten Province, Indonesia (0–20 cm depth).

Soil Parameters	Unit	Value	Category
Textural Grade (pipet)			Silty Clay Loam
Sand	%	16	
Silt	%	45	
Clay	%	40	
pH (1:5, H_2O)		4.56	Acid
Organic matter			
C (Walkley and Black)	%	2.13	Medium
N (Kjeldahl)	%	0.18	Low
C/N	%	12	Medium
Extractant (HCl 25%)			
P_2O_5	mg·100 g^{-1}	14	Very low
K_2O	mg·100 g^{-1}	5	Low
P-Bray 1	ppm	2	Very low
Cation Exchangeable value (NH4-Acetate 1N, pH 7)			
Ca	$cmol_{(+)}$ kg^{-1}	2.61	Low
Mg	$cmol_{(+)}$ kg^{-1}	1.03	Low
K	$cmol_{(+)}$ kg^{-1}	0.07	Very low
Na	$cmol_{(+)}$ kg^{-1}	0.1	Low
CEC	$cmol_{(+)}$ kg^{-1}	12.21	Low
Base saturation	%	31	Low
Exchangeable (KCl 1 M)			
Al^{3+}	$cmol_{(+)}$ kg^{-1}	4.01	
H^+	$cmol_{(+)}$ kg^{-1}	0.14	
Al saturation	%	32.84	High
Heavy Metal (ppm)			
Cu	ppm	14.00	
Cd	ppm	1.25	
Pb	ppm	3.35	

Table 2. Treatment Effects on Soil Chemical Characteristics and Soil Cu Content.

No	Treatments *	pH **	CEC (cmol(+)kg^{-1}) **	Soil Cu (mg·kg^{-1}) **
1	LF0 C0 I0	4.7 fg	14.53 ab	338.5 ab
2	LF0 C0 IA	4.5 g	14.48 ab	332.0 ab
3	LF0 C0 IB	4.6 fg	14.31 ab	334.5 ab
4	LF0 C4 I0	4.9 de	15.98 ab	347.5 ab
5	LF0 C4 IA	4.8 ef	11.87 b	396.5 a
6	LF0 C4 IB	4.8 ef	16.26 ab	366.0 ab
7	LF1 C0 I0	5.0 cd	14.78 ab	372.5 ab
8	LF1 C0 IA	5.1 cd	14.99 ab	402.0 a
9	LF1 C0 IB	5.2 c	15.77 ab	377.5 ab
10	LF1 C4 I0	5.5 b	18.54 a	294.5 b
11	LF1 C4 IA	5.8 a	17.10 ab	341.5 ab
12	LF1 C4 IB	5.6 b	17.06 ab	390.5 ab

* LF0 = without liming and inorganic fertilize, LF1 = with liming and inorganic fertilizer; C0 = 0% compost, C4 = 4% compost; I0 = No inoculation, IA = Inoculant consortium A, IB = Inoculant consortium B. ** Means in a column followed by the same letter are not significantly different at 5% level by DMRT.

3.2. Effect of Treatments on Plant Growth Parameters

There were significant differences between treatments on growth parameters, i.e., plant height, leaf number, stem diameter, shoots, and roots DW (dry weight) as presented in Table 3. The value of plant height, leaf number, stem diameter, shoots, and roots DW were about 10.45–35.25 cm, 4.67–14.67, 0.15–0.57 cm, 0.38–3.62 g, and 0.19–1.40 g, respectively. Treatments 12 (LF1C4IB) and 6 (LF0C4IB) showed the highest value of a seedling growth parameters, i.e., plant height, leaf number, stem diameter, shoots, and root DW (dry weight) in amount of 35.25 and 35 cm; 12.83 and 12.5 leaves; 0.5 and 0.47 cm; 3.62 and 3.08 g, and 1.27 and 1.38 g, respectively.

Table 3. Treatments Effects on Growth Parameters of "Narra" Seedlings.

No	Treatments *	Height (cm) **	Leaf Number **	Stem Diameter (cm) **	Shoot DW (g) **	Root DW (g) **
1	LF0 C0 I0	10.25 d	4. 67 b	0.18 c	0.38 d	0.21 c
2	LF0 C0 IA	10.50 cd	7.17 b	0.15 c	0.43 d	0.38 bc
3	LF0 C0 IB	10.58 cd	5.33 b	0.20 c	0.50 d	0.19 c
4	LF0 C4 I0	23.42 b	11.50 a	0.45 a	2.38 bc	1.10 ab
5	LF0 C4 IA	23.08 b	13.33 a	0.42 ab	2.00 bc	0.92 abc
6	LF0 C4 IB	35.00 a	12.50 a	0.47 a	3.08 ab	1.38 a
7	LF1 C0 I0	11.42 cd	5. 33 b	0.25 c	0.58 d	0.33 bc
8	LF1 C0IA	13.33 bcd	6.83 b	0.27 bc	0.58 d	0.35 bc
9	LF1 C0 IB	13.17 bcd	5.50 b	0.25 c	0.60 d	0.32 bc
10	LF1 C4 I0	18.42 bcd	11.50 a	0.42 ab	1.38 cd	0.63 abc
11	LF1 C4 IA	21.00 bc	14.67 a	0.57 a	2.37 bc	1.40 a
12	LF1 C4 IB	35.25 a	12.83 a	0.50 a	3.62 a	1.27 a

* LF0 = without liming and inorganic fertilizer, LF1 = with liming and inorganic fertilizer; C0 = 0% compost, C4 = 4% compost; I0 = No inoculation, IA = Inoculant consortium A, IB = Inoculant consortium B, DW = dry weight. ** Means in a column followed the same letter are not significantly different at 5% level by DMRT.

3.3. Cu Accumulation in Narra Plants

There were significant differences among all treatments on shoot and root Cu content, as well as plant Cu uptake (Table 4). Cu accumulation data from all treatments ranged from 35.0 to 95.0 mg/kg, 114 to 155 mg/kg, and 61 to 320 µg/plant DW, respectively.

Table 4. Treatment Effects on Cu Accumulation in "Narra" Seedlings.

No	Treatments *	Shoot Cu (ppm) **	Root Cu (ppm) **	Seedling Cu Uptake (µg/plant DW) **
1	LF0 C0 I0	95.00 a	148.50 a	67.36 d
2	LF0 C0 IA	75.50 b	114.00 b	76.11 cd
3	LF0 C0 IB	46.50 ef	155.00 a	52.88 d
4	LF0 C4 I0	39.50 g	143.00 a	250.89 ab
5	LF0 C4 IA	50.00 d	124.50 b	214.04 abc
6	LF0 C4 IB	47.00 e	127.50 b	320.95 a
7	LF1 C0 I0	56.50 c	123.50 b	74.13 cd
8	LF1 C0 IA	57.50 c	119.00 b	75.23 cd
9	LF1 C0 IB	36.50 h	127.00 b	61.98 d
10	LF1 C4 I0	45.50 f	124.00 b	136.97 bcd
11	LF1 C4 IA	36.00 hi	126.50 b	262.48 ab
12	LF1 C4 IB	35.00 hi	119.50 b	277.24 ab

* LF0 = without liming and inorganic fertilizer, LF1 = with liming and inorganic fertilizer; C0 = 0% compost, LF1 = 4% compost; I0 = No inoculation, IA = Inoculant consortium A, IB = Inoculant consortium B. 260, DW = dry weight, ** means in a column followed by the same letter are not significantly different at 5% level by DMRT.

3.4. Plant Performance

Growth performance of "Narra" is presented in Figure 1, better growth performance was showed by the plants which received treatment of 4% compost without or with inoculation and treatment combination of LF1 + 4% without or with inoculation compared to other treatments (LF0 with or without inoculation and LF1 with or without inoculation).

Figure 1. Plant performance of 13 week old "Narra" seedlings planted in soil 445 ppm Cu with combination treatments of inoculant consortia (I0, IA, IB), without or with lime fertilizers (LF0 or LF1), and composts (C0%, C4%).

3.5. Cu Bioaccumulation and Translocation

To better understand the effects of soil characteristics with plant growth and plant Cu accumulation, a correlation test was conducted. The analysis results are presented in Table 5; soil pH had significant positive correlation with CEC (r = 0.374), and seedling growth parameter, i.e., leaf number (r = 0.436), stem diameter 0.583), shoots (r = 410) and roots dry weight (r = 0.364). Significant negative correlation between soil pH and shoot Cu content (r = −0.593), as well as between Cu plant uptake and shoot Cu content (r = −0.441 were observed.

Table 5. Pearson correlation coefficient values (R) among Soil Properties, as well as Seedling Growth against Soil Properties and Cu Accumulation.

Parameters	Seedling Cu Uptake (µg/plant)	pH	CEC	Soil Cu (mg/kg)
Height (cm)	0.896 **	-	-	-
Leaf Number	0.665 **	0.436 **	-	-
Stem Diameter (cm)	0.865 **	0.583 **	-	-
Shoot DW (g)	0.925 **	0.410 *	-	-
Root DW (g)	0.971 **	0.364 *	-	-
Shoot Cu (ppm)	−0.441 **	−0.593 **	-	-
Root Cu (ppm)	-	-	-	-
Seedling Cu uptake(µg/plant)	1	-	-	-
pH	-	1	0.374 *	-
CEC	-	0.374 *	1	-
Soil Cu (mg/kg)	-	-	-	1

** Correlation is significant at the 0.01 level (2-tailed); * Correlation is significant at the 0.05 level (2-tailed).

4. Discussion

4.1. Soil Characteristics Used in the Greenhouse Experiment

The effects of treatment on the chemical characteristics of the soil and soil Cu content are presented in Table 2. All treatments resulted to significant differences in soil pH, CEC, and Cu content. The treatment LF or compost per se, as well as, the combination of LF1 and 4% compost were able to increase soil pH about 4.8 up to 5.8. The treatment which affected the increase in soil pH from highest to lowest were as follows: treatment combination of LF1 and 4% compost, a single treatment of LF1, and a single treatment of 4% compost. In the first two treatments, inoculation of formula A or B resulted in a higher increase in soil pH compared to those without inoculation.

The effect of the treatment on the CEC value revealed that combination treatment of LF1 and 4% compost, as well as single treatment of LF1 or compost, gave a relatively higher CEC value than those of without LF (LF0) or 0% compost. The CEC value of all treatments after the experiment ranged from 11.87 to 18.54 $cmol_{(+)}$ kg^{-1}. The pH-dependent negative charge of clay and organic colloid was a contributor of pH-dependent soil CEC and would have high negative charge density by deprotonation of carboxyl and phenolic functional groups, which would occur at pH more than 5.5. The negative charge of soil and organic colloid is warehouse storage of essential cation elements for plant growth.

The medium of organic C content (2.13%) in Kentrong soil did not appear to be sufficient to effect higher soil CEC value due to acidic soil pH (from 4.5 to 4.7). There was a weak positive correlation between pH and CEC (r = 0.374) which means increasing soil pH would increase soil CEC (Table 1). Soil and soil organic matter have a variable charge that is pH-dependent so that they are important in cation exchange capacity. As pH increases, the degree of negative charge increases due to the deprotonation or dissociation of H^+ from functional groups. The major acidic functional groups are carboxyls, quinones that may dissociate as readily as carboxyl groups, phenolic OH groups, and enols. Since carboxyl and phenolic groups can deprotonate at pH's common in many soils, they are major contributors to the negative charge of soils. It has been estimated that up to 55% of the CEC from SOM is due to carboxyl groups while about 30% of the CEC of SOM up to pH 7 is due to the quinone, phenolic, and enolic groups [32]. Aside from hydrogen and aluminum ion toxicity in acid soil, the toxicity of Mn and added Cu became a constraint to plant and microorganism growth.

Liming and compost application were needed to improve the fertility of this soil. In agricultural practice, liming is carried out for reducing soil acidity. Lime in soil hydrolyzes to release basic conjugate such as carbonate (CO_3^{2-}), hydroxide (OH^-), and silicate (SiO_2^-) and Ca^{2+}/and Mg^{2+}. The basic conjugate could react with H^+ to form weak acid such as water while Ca^{2+}/and Mg^{2+} replace H^+ and Al^{3+} on exchange site of colloidal complex of clay or humus [33]. Liming effects to soil fertility, namely to reduce soil H^+ concentration and soil acidity; to increase the availability of nutrients, particularly Ca, Mg, K, P, and Mo; liming reduces the toxicity of Al, Fe and Mn; to stimulate the activities of the heterotrophic soil organism or those responsible for the decomposition of organic matter with subsequent mineralization of nitrogen; to enhance the symbiotic nitrogen fixation of rhizobium in legume; and to stimulate soil granulation [33–37].

Efficiency of liming to soil pH increase is depending on type and amount of lime, soil properties (pH, SOM, CEC, and clay), management pattern, and crop types [33–37]. He et al., 2021 [38] in their experiment found that when compared to their individual higher soil background values, the addition of limes had a significantly larger impact on soil pH when there was a low background value of soil pH, soil organic matter (SOM), CEC, and clay. This suggests that liming to raise soil pH may be more effective when there is a low background value of soil pH, SOM, CEC, and clay. The soil pH buffering capacity may play a role in how other soil characteristics, excluding the soil background pH, affect the soil pH. The ability of soil pH to maintain a generally constant level after the introduction of alkaline or acidic substances varies depending on the type of soil. The capacity of soils to buffer pH is produced by the precipitation/dissolution and deprotonation/protonation of minerals with varying charges and SOM. The addition of limes to soils with high levels of SOM, CEC, and clay will have a less impact on the soil pH than that would be seen with soils with lower levels of those three components [38].

In this study, liming increased the soil pH so that it could increase soil CEC. However, the increase in pH and CEC were produced by combination treatment of LF1 + 4% compost in ranges 5.5 to 5.8 and 17.06 to 18.54, respectively. The value of pH increase included in pH of 5.5–6.5, which most plants grow well at this pH range, caused liming treatment on acidic soil while the value of CEC was lower than the lowest soil CEC value according to Buni [35] (19.18 $cmol_{(+)}$ kg^{-1}), and the highest (33.34 $cmol_{(+)}$ kg^{-1}).

Compost as a single treatment or as a combination treatment with LF1 was able to improve soil pH and soil CE. The commercial compost which was used in this study was made from a blend of natural and nontoxic sieved grass clippings, palm fronds, green coconut husks, and dried animal manure which have completely decayed following natural decomposition in the forest and it has a pH of 6.3, C/N ratio 22, OM content 19%. The increase in soil pH by compost addition as stated by [39] is mainly due to the addition of basic cations, ammonification, and production of NH_3 during decomposition of the added compost. Additionally, in soils modified with compost, adsorption of H+ ions, the establishment of reducing conditions as a result of increased microbial activity, and the displacement of hydroxyls from sesquioxide surface by organic anion can all contribute to pH increase. It is similar to the report of Sulok et al. [40] that compost has a liming effect due to its high content of calcium, magnesium, sodium, and potassium, and when organic matter decomposes these base cations are released.

The mobility and availability of heavy metals in the soil are commonly low, especially when the soil is high in pH, clay and organic matter [41–43]. Copper heavy metal has a strong affinity for organic matter; especially for dissolved organic matter which is a more important determinant of Cu solubility and bioavailability than pH [43]. As stated by [42], OM buffers the concentrations of cations in the soil solution due to its high cation exchange capacity, and its incorporation into soils can reduce concentrations of HMs in the soil solution, thereby preventing their uptake by roots and their leaching into groundwater. After the experiment, the soil total Cu content was about 294.5–402.0 $mg \cdot kg^{-1}$ where treatment 5 (LF0C4IA) and 8 (LF1C0IA) showed significantly higher soil Cu concentration at the end of the experiment, i.e., 396.5 and 402.0 $mg \cdot kg^{-1}$, respectively. Before the experiment, all soils were created to have final total Cu content of about 445 mg/kg. However, total soil Cu content from composite soil samples before treatment was about 378 $mg \cdot kg^{-1}$. The change in soil total Cu content before and after the experiment could not be discussed since this factor was not measured in the individual soil pot unit before the experiment. Overall, soil Cu concentration up to the end of the experiment relatively the same values. Soil total Cu contents in each pot unit showed as expected values, throughout the experiment of all soils were in a situation of Cu contamination. According to Liu et al. [44] the regulatory concentrations of Cu in agricultural soils ($mg \cdot kg^{-1}$) of different countries are as follows: $\leq$100, Australia; $\leq$63, Canada; $\leq$50 (pH < 6.5), $\leq$100 (pH $\geq$ 6.5, China, Mainland); $\leq$200, China, Taiwan; $\leq$150, European union; <125 in paddy soil Japan; $\leq$270, USA.

4.2. Effect of Treatments on Plant Growth Parameters

The treatment combination of LF1 + 4% compost, as well as compost treatment alone was able to increase plant height, leaf number, stem diameter, shoots, and roots DW. Of both treatments, inoculant consortium A application was able to enhance leaf number, stem diameter, and roots DW (14.67, 0.57 cm, and 1.40 g) while inoculant consortium B was able to enhance plant height and shoot DW (5.25 cm and 3.63 g).

The application of liming and in organic fertilizer (LF1) could not improve "Narra" growth. The seedling performance of this treatment (treatments 7, 8, 9) were not different from the seedlings which did not receive lime and inorganic fertilizer (treatment 1, 2, 3). The addition of 4% compost or combination of LF1 and 4% compost could improve seedling growth which showed better performance compared to those treated without LF and compost (treatments 1, 2, 3) or LF1 only (treatments 7, 8, 9). In the former treatment, the inoculated seedling had better performance than the uninoculated treatment. It seemed that the inoculant has not yet optimally expressed the seedling growth promotion traits on "Narra" planted in Kentrong soil which had low soil pH and only 2.13% organic carbon content. Inoculant requires lime and additional organic matter in the soil to affect the growth of "Narra" seedlings (Figure 1).

The effects of liming to seedling growth as described previously due to its effects to the increase in soil pH and soil CEC. These results were confirmed as in a significant positive correlation between soil pH and growth parameters such as leaf number (0.436), stem diameter (0.583), shoot DW (0.410), and root DW (0.364) (Table 4).

An increase in soil pH would increase "Narra" tree growth through increasing availability of soil nutrients (N, Ca, Mg, P, K, S, Mo), decreasing solubility or toxicity of heavy metal (A^{3+}, Cu^{2+}, Mn^{2+}, Fe^{2+}), and improving the suitable environment for inoculated rhizobacteria to grow and express plant growth promotion on "Narra" seedlings. Moreover, as has been stated previously, the increase pH would promote decomposition activity of organic material which is nutrient such as nitrogen was released and available for plant uptake.

This corroborates the research result of [45] who reported that poultry litter addition to soil can enhance soil pH, CEC, and exchangeable cation, except K. The significant increase in soil pH following the application of poultry litter results from the reduction in exchangeable aluminum of organic colloids. Similarly with another research result those soils with a high CEC are more produce high availability and minimal leaching of K, Ca, Mg, Na and other cations [39].

4.3. Cu Accumulation in Narra Plants

The highest Shoot Cu content was obtained from treatment 1, i.e., without application of liming, inorganic fertilizer, and inoculation (LF0C0I0) while the highest root Cu content was exhibited by treatment 3 (LF0C0IB), i.e., 155.0 g which was similar to those of treatment 1 (LF0 C0 I0) and 4 (LF0 C4 I0) in the amount of 148.5 and 143.0 mg/kg, respectively. Significant negative correlation between soil pH and Shoot Cu content (r = −0.593) (Table 4) occurred which means an increase in soil pH would decrease shoot Cu content. Plant Cu uptake had a significant negative correlation with shoot Cu content (r = −0.441) (Table 4). An increase in plant Cu uptake decreased shoot Cu content. According to [40], reduced Cu contents in plants with the presence of increased supply of some nutrients are often related to secondary effects such as Cu dilution because of enhanced growth rates of the plant where its growth rate is faster than Cu uptake.

4.4. Plant Performance

The organic carbon in Kentrong soil has decomposed as a material similar to hummus in organic soil, which plays a role in the process of nutrient exchange; the formation of aggregates between organic substances and mineral particles, and the immobilization of toxic compounds. The additional 4% compost in this study was as a source of labile organic matter, the organic compound which could be used for rhizobacterial growth and functional activity. After lime addition, soil pH increased close to neutral pH which was suitable for bacterial growth with added organic matter as nutrient sources, eventually, they could express their plant promotion characteristics on "Narra" plant growth.

Inoculant B was able to contribute to better performance of seedling growth on soil with 4% compost or combination LF1 + 4% compost. ACC deaminase attributed to *Pseudomonas synxantha* (PbSM 2.1) was a factor for survival of the "Narra" seedlings on the degraded Kentrong soil. ACC deaminase is an enzyme which could breakdown ACC to ketoglutaric acid and ammonium so that it could not be used as precursor of ethylene formation in plant. Excessive ethylene is usually produced in plant growing in soil with abiotic or biotic stress where excessive presence of this compound could cause senescence impact to the plant. Therefore, ACC deaminase facilitates plant growth by IAA and other plant growth promotion compounds of the inoculant in abiotic stress situation as described by [10].

In this study, factors which were expected as constraints for seedling growth, i.e., Cu and Al heavy metal. However, kinetics and isotherm of the absorption of these metal ions onto Kentrong soil was not evaluated so that it could not determine whether delayed seedling growth of treatment 1, 2, and 3 (Figure 1) due to Cu or Al, and or due to other factors which were associated with soil acidity. Similarly, the available Cu and Al in Kentrong soil was not measured so that it could not predict the toxicity of the Cu and Al in Kentrong soil toward the growth of the seedling.

4.5. Cu Bioaccumulation and Translocation

Metal which are taken up by plants do not accumulate quickly in the environment, but accumulate in plants, resulting in heavy metal accumulation. Plant metal extraction, also known as phytoextraction, necessitates the transfer of heavy metals to an easy-to-harvest region of biomass, particularly the upper part [6,43,46]. There is significant positive correlation among growth variable, i.e., plant height, number of leaves, stem diameter, shoot dry weight, and root dry weight with plant Cu uptake of about $r = 0.896$, 0.665, 0.865, 0.925, and 0.971, respectively. However, plant Cu uptake was not affected by soil Cu concentration soil pH, and CEC as shown in Table 5. Soil pH indicated positive correlation with plant growth parameter, namely number of leaves ($r = 0.436$), stem diameter ($r = 0.583$), shoot dry weight ($r = 0.410$), and root dry weight ($r = 0.364$). Interestingly, shoot Cu content indicated negative correlation with plant Cu uptake ($r = -0.441$) and soil pH ($r = -0.593$). Increasing the shoot Cu content due to a decrease in soil pH would cause a decrease in plant Cu uptake and (Table 5). Plant biomass when subjected to a safe and controlled procedure will result in the release of metals contained therein. Tolerance plants limit metal transmission from soil to roots and from roots to shoots, resulting in tiny metal accumulations in their biomass. Hyperaccumulation plants, on the other hand, actively take up and transmit metals to aboveground biomass. Plants with lower Cu concentrations in their roots than in the soil (low bioaccumulation factor/BCF) and shoot parts with lower Cu concentrations than in their roots (low translocation factor/TF) are good candidates for phytoextraction [43,46].

The TF and BCF values of "Narra" plants grown in 445 ppm Cu-polluted soil were less than one in this study. The "Narra" plants were not acceptable for phytoextraction of Cu in the metal contaminated soil, even with the addition of lime + inorganic fertilizer, 4% compost, and/or inoculation. From Table 5, we can see that the decrease in pH caused an increase in the shoot Cu content but the Cu uptake of plants decreased. This shows that the tolerance mechanism of plants was root-based. Shoot parts of the plant shows limitation in Cu accumulation which is associated with its tolerance mechanism. Similarly, in 9 month old tropical plants Khaya ivorensis and Cedrela fissile, it was reported that these plants showed BCF and TF factor values of <1 and most of the Cu was compartmentalized by the root [6]. However, this conclusion may be premature; more research based on data analysis from mature Narra trees is needed. Furthermore, it is projected that mature trees would accumulate more copper. More or less from this study it was known that "Narra" will be able to grow and become a Phyto stabilizer in Cu-contaminated soil with the help of fertilizer or 4% compost combined with inoculation. According to [47], "Narra" is a suitable plant for phytoremediation in terms of Phyto stabilization, in which heavy metals are locked or sequestered over time and limited to a single location. "Narra" is a semi deciduous plant and is suitable for phyto stabilization of ex-mining land. Fallen leaves with less Cu content would improve soil fertility through increase in soil organic carbon.

Several report about effects inoculation of PGPR on halophyte plant growth and metal bioaccumulation have been provided. Metal-resistant PGPR (*Pantoea agglomerants* RSO6 and RSO7, and *Bacillus Aryabhata* RSO25) contribute to alleviate metal stress on wheat plant. After inoculation, the oxidative stress index (OSI) reduced by between 50% and 75%, phenylalanine ammonium lyase (PAL), which is involved in secondary metabolism and/or lignin synthesis, plays a significant role in managing metal stress in this halophyte when it is inoculated with the proper PGPR. Metal tolerant (1400 µg mL^{-1} Cu, 1000 µg mL^{-1} Cd, and (1000 µg mL^{-1} Cr) bacterium isolated from chilli rhizosphere (*Pseudomonas aeruginosa*), had multiple plant growth promoting biomolecules in the presence and absence of metals. Strain CPSB1 solubilized P at 400 µg mL^{-1} of Cd, Cr and Cu. The strain was positive for indole-3-acetic acid (IAA), siderophores, hydrogen cyanide (HCN), ammonia (NH$_3$) and 1-aminocyclopropane-1-carboxylate (ACC) deaminase when grown with/without metals. The phytotoxic effects on wheat increased with increasing Cd, Cr and Cu rates. The *P. aeruginosa* CPSB1 inoculated wheat in contrast had better growth and yields under Cu, Cd and Cr stress. The root dry biomass of inoculated plants was enhanced by 44, 28 and 48% at 2007 mg Cu kg^{-1}, 36 mg Cd kg^{-1} and 204 mg Cr kg^{-1}, respectively. The bioinoculant enhanced number of spikes, grain and straw yields by 25, 17 and 12%, respectively. *Pseudomonas aeruginosa* CPSB1 significantly declined the levels of catalase (CAT), glutathione reductase (GR) and superoxide dismutase (SOD), proline and malondialdehyde (MDA), and reduced metal uptake by wheat. Single inoculation of PGPR (*Bacillus cereus* and *Pseudomonas moraviensis*) decreased 50% of Co, Ni, Cr and Mn concentrations in the rhizosphere soil. Co-inoculation of PGPR and biofertilizer treatment further augmented the decreases by 15% in Co, Ni, Cr and Mn over single inoculation except Pb and Co where decreases were 40% and 77%, respectively. The maximum decrease in biological concentration factor (BCF) was observed for Cd, Co, Cr, and Mn. *P. moraviensis* inoculation decreases the biological accumulation coefficient (BAC) as well as translocation factor (TF) for Cd, Cr, Cu Mn, and Ni. The PGPR inoculation minimized the deleterious effects of heavy metals, and the addition of carriers further assisted the PGPR [48].

5. Conclusions

Inoculation of the contribution of plant growth promoting rhizobacteria showed significant effects on seedling growth when incorporated with lime, inorganic fertilizer, and compost. Future observations on the improvement of efficiency in organic matter use should be considered in phytoremediation work. Based on the results of this study, "Narra" could grow in Cu-contaminated soil containing 445 ppm Cu when compost at 4% level is added, indicating that the tree could take up the HM thereby reducing Cu in the soil. Moreover, for phytoremediation of Cu contaminated soil using "Narra", the following rhizobacteria bacteria CuNFBM 4.1, MGR and PbSM 2.1 in combination with compost is recommended.

Author Contributions: I.F.D., V.C.C., A.K.R., E.S.P., N.M.C., D.N.S., K.M., S., I., H.P., A.A., K. and N.R., performed research and analyzed data, E.Y. and I.F.D., wrote the original manuscript and draft preparation, E.Y., wrote, reviewed, and edited, I.F.D., V.C.C., A.K.R., E.S.P., N.M.C., D.N.S., K.M., S., I., H.P., A.A., K. and N.R., edited, E.Y., I.F.D., V.C.C., A.K.R., E.S.P., N.M.C., D.N.S., K.M., S., I., H.P., A.A., K. and N.R. All authors have read and agreed to the published version of the manuscript.

Funding: This research funded by the Indonesian Agency for Agricultural Research and Development.

Institutional Review Board Statement: Not applicable.

Data Availability Statement: The data presented in this study are available upon request from the corresponding author. The data are not publicly available yet but will be in due course.

Acknowledgments: The authors wish to thank the Indonesian Agency for Agriculture Research and Development for funding under the project DIPA.

Conflicts of Interest: The authors declare no conflict of interest.

References

1. Raymundo, A.K. Overview of Bioremediation Technologies for Mine Industry. In *A Potential Strategy for Mining Wastes Management*; NAST Monograph Series; NAST: Taguig City, Philippines, 2006; pp. 17–22.
2. Masciandaro, G.; Macci, C.; Peruzzi, E.; Ceccanti, B.; Doni, S. Organic matter-microorganism-plant in soil bioremediation: A synergic approach. *Rev. Environ. Sci. Biotechnol.* **2013**, *12*, 399–419. [CrossRef]
3. Seshadri, B.; Bolan, N.S.; Naidu, R. Rhizosphere-induced heavy metal(loid) transformation in relation to bioavailability and remediation. *J. Soil Sci. Plant Nutr.* **2015**, *15*, 524–548. [CrossRef]
4. Gasco, G.; Alvarez, M.L.; Paz-Ferreiro, J.; Endez, A.M. Combining phytoextraction by Brassica napus and biochar amendment for the remediation of a mining soil in Riotinto. *Chemosphere* **2019**, *231*, 562–570. [CrossRef] [PubMed]
5. El-Mahrouk, E.S.; Eisa, E.A.; Ali, H.M.; Hegazy, M.A.; Abd El-Gayed, M.E. Populus nigra as a Phytoremediator for Cd, Cu, and Pb in Contaminated Soil. *BioResources* **2020**, *15*, 869–893. [CrossRef]
6. Covre, W.P.; Pereira, W.V.D.S.; Gonçalves, D.A.M.; Teixeira, O.M.M.; Amarante, C.B.D.; Fernandes, A.R. Phytoremediation potential of Khaya ivorensis and Cedrela fissilis in copper contaminated soil. *J. Environ. Manag.* **2020**, *268*, 10733. [CrossRef] [PubMed]
7. Adiloğlu, S.; Açikgöz, F.E.; Gürgan, M. Use of phytoremediation for pollution removal of hexavalent chromium-contaminated acid agricultural soils. *Glob. NEST J.* **2021**, *23*, 400–406.
8. Kushwaha, A.; Rani, R.; Kumar, S.; Gautam, A. Heavy metal detoxification and tolerance mechanisms in plants: Implications for phytoremediation Anamika Gautam. *Environ. Rev.* **2015**, *24*, 39–51. [CrossRef]
9. He, L.Y.; Zhang, Y.F.; Ma, Y.F.; Su, L.N.; Chen, Z.J.; Wang, Q.Y.; Sheng, X.F. Characterization of copper-resistant bacteria and assessment of bacterial communities in rhizosphere soils of copper-tolerant plants. *A Soil Ecol.* **2010**, *44*, 49–55. [CrossRef]
10. Glick, B.R. Bacteria with ACC deaminase can promote plant growth and help to feed the world. *Microbiol. Res.* **2014**, *169*, 30–39. [CrossRef]
11. Angus, A.A.; Hirsch, A.M. Biofilm formation in the rhizosphere: Multispecies interactions and implications for plant growth. *Mol. Microb. Ecol. Rhizos.* **2013**, *1*, 701–712.
12. Abou-Shanab, R.A.I.; Angle, J.S.; Chaney, R.I. Bacterial inoculants affecting nickel uptake by Alyssum murale from low, moderate and high Ni soils. *Soil Biol. Biochem.* **2016**, *38*, 2882–2889. [CrossRef]
13. Jiang, C.; Sheng, X.; Qian, M.; Wang, X. Isolation and characterization of a heavy metal-resistant Burkholderia sp. from heavy metal-contaminated paddy field soil and its potential in promoting plant growth and heavy metal accumulation in metal-polluted soil. *Chemosphere* **2008**, *72*, 157–164. [CrossRef] [PubMed]
14. Ma, Y.; Oliveira, R.S.; Freitas, H.; Zhang, C. Biochemical and Molecular Mechanisms of Plant-Microbe-Metal Interactions: Relevance for Phytoremediation. *Front. Plant Sci.* **2016**, *7*, 918. [CrossRef] [PubMed]
15. Yuniarti, E.; Dalmacio, I.F.; Paterno, E.S. Plant Growth-Promoting Potency of Heavy Metal Resistant Rhizobacteria from Gold and Copper Mine. *Int. J. Agric. Innov. Res.* **2019**, *7*, 560–566.
16. Yuniarti, E.; Dalmacio, I.F.; Paterno, E.S. Rhizobacteria Resisten Logam Berat Asal Tambang Emas Pongkor Indonesia dan Tambang Tembaga Marinduque Filipina. *AGRIC* **2019**, *31*, 75–88. [CrossRef]
17. Barstow, M. Erocarpus Indicus. The IUCN Red List of Threatened Species. 2018. Available online: https://www.iucnredlist.org/species/33241/2835450 (accessed on 10 June 2022). [CrossRef]
18. Orwa, C.; Mutua, A.; Kindt, R.; Jamnadass, R.; Anthony, S. Agroforestree Database: A Tree Reference and Selection Guide Version 4.0. Available online: http://www.worldagroforestry.org/sites/treedbs/treedatabases.asp (accessed on 8 June 2022).
19. Shi, Y.; Zhang, L.; Zhao, M.; Wang, G. Inhibitory effects of aqueous extracts from leaves of commontropical green plants on urea hydrolysis in soils. *Adv. Mater. Res.* **2014**, *1010–1012*, 614–617. [CrossRef]
20. Anggono, W.; Sutrisno, S.; Suprianto, F.; Evander, J.; Gotama, G. Biomass Briquette Investigation from Pterocarpus Indicus Twigs Waste as an Alternative Renewable Energy. *Int. J. Renew. Energy Res.* **2018**, *8*, 10–12.
21. Nurmila, N.; Sinay, H.; Watuguly, T. Identifikasi dan Analisis Kadar Flavonoid Ekstrak Getah Angsana (Pterocarpus indicus Willd) Di Dusun Wanath Kecamatan Leihitu Kabupaten Maluku Tengah. *J. Biol. Pendidik. Terap.* **2019**, *5*, 65–71. [CrossRef]
22. Ragasa, G.Y.; De Luna, R.D.; Hofilena, J.G. Antimicrobial terpenoids from Pterocarpus indicus. *Nat. Prod. Res.* **2005**, *19*, 305–309. [CrossRef]
23. Abdurrozak, M.I.; Syafnir, L. Uji Efektivitas Ekstrak Etanol Daun Angsana (*Pterocarpus indicus* Willd) sebagai Biolarvasida terhadap Larva Nyamuk Culex Sp. *J. Ris. Farm.* **2021**, *1*, 33–37. [CrossRef]
24. Rosianty, Y.; Waluyo, E.A.; Himawan, M.S.A. Potential of Carbon Storage in Angsana Plant (*Pterocarpus indicus* Willd) Inlir Barat District, Palembang City. *Sylva J. Ilmu-Ilmu Kehutanaman* **2021**, *10*, 6–11. [CrossRef]
25. Kainama, N.; Matinahoru, J.; Latumahina, F.S. Agroforestry Based Social Forestry on the Island of Ambon. *Plant Cell Biotechnol. Mol. Biol.* **2021**, *22*, 55–63.
26. Kandasamy, N.; Kaliappan, K.; Palanisamy, T. Upcycling sawdust into colorant: Ecofriendly natural dyeing of fabrics with ultrasound assisted dye extract of Pterocarpus indicus Willd. *Ind. Crops Prod.* **2021**, *171*, 113969. [CrossRef]
27. Manurung, M.A.; Mardhiansyah, M.; Sribudiani, E. Pengaruh Lama Perendaman Air Kelapa Terhadap Perkecambahan Semai Angsana (*Pterocarpus indicus* L). *Sylva J. Ilmu-Ilmu Kehutanaman* **2021**, *5*, 7–11. [CrossRef]
28. Mangkoedihardjo, S.; Ratnawati, R.; Alfiant, N. Phytoremediation of Hexavalent Chromium Pol-luted Soil Using Pterocarpus indicus and *Jatropha curcas* L. *World Appl. Sci. J.* **2008**, *4*, 338–342.

29. Suthep, S.; Duangrat, S.; Prayad, P.; Kraichat, T. Phytoremediation of Cadmium by Selected Leguminous Plants in Hydroponics Culture. *RRJEES* **2016**, *4*, 1–7.
30. Eviati, S. *Petunjuk Teknis Edisi 2 Analisis Kimia Tanah, Tanaman, Air, dan Pupuk*; Soil Research Institute: Karnal, Haryana, 2009; pp. 1–234.
31. Rawat, P.; Das, S.; Shankhdhar, D.; Shankhdhar, S.C. Phosphate-Solubilizing Microorganisms: Mechanism and Their Role in Phosphate Solubilization and Uptake. *J. Soil Sci. Plant Nutr.* **2021**, *21*, 49–68. [CrossRef]
32. Havlin, J.L. Fertility. In *Reference Module in Earth Systems and Environmental Sciences*; Elsevier: Amsterdam, The Netherlands, 2013; pp. 1–11.
33. Brady, N.C.; Weil, R.R. *The Nature and Properties of Soils*; Pearson Prentice Hall: Hoboken, NJ, USA, 2008; pp. 1–975.
34. Bolan, N.S.; Adriano, D.C.; Curtin, D. Soil Acidification and Liming Interactions with Nutrient and Heavy Metal Transformation and Bioavailability. *Adv. Agron.* **2003**, *78*, 5–272.
35. Buni, A. Effects of Liming Acidic Soils on Improving Soil Properties and Yield of Haricot Bean. *J. Environ. Anal. Toxicol.* **2014**, *5*, 1–4. [CrossRef]
36. Holland, J.E.; Bennett, A.E.; Newton, A.C.; White, P.J.; McKenzie, B.M.; George, T.S.; Pakeman, R.J.; Bailey, J.S.; Fornara, D.A.; Hayes, R.C. Liming impacts on soils, crops and biodiversity in the UK: A review. *Sci. Total Environ.* **2018**, *610–611*, 316–332. [CrossRef]
37. Ameyu, T. A Review on the Potential Effect of Lime on Soil Properties and Crop Productivity Improvements. *Ethiop. J. Environ. Earth Sci.* **2019**, *9*, 17–23. [CrossRef]
38. He, L.L.; Huang, D.Y.; Zhang, Q.; Zhu, H.-H.; Xu, C.; Li, B.; Zhu, Q.-H. Meta-analysis of the effects of liming on soil pH and cadmium accumulation in crops. *Ecotoxicol. Environ. Saf.* **2021**, *223*, 112691. [CrossRef] [PubMed]
39. Duong, T.T.T. Compost Effect on Soil Properties and Plant Growth. Ph.D. Thesis, The University of Adelaide, Adelaide, Australia, 2013; pp. 1–92.
40. Sulok, K.M.T.; Ahmed, O.H.; Khew, C.Y.; Lai, P.S.; Zehnder, J.A.M.; Wasli, M.E.; Aziz, Z.F.A. Use of organic soil amendments to improve soil health and yield of immature pepper (*Piper nigrum* L.). *Org. Agric.* **2021**, *11*, 145–161. [CrossRef]
41. Kabata-Pendias, A.; Pendias, H. *Trace Elements in Soils and Plants*; CRC Press: London, UK, 2001.
42. White, P. Heavy metal toxicity. In *Plant Stress Physiology*; Shabala, S., Ed.; CPI Group (UK). Ltd. Croydon: London, UK, 2012; pp. 210–237.
43. Yoon, J.; Cao, X.; Zhou, Q.; Ma, L.Q. Accumulation of Pb, Cu, and Zn in native plants growing on acontaminated Florida site. *Sci. Total Environ.* **2006**, *368*, 456–464. [CrossRef] [PubMed]
44. Liu, L.; Li, W.; Song, W.; Guo, M. Remediation techniques for heavy metal-contaminated soils: Principles and applicability. *Sci. Total Environ.* **2018**, *633*, 206–219. [CrossRef] [PubMed]
45. Oreoluwa, A.T.; Yetunde, A.T.; Joseph, U.E.; Chengsen, Z.; Hongyan, W. Effect of Biochar and Poultry Litter Application on Chemical Properties and Nutrient Availability of an Acidic Soil. *Commun. Soil Sci. Plant Anal.* **2020**, *51*, 1670–1679. [CrossRef]
46. Wang, Z.; Liu, X.; Qin, H. Bioconcentration and translocation of heavy metals in the soil-plants system in Machangqing copper mine, Yunnan Province, China. *J. Geochem. Explor.* **2019**, *200*, 159–166. [CrossRef]
47. Kurek, E.; Majewska, M. Microbially Mediated Transformations of Heavy Metals in Rhizosphere. In *Toxicity of Heavy Metals to Legumes and Bioremediation*; Zaidi, A., Ahmad Wani, P., Saghir Khan, M., Eds.; Springer: Berlin/Heidelberg, Germany, 2012; pp. 129–146.
48. Hassan, T.U.; Bano, A.; Naz, I. Alleviation of heavy metals toxicity by the application of plant growth promoting rhizobacteria and effects on wheat grown in saline sodic field. *Int. J. Phytoremediat.* **2019**, *19*, 522–529. [CrossRef]

Article

Bioaccumulation and Mobility of Heavy Metals in the Soil-Plant System and Health Risk Assessment of Vegetables Irrigated by Wastewater

Muhammad Tansar Abbas [1], Mohammad Ahmad Wadaan [2], Hidayat Ullah [1,*], Muhammad Farooq [3], Fozia Fozia [4,*], Ijaz Ahmad [5], Muhammad Farooq Khan [2], Almohannad Baabbad [2] and Zia Ullah [6]

[1] Institute of Chemical Sciences, Gomal University, Dera Ismail Khan 29220, Pakistan; mtansar5@gmail.com
[2] Zoology Department, College of Science, King Saud University, P.O. Box 2455, Riyadh 11451, Saudi Arabia; wadaan@ksu.edu.sa (M.A.W.); fmuhammad@ksu.edu.sa (M.F.K.); almbaabbad@ksu.edu.sa (A.B.)
[3] National Center of Excellence in Physical Chemistry, University of Peshawar, Peshawar 25120, Pakistan; drfarooq@uop.edu.pk
[4] Department of Biochemistry, KMU Institute of Dental Sciences, Kohat 26000, Pakistan
[5] Department of Chemistry, Kohat University of Sciences & Technology, Kohat 26000, Pakistan; drijaz_chem@yahoo.com
[6] College of Professional Studies, Northeastern University, Boston, MA 02115, USA; ziaullah.z@northeastern.edu
* Correspondence: hidayatktk@gu.edu.pk (H.U.); drfoziazeb@yahoo.com (F.F.)

Abstract: Accumulation of heavy metals in soil and vegetables is presently a challenging environmental concern worldwide. The present study was designed to elucidate heavy metals contamination of vegetables irrigated with domestic wastewater and associated health risks. The study area comprises three zones: Kot Addu, Alipur, and Muzaffargarh. A total of 153 samples of wastewater, topsoil, and vegetables were analyzed for physicochemical parameters and concentration levels of eight metal elements (Cu, Fe, Zn, Mn, Pb, Cd, Ni, and Cr) determined through analytical procedures. The outcome of the present investigation reveals that heavy metal concentrations in wastewater, soil, and vegetables irrigated with wastewater were slightly higher than the WHO-suggested limit. The heavy metals concentration observed in vegetables irrigated with wastewater can be ranked in order of Ni > Mn > Cr > Pb > Cu > Fe > Zn > Cd. Transfer factor (TF), daily ingestion of metals (DIM), and health risk index (HRI) were calculated. Spinach exhibited higher values of transfer factor than cabbage, cauliflower, and radish, which were followed by tinda and carrot. Minimum values of HRI were observed for Cr (0.0109) in almost all of the vegetables ingested by adults and children. Cabbage exhibited higher values of HRI for Pb (4.0656) in adults, followed by cadmium (HRI = 2.993). Minimum values of HRI were calculated for Cd (0.0115; child). Cauliflower exhibited higher values of HRI (5.2768) for Pb in children. Pb, HRI values (4.5902) were observed in adults living in Kot Addu. The results exhibited similar trends of HRI in adults and children living in Muzaffargarh and Alipur.

Keywords: heavy metals; vegetables; metals intake; health risk; wastewater-irrigation

1. Introduction

Though often overlooked, water is an imperative gift of nature for numerous lives that depend on it. All the activities of life are performed by water. Only a small fraction (<1%) of fresh water is available for terrestrial ecosystems [1–3]. Approximately 70% of the globally available freshwater is consumed by agricultural activities [4]. Currently, all over the world, the scarcity of water on land is due to numerous factors, including rapid population growth, uneven distribution patterns, abrupt climatic variations, and massive contamination by industrial and agricultural activities [5]. Owing to the shortage of freshwater around the world, most people are becoming more reliant on groundwater, which is characteristically expensive and also of poor quality because of high SAR (sodium adsorption ratio), higher cations exchange capacity, excessive sodium salts of carbonate, bicarbonate, and heavy metals. However, domestic and industrial sewage waste has become an alternative source of water for irrigation intentions. It has been estimated that the annual world wastewater production in arid and semi-arid regions is 1500 km^3, and about 20 million hectors (7% of the total irrigated) of land is irrigated with untreated wastewater for crop production [6,7]. The wastewater is the farmers' second choice for using this water in agricultural drives when there is a freshwater shortage. The quality of wastewater varies in terms of chemical composition and depends on sources of effluent production [8]. Recently, the recycling of wastewater in agriculture has become a widespread practice in regions located in dry arid zones, where scarcity of water is more pronounced [9]. The production of healthy food is a major concern all over the ecosphere [10]. The main source of income in Pakistan's remote villages, particularly in Sothern Punjab, is agriculture and growing vegetable crops commercially. Vegetables and fruits are important constraints for maintaining good health. These are sources of vitamins and important nutrients and support the body's metabolism. WHO suggested that daily intake of vegetables and fruits should be not less than 400 g per person in terms of acquiring numerous micro and macro-nutrients necessary for balanced health stability [11,12]. Unfortunately, in Pakistan, the average daily intake of vegetables is 130 g, which is 66% lower than the defined limit per day by WHO [13]. Besides the importance of vegetables, no information is available about the preferences of particular vegetables by local people and the magnitude of consumption of vegetables at international as well as national and regional levels. In underdeveloped countries, people consume vegetables as a source of food [14]. Humans grow vegetables in sewage-treated land, causing the heavy metals to enter the body. Long-term practices of irrigation through wastewater may lead to the accumulation of heavy metals in various crop plants and agricultural soil. Most vegetable crops are facing the severe problem of heavy metal pollution all over the world. The impacts of heavy metal pollution on vegetable growth and associated health risks are more pronounced in the areas close to industrial sites and the crops irrigated with domestic and industrial effluents [15]. Heavy metals having a non-biodegradable nature, usually known as trace elements, are essential in water sources [16]. Toxic concentrations of all trace metals, both necessary and unnecessary, affect biological processes, destroy cell membranes, and affect the three-dimensional structure of enzymes [16]. Some elements like copper, zinc, and selenium are important for maintaining human metabolic activities at the cellular level in various organs [17]. Arsenic, lead, cadmium, chromium, cobalt, and mercury are assumed to be potentially hazardous and exhibit negative health consequences, if consumed for a long time and ultimately lead to cancerous diseases [18]. Important constituents of vitamins, protein, iron, calcium, and other nutrients are directly obtained from the vegetables [19]. A higher concentration of Zn can cause shock, frits, vomiting, and forehead irritation [20]. The composition of white and red blood cells and severe digestive problems are associated with low concentrations of arsenic, while increased amounts result in black spots on the soles of the feet and brownish skin. The essential activities of enzymes are stimulated when the concentration level of Ni is low, and Ni has a detrimental effect in excessive amounts [21]. The level of health risks posed by wastewater with heavy metals was quantified by the application of different indices, including the transfer factor (TF), daily intake of metals (DIM), and health risk index (HRI). Little information is available about the impacts of heavy metals contamination on vegetables irrigated by sewage wastewater to common people of society. To recognize the health risks caused by vegetables contaminated with excessive concentrations of heavy metals, suitable information is very important. Reference [3] made studies to examine the influence of heavy metals toxicity in vegetable crops: broad bean (*Vacia faba*), durum wheat (*Triticum turgidum*), soft wheat (*Triticum æstivum*), oat (*Avena sativa*), nettle (*Urtica dioica*), broadleaf plantain (*Plantago major*), alfalfa (*Medicago sativa*), and mallow (*Malva sylvestris*), irrigated waste wastewater in Marrakech city, Morocco. Reference [4] selected the three vegetables: lettuce, cabbage, and tomato, for elucidation of the health risks of heavy metals from vegetables grown on soil irrigated with untreated and treated wastewater in Arba Minch, Ethiopia. Reference [22] conducted research work in India to explore the impact of heavy metals accumulation in vegetables including coriander, onion, and tomato. Reference [23] had investigated the impacts of heavy metal pollution on Ethiopian agriculture. Reference [24] revealed the role of wastewater use for the irrigation of vegetation crops and associated health risks. They had selected only two species of vegetables, namely cauliflower and cabbage. However, the data matrix related to heavy metals pollution in underground water, soil, and vegetables had not yet been examined in the area under consideration (Kot Addu, Muzaffargarh and Alipur), Pakistan. Therefore, the current investigations were made to elucidate the heavy metals contamination of vegetables irrigated with domestic sewage wastewater and associated health risks.

2. Materials and Methods

2.1. Study Area

The study area comprises three zones (Figure 1), namely classified as Kot Addu (30.4685° N, 70.9606° E), Alipur (29.3817° N, 70.9131° E), and Muzaffargarh (30.0736° N, 71.1805° E), located along elevation gradient from 103 m (a.s.l) to 125 m (a.s.l). The study was designed to cover the 8249 Km^2 area of district Muzaffargarh, part of the southern Punjab province of Pakistan. The area under consideration is classified as subtropical dunes. The climate is very hot, with a harsh dry season in summers with maximum values recorded for temperature at approximately 54 °C (129 °F) and winter with minimum values recorded for temperature at approximately −1 °C (30 °F). Rainfall concentrates from late July to August, and the area receives an average annual rainfall of 127 mm (5.0 in). Wind erosion is common, characterized by dust storms. According to the census 2017 report, a total of 4,322,009 persons harbor in the area.

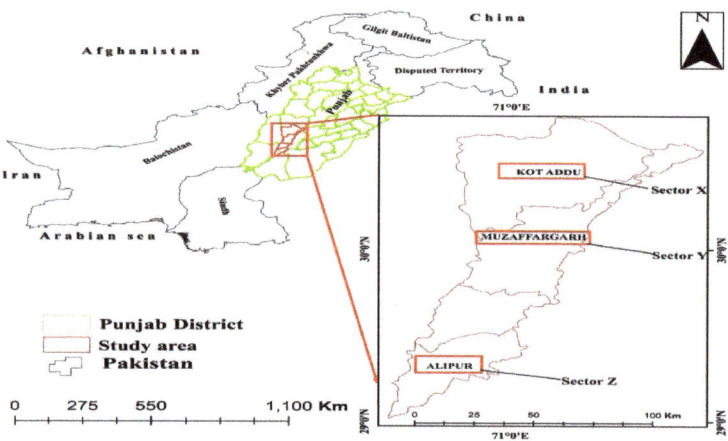

Figure 1. Showing Study area for Sampling: Kot Addu, Muzafargarh, and Alipur.

The study area has been classified into three zones: (i) Kot Addu designated as sector X, (ii) Muzaffargarh designated as sector Y and (iii) Alipur designated as sector Z. In contrast, sector Z was used to collect fresh water, irrigated soil, and vegetables. Samplings were performed during 2021 and 2022. Sectors X and Y are separated by 60 km, which is the longest distance, and 9 km, which is the shortest distance between subsectors of each prescribed sector. From the examining areas, tests of sewage water, soil, and vegetables were gathered. Sector Z, on the other hand, was used to collect freshwater samples along with their irrigated soil and vegetables. Standard Recommended Methods (SRM) were used for each sampling and analysis.

2.2. Field Sampling

Sampling of wastewater was carried out from October 2021 to April 2022, from various disposal sectors and explored to determine the heavy metals contents. Different sampling procedures were applied to evaluate the quality of water used for irrigation of vegetable crops [5]. Sewage wastewater samples were collected in 500 mL plastic bottles from all the sources present in the area under consideration. All the bottles were washed with 5% nitric acid, then treated with deionized water and cleaned twice. A triplicate sample was taken from each sampling site for accuracy. In the study area (sectors X, Y, and Z), composite soil samples were collected from topsoil at 0 to 30 cm depths and stored in plastic bags. Composite samples of broad-leaved vegetables including *Spinacia oleracea* (spinach), *Brassica oleracea var. capitata* (cabbages), *Brassica oleracea var. botrytis* (cauliflower), *Raphanus sativus* (radish), *Brassica rapa* subsp. rapa (turnip), *Praecitrullus fistulosus* (tinda), *Daucus carota* (carrots), *Lactuca sativa* (lettuce), and *Colocasia esculenta* (Colocasia) were collected from the study area irrigated with sewage and partially sewage water. The vegetable samples were cleaned with distilled water to eliminate dust and other pollutant matter, and for specimen drying purposes, 105 °C temperatures were fixed in the oven for 12 h. The dried samples were then pulverized, homogenized, and transferred into clean sample bottles. For analysis, only the edible section of the vegetable was used.

2.3. Chemical Analysis

2.3.1. Determination of Biochemical Composition and Concentration of Trace Metals in Wastewater Samples

The collected water samples were used to determine the pH, electrical conductivity of wastewater, and total dissolved solids (TDS). pH was determined by using a digital pH meter (Jenway 3510, Tokyo, Japan). For additional analysis of the water sample, SRM was used [5,25]. The cation exchange capacity of collected water samples was determined by using a digital EC meter. KCl standard solution was used for calibration. The electrode bulb was washed with deionized water after analyzing each sample [26,27]. Utilizing a movable meter (Jenway 4510 Conductivity/TDS Meter; 230 VAC/UK), the recommended methods for estimating TDS for samples were used [28,29]. The samples of sewage water and unadulterated water were gathered over more than seven days. The sewage and new water processing was finished by blending 50 mL of the samples with 10 mL moved nitric corrosive in a particular sample volume (HNO_3). The digesting process was prolonged in a water bath for up to one hour at eighty degrees Celsius and then filtered. The filtrate was diluted after being filtered using Whatman (no-42) filter sheets and 0.05 L deionized water. We utilized a Perkin Elmer Atomic absorption spectrophotometer (AAS) model 7000 to distinguish the measures of heavy metals such as Cr, Pb, Cd, Cu, Fe, Ni, Zn, and Mn. To guarantee accurate results, standard solutions of each of these metals were employed.

2.3.2. Determination of Biochemical Composition and Concentration of Trace Metals in Soil Samples

The soil samples' pH, EC, and organic matter (OM) were calculated. SRM was followed to perform analysis. The nature of edaphic features, soil chemical reactions (pH), and nutrient cycling between the soils and vegetables are the important dimensions to evaluate the sewage quality in terms of heavy metals contamination. Soil sample extracts were prepared and then pH was determined by using a digital pH meter, following the methods of [29]. Soil electrical conductivity was determined using a digital EC meter by following standard procedure [25,30]. The amount of soil organic matter content was determined using the Walkley–Black method [27,31]. Heavy metal testing was also conducted on the subsurface soil. A 500 mL dried soil sample was mixed with 15 mL of a 5:1:1 solution of HNO_3, H_2SO_4, and $HClO_4$ overnight. At two separate temperatures, the first digestion took place for 30 min at 50 °C. The second digestion was carried out for 180 min at 120 °C. A VELP Scientific digester was employed for indigestion, and it was kept running until transparent solutions were produced. The Whatman filter paper was used for the filtering process (no-42). A total of 50 mL of deionized water was used to dilute the filtrate [13,17]. Perkin Elmer's Atomic absorption spectrophotometer (AAS) model 7000 was used for further analysis.

2.3.3. Determination of Biochemical Composition and Concentration of Trace Metals in Vegetable Samples

The vegetable species under consideration after sampling were subjected to biochemical analysis to estimate the heavy metals (Cr, Pb, Cd, Cu, Fe, Ni, Zn, and Mn) contents by following the standard procedure [27]. After sampling the vegetable specimens from the study area, each specimen was dried for 24 h and completely cleaned with distilled water followed by 1% HCl to remove any unwanted particles. Only a part of the vegetables was used for chemical analysis, following standard procedure [27]. The dried material was crushed into a fine powder and stored in tidy containers that were properly sealed. With 50 mL of deionized water, the filtrate was made more weakened [5]. Perkin Elmer's Atomic absorption spectrophotometer (AAS) model 7000 was used for further analysis.

2.4. Health Risk Assessment

How harmful contaminated vegetables are to one's health is determined by their hazard quotient. The hazard quotient describes the correlation between the reference dosage and the computed dose. The general public views it as safe if the ratio is less than one. The population will be in grave danger if the value is greater than one or the same as the esteem. The researchers were proven to be accurate while using this method [31]. The equation reads as follows:

$$HQ = (W_{plant}) \times (M_{plant})/R_fD \times B$$

(W_{plant}) = Dry weight of a contaminated piece of the plant that is consumed in mg/d.
(M_{plant}) = Heavy metal substance that is tracked down in vegetables in mg/kg.
R_fD = Typical grouping of metal in food (mgd^{-1}).
B = The typical mass of the body is determined in kg.
Adult = 55.9 kg, Child = 32.7 kg.

The conventional values and RfD values for harmful excess heavy metals were developed by the Department of Food and Rural Affairs, the Department of the Environment, and the Institute of Integrated Risk Information.

2.5. Determination of Transfer Factor (TF)

The transfer factor refers to the estimation of the limit of easily heavy metals that can move from soil to vegetative parts of crops. If the value of the transfer factor is greater than one, then the plants showed a strong ability to absorb metals from agricultural soil. The TF was determined by using Equation [32].

$$TF = C\ plant/CAS$$

where C plant = Heavy metals in edible fragments of vegetables
CAS = HMs concentration within agricultural soil (mg/Kg dry weight)

2.6. Daily Dietary Index (DDI)

The daily consumption of dietary diversity indices was determined by the following [33].

2.7. Health Risk Index (HRI)

To determine the "chronic health risks" HRIs of heavy metals contents were measured by using the following Equation [15]. The health risk was determined by utilizing DIM and oral dose. An HRI value of less than one (HRI < 1) is suggested to be safe for health [24].

HRI was determined as:

$$HRI = DIM/RfD$$

3. Statistical Analysis

The data obtained from the lab experiment were statistically analyzed by IBM SPSS software (Version 26) computer program. Analysis of variance (ANOVA) was applied for the exploration of the impact of heavy metals on the quality of vegetables to describe pictorial irrigation water situations [34]. The presence of heavy metals and health risks were statistically evaluated [35].

4. Results and Discussion

4.1. Exploration of Heavy Metals Concentration in Wastewater

The physicochemical attributes of the wastewater collected from the three different sectors (X, Y, and Z) are presented (Table 1). The result of the analysis of variance (ANOVA) described that there were pronounced variations among pH levels of wastewater samples collected from three different sectors (X, Y, Z). The pH values ranged from 8.37 to 7.80 ± 0.28, 8.70 to 7.40 ± 0.65, 7.25 to 7.04 ± 0.11, respectively ($F = 26.96$; $p < 0.000$). The WHO acceptable pH limit for water is 6.5 to 8.8. However, wastewater comprises carbonates of sodium and calcium along with sodium chlorides and sodium Sulphate. Due to the presence of these salts, the pH of wastewater was slightly higher and exhibited a more alkaline nature than the pH of fresh water. The trends found in the chemical reaction (pH) of wastewater were also reported by [36]; they elucidated the physiochemical attributes of wastewater in Bhakkar. The result (Table 1) reveals that there were significant variations exist in terms of EC among the three sectors (F & p-value = 474.06***). EC values of wastewater ranged from 1740 to 1680 ± 30.00 µS/cm, 1710 to 1640 ± 35.00 µSc/m, and 1335 to 1298 ± 18.50 µS/cm, respectively. Similar findings were also reported by [15]. The result (Table 1) suggested that wastewater samples collected from the area under consideration resources were saline. In most of the analyzed wastewater samples, EC values were above the permissible limit (1400 µS/cm) guided by [37]. The exceeded values of EC in wastewater are not surprising, because numerous researchers also reported similar findings [38]. In the wastewater used for irrigation, heavy metals Pb, Cr, Cd, Cu, Zn, Ni, Fe, and Mn had a range of 0.28 to 0.45 ± 0.09, F & p-values = 26.68***; 0.77 to 0.98 ± 0.11, F & p-values = 100.78***; 0.64 to 0.96 ± 0.16, F & p-values 69.72***; 0.32 to 0.48 ± 0.08, 0.14 to 0.42 ± 0.14, F & p-values = 29.74***; 0.29 to 0.47 ± 0.09, F & p-values = 11.25*** 8.14 to 10.35 ± 1.11, F & p-values = 83.82***; 0.3 to 0.52 ± 0.11, F & p-values= 7.32*** (Table 1) in sector X, respectively, and exhibited significant variation for all the variables (Table 1). Similarly, the heavy metals concentrations observed in wastewater collected from sector Y showed significant differences (Table 1) and ranged from 0.22 to 0.45 ± 0.12, 0.68 to 0.96 ± 0.14, 0.52 to 0.77 ± 0.13, 0.35 to 0.40 ± 0.03, 0.20 to 0.27 ± 0.04, 0.25 to 0.37 ± 0.06, 9.65 to 10.42 ± 0.39, 0.31 to 0.43 ± 0.06, respectively. In the freshwater collected from sector Z used for irrigation, heavy metals Pb, Cr, Cd, Cu, Zn, Ni, Fe, and Mn had range of 0.08 to 0.10 ± 0.01, 0.07 to 0.15 ± 0.04, 0.06 to 0.08 ± 0.01, 0.14 to 0.22 ± 0.04, 0.07 to 0.1 ± 0.02, 0.10 to 0.22 ± 0.06, 4.78 to 5.9 ± 0.56, 0.17 to 0.32 ± 0.08, respectively (Table 1), similar heavy metals contamination constraints were reported by [10]. The magnitude of heavy metals contamination in wastewater was ranked in order of Fe> Cr > Mn > Cd > Ni > Cu > Zn > Pb. The results demonstrated that all the heavy metal contents overflowed from the allowed limits good for health suggested by WHO, used for irrigation drives.

Table 1. Physicochemical attributes of wastewater and heavy metals concentration. Samples were collected from three sectors (X, Y, and Z).

Variables		pH	EC (µS/cm)	TDS (mg/L)	Pb (mg/L)	Cr (mg/L)	Cd (mg/L)	Cu (mg/L)	Zn (mg/L)	Ni (mg/L)	Fe (mg/L)	Mn (mg/L)
Sector X		8.07 ± 0.28	1710 ± 30.00	1120.5 ± 24.50	0.36 ± 0.09	0.87 ± 0.11	0.8 ± 0.16	0.4 ± 0.08	0.28 ± 0.14	0.38 ± 0.09	9.245 ± 1.11	0.41 ± 0.11
Min–Max		7.80–8.35	1680.00–1740.00	1096.00–1145.00	0.28–0.45	0.77–0.98	0.64–0.96	0.32–0.48	0.14–0.42	0.29–0.47	8.14–10.35	0.30–0.52

Table 1. Cont.

Variables	pH	EC (μS/cm)	TDS (mg/L)	Pb (mg/L)	Cr (mg/L)	Cd (mg/L)	Cu (mg/L)	Zn (mg/L)	Ni (mg/L)	Fe (mg/L)	Mn (mg/L)
Sector Y	8.05 ± 0.65	1675 ± 35.00	1090 ± 30.00	0.33 ± 0.12	0.82 ± 0.14	0.64 ± 0.13	0.37 ± 0.03	0.23 ± 0.04	0.31 ± 0.06	10.03 ± 0.39	0.37 ± 0.06
Min–Max	7.40–8.70	1640–1710	1060–1120	0.22–0.45	0.68–0.96	0.52–0.77	0.35–040	0.20–0.27	0.25–0.37	9.65–10.42	0.31–0.43
Sector Z	7.145 ± 0.11	1316.5 ± 18.50	695 ± 15.00	0.09 ± 0.01	0.11 ± 0.04	0.07 ± 0.01	0.18 ± 0.04	0.08 ± 0.02	0.16 ± 0.06	5.34 ± 0.56	0.24 ± 0.08
Min–Max	7.04–7.25	1298–1335	680–710	0.08–0.10	0.07–0.15	0.06–0.08	0.14–0.22	0.07–0.10	0.10–0.22	4.78–5.90	0.17–0.32
F & p-value	26.96 ***	474.06 ***	596.23 ***	26.68 ***	100.78 ***	69.72 ***	29.74 ***	6.45 *	11.25 ***	83.82 ***	7.32 ***
WHO, 2016	6.5 8.8	1400	1000	0.1	0.1	0.01	0.2	2	0.2	5	0.2

* $p < 0.05$, *** $p < 0.001$; StD = values are expressed in ±; (i) Kot Addu designated as sector X, (ii) Muzaffargarh designated as sector Y, and (iii) Alipur designated as sector Z.

4.2. Exploration of Heavy Metals Concentration in Soil Irrigated with Wastewater

Biochemical properties of soil samples collected from the study area irrigated with sewage wastewater were analyzed to determine the level of heavy metals' heterogeneity. Results of the ANOVA (Table 2) revealed that there were no significant differences observed among the three sectors for the pH of soil irrigated with wastewater (F & p-values= 0.29^{NS}). pH ranged from 8.11 to 8.18. The values of soil chemical reactions (pH) were within the range as suggested by WHO. The slightly alkaline nature of wastewater-irrigated soil is because sewage waste is comprised of sufficient content of magnesium, calcium, and bicarbonates. The EC of soil samples collected from the area irrigated with wastewater ranged from 457.50 EC (μS/cm) to 466.25 EC (μS/cm) and exhibited significant differences (F & p-values = 50.49***). The results demonstrated that the EC (μS/cm) of the wastewater was not below the WHO's limits (1400 μS/cm), hence the EC of wastewater-irrigated soil was not acceptable for irrigation drives. These findings in terms of electrical conductivity of soil were similar to the outcomes of [39]. The soil was enriched with decomposable ingredients, resulting in considerable content of organic matter (%) ranging from 1.375 to 1.4725. Our results are in line parallel with the findings reported by [26], who explored the organic matter of soil in the D.G. Khan District. Heavy metals Pb, Cr, Cd, Cu, Zn, Ni, Fe, and Mn in soil irrigated with wastewater ranged from 74.50, 63.00, 3.20, 35.40, 31.50, 78.00, 220.00, and 310.00 mg/Kg in sector X, 70.75, 56.00, 3.35, 38.00, 33.00, 85.50, 230.00, and 365.00 values were observed in sector Y, respectively. Similarly, 46.25, 34.25, 1.25, 16.75, 15.75, 36.50, 192.50, and 134.25 mg/Kg were observed in sector Z, accordingly. The results followed the permissible limits for heavy metals described by WHO. The heavy metals contamination in soil irrigated with sewage wastewater can rank in the following order: Mn > Fe > Ni > Pb > Cr > Cu > Zn > Cd (Table 2).

Table 2. Heavy metals accumulation in soil irrigated with wastewater. Samples were collected from three sectors (X, Y, and Z).

Variables	pH	EC (μS/cm)	O.M (%)	Pb (mg/L)	Cr (mg/L)	Cd (mg/L)	Cu (mg/L)	Zn (mg/L)	Ni (mg/L)	Fe (mg/L)	Mn (mg/L)
Sector X	8.18 ± 0.28	457.5 ± 22.50	1.37 ± 0.02	74.5 ± 6.50	63 ± 6.00	3.2 ± 0.30	35.4 ± 1.60	31.5 ± 2.50	78 ± 6.00	220 ± 25.00	310 ± 30.00
Min–Max	7.90–8.47	335.00–480.00	1.35–1.40	68.00–81.00	57.00–69.00	2.90–3.50	33.80–37.00	29.00–34.00	72.00–84.00	195.00–245.00	280.00–340.00
Sector Y	8.11 ± 0.24	466.25 ± 14.98	1.47 ± 0.12	70.75 ± 7.98	56 ± 5.51	3.35 ± 0.64	38 ± 4.98	33 ± 6.01	85.5 ± 7.49	230 ± 39.98	365 ± 45.03
Min–Max	7.80–8.30	450.00–480.00	1.35–1.60	62.00–78.00	49.00–60.00	2.60–3.90	32.00–42.00	27.00–39.00	78.00–93.00	180–260	310–400
Sector Z	8.07 ± 0.35	303.75 ± 32.49	0.85 ± 0.44	46.2 ± 8.49	34.25 ± 3.99	1.25 ± 0.39	16.75 ± 4.49	15.75 ± 2.98	36.5 ± 5.49	192.5 ± 22.49	134.25 ± 12.98
Min–Max	7.73–8.07	275.00–303.75	0.40–0.85	38.00	46.25	31.00–34.25	0.90–1.25	13.00–16.75	13.00–15.75	30.00–36.50	175.00–192.50

Table 2. *Cont.*

Variables	pH	EC (µS/cm)	O.M (%)	Pb (mg/L)	Cr (mg/L)	Cd (mg/L)	Cu (mg/L)	Zn (mg/L)	Ni (mg/L)	Fe (mg/L)	Mn (mg/L)
F & p-value	0.29 NS	50.49 ***	10.60 ***	17.41 ***	62.34 ***	49.31 ***	32.22 ***	22.65 ***	78.85 ***	2.36 NS	82.82 ***
WHO, 2007/2016	—	—	—	60	100	1	100	200	67.9	150	500

*** $p < 0.001$; NS: Non-significant; StD = values are expressed in ±; (i) Kot Addu designated as sector X, (ii) Muzaffargarh designated as sector Y, and (iii) Alipur designated as sector Z.

4.3. Exploration of Heavy Metals in Vegetables Irrigated with Wastewater

In the present investigation, determination of different levels of heavy metals accumulated in vegetables and crop plants, irrigated with municipal sewage waste, industrial waste was the major concern and compared with the permissible limits set by WHO/FAO. Heavy metals Cu, Fe, Zn, Mn, Pb, Cd, Ni, and Cr in vegetables irrigated with wastewater from sector X also showed significant differences in almost all the variables (Table 3), and all the metals ranged from 9 to 29, (F & p-values = 20.48***) 6.3 to 26.75, (F & p-values 4.89**); 7.50 to 22.30, (F & p-values = 4.46**); 10.30–38.40, (F & p-values = 19.67***); 6.00 to 35.00, (F & p-values = 10.57***); 14.30 to 42.50 (F & p-values = 13.26***); 14.30 to 28.20, (F & p-values = 3.36**), and 4.30 to 38.00 mg/Kg (F & p-values = 1.59*) and those from sector Y wastewater-irrigated vegetables 11.30 to 23.40, 5.70 to 26.40, 6.78 to 17.75, 11.45 to 41.38, 7.40 to 38.00, 2.9 to 8.48, 14.90 to 48.84 and 10.47 to 34.00 mg/Kg, respectively. Cu, Fe, Zn, Mn, Pb, Cd, Ni, and Cr in vegetables irrigated with fresh water from the sector ranged from 6.95 to 15.80, 2.45 to 48.95, 2.96 to 14.32, 3.40 to 22.45, 4.31 to 10.35, 0.90 to 3.70, 4.38 to 29.92, and 4.37 to 27.40 mg/Kg, respectively. The result (Table 3) reveals that heavy metal elements, Mn, Pb, Cd, Ni, and Cr had overriding values in vegetables and exhibited exceeded levels suggested by the World Health Organization (WHO). These results are similar to the findings of [10]. The magnitude of contamination of vegetables irrigated with wastewater can be ranked: Ni > Mn > Cr > Fe> Pb > Cu > Zn > Cd. It can be concluded that nickel, manganese, and chromium (mg/Kg) contents were higher in all the vegetables under consideration. The present investigation reveals that if local farmers continue the practice of using sewage wastewater and untreated municipal industrial waste for the irrigation of vegetables may lead to the accumulation of heavy metals, particularly Cd, Cr, and Pb may occur from water to plants and livestock. These results are congruent with the findings of [40,41]. In the present research, nickel concentration was found in an excessive amount in vegetables irrigated with sewage wastewater, soil, and water samples subjected to chemical analysis (Table 3). This is because the study area is a dry subtropical sand dunes desert zone facing the regular windblown dust storm derived from the weathering of rocks and soil from the Suleman Mountain range, Dera Ghazi Khan.

The presence of nickel in the air also accumulates from the combustion of coal, fuel, petrol, and diesel by vehicles and thermal power plants occurring in nearby areas of the study area. Similar findings in terms of nickel sources were also reported by [13]. The higher level of nickel in vegetables and other fruit stuff poses severe health risks (Table 3), and these findings were in parallel to the results of [5]. In the soil, nickel accumulation is sustained by earthworms and micro-organisms [42]. Many other environmental factors also cause the accumulation of excessive heavy metals [43]. Another cause of the higher content of nickel in vegetation crop plants is due to nickel being preferred by all angiospermic plant species as it plays an imperative role in the conversion of urea into ammonium [43]. Chromium exhibited an excessive level in the present investigation (Table 3) than the suitable amount for health suggested by WHO. Many industries have been established in the area under consideration and producing chromium pollution in water and soil. The effluent discharge from these industries contains much Cr content in its composition [26]. Besides these, Zinc, Pb, Cd, and Mn should be further scrutinized in derived products from the study crops and livestock, particularly meat, milk, and leafy vegetables to determine considerable food safety problems. Besides these, wastewater reuse may cause environmental risks along with health diseases in humans [36].

Table 3. Heavy metals contents in vegetables irrigated with wastewater. Samples were collected from three sectors (X, Y, and Z).

Variables	Pb (mg/L)	Cr (mg/L)	Cd (mg/L)	Cu (mg/L)	Zn (mg/L)	Ni (mg/L)	Fe (mg/L)	Mn (mg/L)
Sector X	20.5 ± 14.50	21.15 ± 16.85	4.53 ± 3.23	19 ± 10.00	14.9 ± 7.40	28.4 ± 14.10	16.53 ± 10.23	24.35 ± 14.05
Min–Max	6.00–35.00	4.30–38.00	14.30–42.50	9.00–29.00	7.50–22.30	14.30–28.20	6.30–20.45	10.30–38.40
Sector Y	22.70 ± 6.05	22.24 ± 10.35	5.69 ± 5.49	17.35 ± 14.97	12.27 ± 15.30	31.87 ± 2.79	16.05 ± 16.97	26.42 ± 11.77
Min–Max	7.40–38.00	10.47–34.00	2.90–8.48	11.30–23.40	6.78–17.75	14.90–48.84	5.70–26.40	11.45–41.38
Sector Z	7.33 ± 4.43	15.89 ± 23.25	2.35 ± 5.68	11.38 ± 9.53	8.64 ± 3.02	17.15 ± 1.36	25.7 ± 12.77	12.93 ± 11.52
Min–Max	4.31–10.35	4.37–27.40	0.99–3.70	6.95–15.80	2.96–14.32	4.38–29.92	2.45–48.95	3.40–22.45
F & p-value	10.57 ***	1.59 NS	13.26 ***	20.48 ***	4.46 **	3.36 **	4.89 **	19.67 ***
WHO, 2007/2016	0.3	2.3	0.2	73	99	0.1	425	02

** $p < 0.01$, *** $p < 0.001$; NS: non-significant; StD = values are expressed in ±; (i) Kot Addu designated as sector X, (ii) Muzaffargarh designated as sector Y, and (iii) Alipur designated as sector Z.

Pearson Correlation Analysis of Heavy Metals in Vegetables Irrigated with Wastewater

Selected species of vegetables irrigated with wastewater were subjected to physio-chemical analysis. The outcomes of the chemical analysis of vegetables were then used for statistical analysis to determine the interrelationship of heavy metals and delimit the major sources contributing to the heavy metals in wastewater used for irrigation and vegetable species accumulate excessive amounts of heavy metals in their vegetative parts. Pearson's correlation was performed by using SPSS software (Version 26) computer program. The results (Table 4) show that if the magnitude of one heavy metal concentration increases in vegetable vegetative parts, then the other concentrations will also increase accordingly and vice versa. The greater "r—value; $p < 0.001$" exhibited a stronger correlation of the two heavy metals to each other. Among the seven types of heavy metals under consideration, Cu exhibited a strong positive correlation with Pb (r = 0.688; $p < 0.001$), and Cu showed a strong positive correlation with Cr (r = 0.605; $p < 0.001$). Similarly, there were highly significant correlations accounted for Zn and Ni (r = 0.638; $p < 0.001$) followed by Zn and Pb (r = 0.458; $p < 0.001$). Besides these, lower values of correlation were observed between Zn and Cd (r = 0.060; $p < 0.002$; Table 4). The results demonstrated that maximum values of the correlation coefficient suggest the common source of similar metal accumulation in vegetables by absorbing wastewater.

Table 4. Representing the Pearson's correlation matrix of heavy metal elements in wastewater collected from three different sectors (X, Y, and Z).

Metals	Cu	Fe	Zn	Mn	Pb	Cd	Ni
Fe	0.028						
	0.889						
Zn	0.275	0.154					
	0.164	0.443					
Mn	0.283	−0.333	0.226				
	0.153	0.090	0.257				
Pb	0.688	0.120	0.458	0.081			
	0.000	0.552	0.016	0.687			
Cd	0.334	-0.108	0.060	0.566	0.049		
	0.590	0.767	0.002	0.808	0.764		
Ni	0.153	0.010	0.638	0.351	0.098	0.251	
	0.447	0.959	0.000	0.072	0.625	0.206	
Cr	0.605	0.070	0.172	0.154	0.469	0.171	−0.064
	0.001	0.728	0.390	0.443	0.014	0.394	0.753

4.4. Taxation of Heavy Metal Transfer Factor (HMTF)

Transfer factor (TF) is the quotient that exhibited the potential of plants to absorb the metal contents from the soil, which is enriched with metal pollutants. The result (Table 5) demonstrated the heavy metal transfer factor [43]. Heavy metals associated health risks were determined by transfer factor (TF), consuming the vegetables irrigated with wastewater. The nine vegetables exhibited significantly different transfer factors among the three sector types, as all the values of heavy metals transfer factors vary (Table 5). The F-value indicated a highly significant variance in terms of lead (F = 2.49*; Table 5). The lead (TF) ranged from 0.24 to 0.47 (Table 5) in spinach. Chromium also exhibited significant differences in all three sectors (8.84**). All the vegetables exhibited significant differences. Spinach showed range (0.43 to 0.82), cabbage exhibited (TF) for Cr ranged (1.76 to 2.96), cauliflower (0.41 to 0.61), radish (0.07–0.13), turnip (0.17–0.23), tinda (0.17 to 0.35), carrot (0.37 to 0.64), and lettuce (0.41 to 0.70). Ferric (TF) played an overriding role in overall analysis variance, as much of a significant difference exists (12.99***; Table 5). The magnitude of the effect of sector types can be ranked in terms of difference among the nine types of vegetables that accumulate heavy metals from the soil irrigated by wastewater enriched with heavy metals and showed the potential of transfer factor (Fe = 12.99*** > Cr = 8.84*** > Cu = 8.25*** > Mn = 4.67*** > Zn = 2.82* > Cd = 2.78* > Pb = 2.63*; Table 5). Spinach exhibited higher values of variance in terms of Fe and ranged (0.07–0.11). TF of cabbage was (0.08 to 0.12) for Fe, cauliflower showed range (Fe = 0.05 to 0.06), radish (0.13 to 0.15), tinda (Fe = 0.02 to 0.04), carrot (0.01 to 0.10), and lettuce also exhibited similar values of TF in term of Fe accumulation to carrot (0.01 to 0.11), and Colocasia showed the transfer factor values for Fe accumulation was (0.02 to 0.03; Table 5). Similar findings in terms of transfer factor were reported by [24,26] in broad-leaved perennial vegetables.

Table 5. Comparison of transfer factor (TF) of selected heavy metals accumulated in vegetables irrigated with wastewater.

Vegetables	Sectors	Pb	Cr	Cd	Cu	Zn	Ni	Fe	Mn
Spinach	X	0.37	0.43	1.05	0.76	0.56	0.54	0.11	0.11
	Y	0.45	0.52	1.64	0.59	0.51	0.45	0.07	0.10
	Z	0.24	0.82	2.00	0.94	0.91	0.90	0.07	0.16
Cabbage	X	0.42	0.53	1.76	0.65	0.71	0.41	0.12	0.03
	Y	0.54	0.55	1.82	0.58	0.43	0.38	0.11	0.03
	Z	0.18	0.80	2.96	0.62	0.19	0.12	0.08	0.09
Cauliflower	X	0.48	0.61	0.40	0.82	0.30	0.18	0.06	0.07
	Y	0.47	0.61	0.81	0.62	0.35	0.17	0.05	0.06
	Z	0.09	0.41	0.72	0.83	0.52	0.57	0.05	0.03
Radish	X	0.38	0.07	0.74	0.42	0.38	0.40	0.15	0.04
	Y	0.39	0.09	1.01	0.30	0.41	0.35	0.14	0.04
	Z	0.21	0.13	1.40	0.67	0.69	0.79	0.13	0.05
Turnip	X	0.10	0.17	1.33	0.31	0.35	0.31	0.03	0.07
	Y	0.16	0.24	2.48	0.34	0.30	0.20	0.06	0.11
	Z	0.18	0.35	0.79	0.41	0.50	0.55	0.04	0.15
Tinda	X	0.18	0.19	0.77	0.35	0.46	0.39	0.03	0.10
	Y	0.24	0.19	0.87	0.39	0.54	0.44	0.02	0.09
	Z	0.22	0.39	0.98	0.53	0.81	0.67	0.03	0.17
Carrot	X	0.20	0.37	2.30	0.49	0.24	0.28	0.09	0.10
	Y	0.45	0.44	1.28	0.57	0.25	0.26	0.10	0.10
	Z	0.16	0.64	1.46	0.44	0.34	0.59	0.01	0.07
Lettuce	X	0.15	0.41	2.11	0.34	0.46	0.48	0.11	0.12
	Y	0.20	0.46	1.70	0.36	0.45	0.57	0.11	0.10
	Z	0.22	0.70	1.19	0.59	0.95	0.76	0.07	0.12

Table 5. Cont.

Vegetables	Sectors	Pb	Cr	Cd	Cu	Zn	Ni	Fe	Mn
Colocasia	X	0.08	0.26	2.39	0.59	0.25	0.40	0.03	0.09
	Y	0.10	0.32	2.53	0.60	0.21	0.38	0.03	0.08
	Z	0.17	0.70	1.43	0.68	0.25	0.82	0.02	0.08
F & p-Value		2.63 *	8.84 ***	2.78 *	8.25 ***	2.82 *	2.79 *	12.99 ***	4.67 **

* $p < 0.05$, ** $p < 0.01$, *** $p < 0.001$.

4.5. Daily Intake of Metals (DIM) and Health Risk Index (HRI)

The result (Table 6) described the daily intake of metals by humans and the health risk index. The present research work was conducted to determine human exposure to various levels of heavy metals contaminated vegetables and to elucidate pollution-related hazards to human health. Vegetables are the major sources of food in underdeveloped countries. In Pakistan, vegetables are grown in soil that is irrigated with wastewater, consequently, people consume higher values of heavy metals daily [44]. However, the metals consumption by vegetables irrigated with underground water is less than those grown in soil irrigated with sewage wastewater. The availability of such secondary data provides information about health risks from metal consumption. The daily intake of metals in Kot Addu (sector X), heavy metals such as Pb, Cr, Cd, Ni, Zn, Cu, Fe, and Mn for adults ranged from 0.0058 to 0.783 and for children ranged from 0.0013 to 0.516. The daily intake of metals (Pb, Cr, Cd, Ni, Zn, Cu, Fe, and Mn) in Muzaffargarh (sector Y) ranged from 0.0092 to 0.989 for adults and children ranging from 0.0096 to 0.0.698. Sector Z exhibited lower values of daily intake of heavy metals for adults and children than sector X and sector Y (Table 6) in terms of values mean across nine vegetable types (spinach, cabbage, cauliflower, radish, turnip, tinda, carrot, lettuce, and colocasia). The health risk index was calculated in terms of vegetable consumption. Spinach grown in sector X (Kot Addu) exhibited maximum values of HRI for Pb (3.541) and is followed by cadmium (1.7836), Ni (1.1148), and Mn (0.1326) for adults. Minimum values of HRI were observed for Cr (0.0094) in sector X (Kot Addu). The results (Table 6) exhibited that there were higher values of HRI (4.070) for Pb observed, taken by children living in Kot Addu followed by cadmium (2.0504). Minimum values of HRI were observed for Cr (0.0109). Cabbage also showed parallel values of HRI for Pb (4.0656) in adults, followed by cadmium (HRI= 2.993; Table 6). Minimum values of HRI were calculated Cd (0.0115; Table 5; Child). Cauliflower exhibited higher values of HRI (5.2768) for Pb in children. Pb HRI values of 4.5902 were observed in adults (sector X; Table 6). The results (Table 6) exhibited similar trends of HRI in adults and children living in sector Y and sector Z. However, the magnitude of HRI for sector Z was less than for sector X and sector Y. HRI values of radish were also higher in children (4.2214). Among the vegetables, turnip (HRI; Pb = 0.9180) and tinda (1.7049) exhibited minimum values HRI for adults, and similar trends were recorded for children living in sector X (Table 6). HRI values for carrots in terms of Cd were 3.9135 in adults, which was maximum than the summit. Among the children living in sector X, higher values of HRI were calculated for cadmium (4.4988). Minimum values of HRI were observed for ferric in children (Table 6). Colocasia exhibited the minimum values of HRI for all heavy metals contaminants in adults and children than other vegetables discussed above cultivated in sector x irrigated with wastewater. The HRI value for lead was 0.7869 in adults, Cr (HRI = 0.0057), and cadmium exhibited maximum values of HRI observed in Colocasia (Table 6). Similarly, copper (HRI = 0.2754) in adults was observed. All the remaining heavy metals showed minimum values of HRI by consuming Colocasia irrigated with wastewater in sector X (Table 6). Mirror reflecting to sector X, trends were attributed by sector Y and sector Z for the heavy metals in terms of daily intake and HIR (Tables 6–8).

Table 6. HRI and DIM for adults and children consuming different vegetables of sector X.

Vegetables	Factor	Lead (Pb)	Chromium (Cr)	Cadmium (Cd)	Copper (Cu)	Zinc (Zn)	Nickel (Ni)	Ferric (Fe)	Manganese (Mn)
\multicolumn{10}{c}{Sector X (Kot Addu)}
\multicolumn{10}{c}{Adult}
Spinach	DIM	0.0142	0.0142	0.0018	0.0142	0.0092	0.0223	0.0121	0.0186
	HRI	3.5410	0.0094	1.7836	0.3541	0.0306	1.1148	0.0402	0.1326
\multicolumn{10}{c}{Child}
Spinach	DIM	0.0163	0.0163	0.0021	0.0163	0.0106	0.0256	0.0139	0.0213
	HRI	4.070	0.0109	2.0504	0.4071	0.0352	1.2815	0.0462	0.1525
\multicolumn{10}{c}{Adult}
Cabbage	DIM	0.0163	0.0173	0.0030	0.0121	0.0117	0.0172	0.0139	0.0054
	HRI	4.0656	0.0115	2.9902	0.3016	0.0390	0.8577	0.0463	0.0386
\multicolumn{10}{c}{Child}
Cabbage	DIM	0.0187	0.0199	0.0034	0.0139	0.0134	0.0197	0.0160	0.0062
	HRI	4.6737	0.0133	3.4374	0.3468	0.0448	0.9860	0.0533	0.0444
\multicolumn{10}{c}{Adult}
Cauliflower	DIM	0.0184	0.0199	0.0007	0.0152	0.0050	0.0075	0.0081	0.0114
	HRI	4.5902	0.0133	0.6820	0.3803	0.0165	0.3751	0.0269	0.0813
\multicolumn{10}{c}{Child}
Cauliflower	DIM	0.0211	0.0229	0.0008	0.0175	0.0057	0.0086	0.0234	0.0114
	HRI	5.2768	0.0153	0.7840	0.4372	0.0190	0.4312	0.0269	0.0935
\multicolumn{10}{c}{Adult}
Radish	DIM	0.0147	0.0023	0.0013	0.0079	0.0062	0.0166	0.0163	0.0071
	HRI	3.6722	0.0015	1.2590	0.1967	0.0206	0.8289	0.0542	0.0510
\multicolumn{10}{c}{Child}
Radish	DIM	0.0169	0.0026	0.0014	0.0090	0.0071	0.0191	0.0187	0.0082
	HRI	4.2214	0.0017	1.4473	0.2261	0.0237	0.9528	0.0623	0.0586
\multicolumn{10}{c}{Adult}
Turnip	DIM	0.0037	0.0056	0.0023	0.0058	0.0057	0.0131	0.0039	0.0118
	HRI	0.9180	0.0038	2.2558	0.1443	0.0190	0.6531	0.0129	0.0839
\multicolumn{10}{c}{Child}
Turnip	DIM	0.0042	0.0065	0.0026	0.0066	0.0065	0.6531	0.0045	0.0135
	HRI	1.0554	0.0043	2.5931	0.1658	0.0218	0.7508	0.0149	0.0965
\multicolumn{10}{c}{Adult}
Tinda	DIM	0.0068	0.0061	0.0013	0.0066	0.0075	0.0163	0.0033	0.0158
	HRI	1.7049	0.0041	1.3115	0.1639	0.0250	0.8157	0.0110	0.1132
\multicolumn{10}{c}{Child}
Tinda	DIM	0.0078	0.0071	0.0015	0.0075	0.0086	0.0188	0.0038	0.0182
	HRI	1.9599	0.0047	1.5076	0.1885	0.0287	0.9378	0.0127	0.1301
\multicolumn{10}{c}{Adult}
Carrot	DIM	0.0079	0.0120	0.0039	0.0091	0.0039	0.0118	0.0103	0.0166
	HRI	1.9672	0.0080	3.9135	0.2282	0.0131	0.5902	0.0344	0.1188
\multicolumn{10}{c}{Child}
Carrot	DIM	0.0090	0.0138	0.0045	0.0105	0.0045	0.0136	0.0119	0.0191
	HRI	2.2615	0.0092	4.4988	0.2623	0.0151	0.6784	0.0396	0.1365

Table 6. *Cont.*

					Sector X (Kot Addu)				
					Adult				
Vegetables	Factor	Lead (Pb)	Chromium (Cr)	Cadmium (Cd)	Copper (Cu)	Zinc (Zn)	Nickel (Ni)	Ferric (Fe)	Manganese (Mn)
					Adult				
Lettuce	DIM	0.0058	0.0133	0.0102	0.0062	0.0075	0.0199	0.0118	0.0201
	HRI	1.4426	0.0089	0.004	0.0072	0.0250	0.9967	0.0393	0.1439
					Child				
Lettuce	DIM	0.0066	0.0153	0.0036	0.1561	0.0086	0.0229	0.0136	0.0232
	HRI	1.6584	0.0102	3.5935	0.1794	0.0287	1.1458	0.0452	0.1654
					Adult				
Colocasia	DIM	0.0031	0.0086	0.0047	0.0110	0.0041	0.0165	0.0032	0.0146
	HRI	0.7869	0.0057	4.6737	0.2754	0.0138	0.8236	0.0107	0.1045
					Child				
Colocasia	DIM	0.0036	0.0098	4.0656	0.0127	0.0048	0.0189	0.0037	0.0168
	HRI	0.9046	0.0066	0.0110	0.3166	0.0159	0.9468	0.0123	0.1202

Table 7. HRI and DIM for adults and children consuming different vegetables of sector Y (Muzaffargarh).

					Sector Y (Muzaffargarh)				
					Adult				
Vegetables	Factor	Cu	Fe	Zn	Mn	Pb	Cd	Ni	Cr
Spinach	DIM	0.0117	0.0089	0.0088	0.0196	0.0168	0.0029	0.0201	0.0152
	HRI	0.2925	0.0297	0.0294	0.1401	4.1968	2.8853	1.0072	0.0101
					Child				
Spinach	DIM	0.0134	0.0103	0.0101	0.0226	0.0193	0.0033	0.0232	0.0175
	HRI	0.3362	0.0342	0.0338	0.1611	4.8245	3.3168	1.1579	0.0117
					Adult				
Cabbage	DIM	0.0115	0.0127	0.0075	0.006	0.0199	0.0032	0.017	0.0163
	HRI	0.2872	0.0425	0.0251	0.0429	4.9837	3.2	0.8525	0.0108
					Child				
Cabbage	DIM	0.0132	0.0147	0.0087	0.0069	0.0229	0.0037	0.0196	0.0187
	HRI	0.3302	0.0488	0.0288	0.0493	5.7291	3.6787	0.98	0.0125
					Adult				
Cauliflower	DIM	0.0123	0.0066	0.0061	0.0119	0.0173	0.0014	0.0078	0.0178
	HRI	0.3069	0.0221	0.0204	0.0852	4.3279	1.4164	0.3908	0.0119
					Child				
Cauliflower	DIM	0.0141	0.0076	0.007	0.0137	0.0199	0.0016	0.009	0.0205
	HRI	0.3528	0.0254	0.0234	0.0979	4.9752	1.6283	0.4493	0.0137
					Adult				
Radish	DIM	0.0059	0.017	0.0071	0.0072	0.0143	0.0018	0.0157	0.0027
	HRI	0.1482	0.0567	0.0236	0.0515	3.5804	1.7836	0.7869	0.0018
					Child				
Radish	DIM	0.0068	0.0195	0.0081	0.0083	0.0165	0.0021	0.0181	0.0031
	HRI	0.1704	0.0651	0.0271	0.0592	4.1159	2.0504	0.9046	0.0021

Table 7. *Cont.*

		\multicolumn{8}{c}{Sector Y (Muzaffargarh)}
					Adult				
Vegetables	Factor	Cu	Fe	Zn	Mn	Pb	Cd	Ni	Cr
					Adult				
Turnip	DIM	0.0068	0.0077	0.0051	0.0217	0.006	0.0044	0.0088	0.0069
	HRI	0.1692	0.0257	0.017	0.1551	1.5082	4.3542	0.4394	0.0046
					Child				
Turnip	DIM	0.0078	0.0089	0.0059	0.025	0.0069	0.005	0.0101	0.008
	HRI	0.1945	0.0295	0.0196	0.1782	1.7338	5.0054	0.5051	0.0053
					Adult				
Tinda	DIM	0.0077	0.003	0.0093	0.0167	0.0088	0.0015	0.0199	0.0055
	HRI	0.1928	0.01	0.031	0.1191	2.1902	1.5213	0.9954	0.0037
					Child				
Tinda	DIM	0.0089	0.0034	0.0107	0.0192	0.0101	0.0017	0.0229	0.0063
	HRI	0.2216	0.0115	0.0357	0.1369	2.5178	1.7489	1.1443	0.0042
					Adult				
Carrot	DIM	0.0114	0.0123	0.0042	0.0183	0.0167	0.0023	0.0115	0.013
	HRI	0.2846	0.0411	0.0142	0.1307	4.1706	2.2558	0.5744	0.0087
					Child				
Carrot	DIM	0.0131	0.0142	0.0049	0.021	0.0192	0.0026	0.0132	0.0149
	HRI	0.3272	0.0472	0.0163	0.1503	4.7943	2.5931	0.6603	0.01
					Adult				
Lettuce	DIM	0.0073	0.0138	0.0078	0.0196	0.0075	0.003	0.0256	0.0136
	HRI	0.1818	0.0462	0.0261	0.1403	1.8754	2.9902	1.2811	0.0091
					Child				
Lettuce	DIM	0.0084	0.0159	0.009	0.0226	0.0086	0.0034	0.0295	0.0156
	HRI	0.209	0.0531	0.03	0.1613	2.1559	3.4374	1.4727	0.0104
					Adult				
Colocasia	DIM	0.0119	0.0041	0.0036	0.0162	0.0039	0.0044	0.0173	0.0093
	HRI	0.2977	0.0138	0.0119	0.1158	0.9705	4.4486	0.863	0.0062
					Child				
Colocasia	DIM	0.0137	0.0048	0.0041	0.0186	0.0045	0.0051	0.0198	0.0107
	HRI	0.3422	0.0159	0.0136	0.1331	1.1157	5.1139	0.992	0.0072

Table 8. HRI and DIM for adults and children consuming different vegetables of sector Z (Alipur).

		\multicolumn{8}{c}{Sector Z (Alipur)}
					Adult				
Vegetables		Cu	Fe	Zn	Mn	Pb	Cd	Ni	Cr
Spinach	DIM	0.0083	0.0069	0.0075	0.0113	0.0058	0.0013	0.0172	0.0147
	HRI	0.2072	0.0231	0.025	0.0805	1.4426	1.3115	0.8577	0.0098
					Child				
Spinach	DIM	0.0095	0.008	0.0086	0.0129	0.0066	0.0015	0.0197	0.0169
	HRI	0.2382	0.0265	0.0288	0.0925	1.6584	1.5076	0.986	0.0113

Table 8. Cont.

Vegetables		Cu	Fe	Zn	Mn	Pb	Cd	Ni	Cr
		\multicolumn{8}{c}{Sector Z (Alipur)}
					Adult				
Cabbage	DIM	0.0055	0.0082	0.0016	0.0061	1.6584	0.0019	0.0023	0.0144
	HRI	0.1371	0.0275	0.0052	0.0438	1.1148	1.941	0.1149	0.0096
					Child				
Cabbage	DIM	0.0063	0.0095	0.0018	0.0071	0.0051	0.0022	0.0026	0.0165
	HRI	0.1575	0.0316	0.006	0.0504	1.2815	2.2313	0.1321	0.011
					Adult				
Cauliflower	DIM	0.0073	0.0047	0.0043	0.0018	0.0023	0.0005	0.0108	0.0073
	HRI	0.183	0.0157	0.0142	0.0127	0.5653	0.4721	0.5414	0.0049
					Child				
Cauliflower	DIM	0.0084	0.0054	0.0049	0.0021	0.0026	0.0005	0.0124	0.0084
	HRI	0.2103	0.018	0.0164	0.0146	0.6498	0.5428	0.6224	0.0056
					Adult				
Radish	DIM	0.0059	0.0128	0.0057	0.0036	0.0052	0.0009	0.0151	0.0023
	HRI	0.1478	0.0428	0.0191	0.0259	1.3010	0.918	0.7541	0.0015
					Child				
Radish	DIM	0.0068	0.0147	0.0066	0.0042	0.0060	0.0011	0.0173	0.0026
	HRI	0.1699	0.0491	0.022	0.0297	1.4956	1.0554	0.8669	0.0018
					Adult				
Turnip	DIM	0.0036	0.0039	0.0041	0.0105	0.0044	0.0005	0.0104	0.0062
	HRI	0.0911	0.0131	0.0137	0.0748	1.0899	0.5194	0.5225	0.0042
					Child				
Turnip	DIM	0.0042	0.0045	0.0047	0.012	0.0050	0.0006	0.012	0.0072
	HRI	0.1048	0.0151	0.0157	0.0859	1.2529	0.597	0.6006	0.0048
					Adult				
Tinda	DIM	0.0047	0.003	0.0067	0.0118	0.0054	0.0006	0.0128	0.0071
	HRI	0.1167	0.0101	0.0223	0.0841	1.3574	0.6453	0.6379	0.0047
					Child				
Tinda	DIM	0.0054	0.0035	0.0077	0.0135	0.0062	0.0007	0.0147	0.0081
	HRI	0.1342	0.0116	0.0256	0.0967	1.5604	0.7418	0.7333	0.0054
					Adult				
Carrot	DIM	0.0977	0.0013	0.0028	0.0047	0.0039	0.001	0.0112	0.0115
	HRI	0.0045	0.0043	0.0093	0.0333	0.9758	0.9548	0.5613	0.0077
					Child				
Carrot	DIM	0.0045	0.0015	0.0032	0.0054	0.0045	0.0011	0.0129	0.0132
	HRI	0.1123	0.0049	0.0107	0.0383	1.1217	1.0976	0.6453	0.0088
					Adult				
Lettuce	DIM	0.0052	0.0073	0.0078	0.0086	0.0052	0.0008	0.0146	0.0126
	HRI	0.1288	0.0243	0.0261	0.0611	1.3049	0.7817	0.7279	0.0084
					Child				
Lettuce	DIM	0.0059	0.0084	0.009	0.0098	0.0060	0.0009	0.0167	0.0144
	HRI	0.1481	0.028	0.03	0.0702	1.5001	0.8986	0.8367	0.0096

Table 8. Cont.

Vegetables		Cu	Fe	Zn	Mn	Pb	Cd	Ni	Cr
Sector Z (Alipur)									
Adult									
Colocasia	DIM	0.006	0.0024	0.0021	0.0059	0.0040	0.0009	0.0157	0.0125
	HRI	0.1489	0.008	0.0069	0.042	1.0125	0.939	0.7848	0.0083
Child									
Colocasia	DIM	0.0068	0.0028	0.0024	0.0068	0.0047	0.0011	0.018	0.0144
	HRI	0.1711	0.0092	0.008	0.0482	1.1639	1.0795	0.9022	0.0096

Besides these, Zinc, Pb, Cd, and Mn should be further scrutinized in derived products from the study crops and livestock, particularly meat, milk, and leafy vegetables to determine considerable food safety problems. Besides these, wastewater reuse may cause environmental risks along with health diseases in humans [24]. The results (Table 3) exhibited that observed heavy metals values are above the suggested permissible limit by WHO and described that the human population living in sector X (Kot addu), Sector-Addu), sector Y (Muzaffargarh), and sector Z (Alipur) are using low-quality vegetables. The results for DIM for the ingestion pathway in three study sites are shown in (Tables 6–8) for both children and adults. The results showed that these values were slightly above the reference dose as recommended by WHO or other international bodies. Our results are in agreement with the report of [45]. The HIR indices for the heavy metal consumption by vegetables in the study areas can be ranked in the order cadmium > lead concentration > nickel > chromium for adults, and values of lead were greater for the children living in three study sites and were similar to the finding of [10]. Results of the present investigation demonstrated that Cd, Zn, Fe, Mn, and nickel may cause cancer in humans and could be a major health risk. These findings are in line parallel to [39].

5. Conclusions

In the present investigation, the analysis described that soil irrigated with municipal and industrial wastewater had alterations in almost all of the physicochemical parameters. Vegetable crops irrigated with wastewater have considerable amounts of heavy metals, some of which could be harmful to adults' and children's health. These heavy metal transfers from soil to vegetables are significantly driven by soil chemical reactions (pH) and soil cations exchange capacity. Owing to the problem of the bulk of sewage, industrial and domestic wastewater disposal, there is a great problem in Pakistan. It is recommended from the present research work outcomes that wastewater should be pretreated before being used for irrigation of vegetable crops. However, as sources of organic matter, sewage wastewater uses cannot be neglected. This study highlights the primary health issues caused by the use of sewage wastewater for irrigation of vegetable crops and demark the drastic effects of heavy metals in soils and water as well as in vegetables and human populations consuming such poor foodstuff, but also suggests that industrial waste could be used in the best way at commercial level for the country's economic improvement. In Pakistan, the demand for water and protein sources is very high. It has been estimated that more than 12 billion rupees are paid for water resources, and more than 35 million rupees are paid internationally for protein sources (lysine and glutamic acid), which affect the economy of Pakistan. However, the findings of the present research work will be significantly useful in terms of balancing the country's payment along with creating an opportunity for the fermentation industry in Pakistan by using sewage wastewater and some micro-organisms (*Corny bacterium Glutamicum*). By this, we can synthesize amino acids like lysine and glutamic acid at the local scale and can pretreat the sewage waste for irrigation purposes. Therefore, the present study was organized to focus the attention of scientists, health concerns, and government waste management authorities to develop local processes and guidelines to enhance lysine and glutamic acid production by using industrial effluent and making wastewater suitable for irrigation drives. This research also reveals that the potential human health risks involved in the consumption of vegetables grown with untreated wastewater should not be neglected. As a result, monitoring the wastewater quality used for growing vegetables is crucial to reduce human health risk.

Author Contributions: Conceptualization, H.U.; Data curation, M.T.A.; Formal analysis, M.F. and I.A.; Funding acquisition, M.A.W., M.F.K. and A.B.; Investigation, M.T.A. and F.F.; Methodology, M.T.A.; Resources, H.U., F.F. and Z.U.; Software, M.F.; Supervision, H.U.; Validation, I.A.; Writing—original draft, M.T.A.; Writing—review and editing, H.U., F.F., I.A., M.A.W., M.F.K., A.B. and Z.U. All authors have read and agreed to the published version of the manuscript.

Funding: This research received no external funding.

Institutional Review Board Statement: Not applicable.

Informed Consent Statement: Not applicable.

Data Availability Statement: All the relevant data are provided in the article.

Acknowledgments: The authors express their sincere appreciation to the Research Supporting Project number (RSP2023R466), King Saud University, Riyadh, Saudi Arabia.

Conflicts of Interest: The authors declare no conflict of interest.

References

1. Gleick, P.H. Water Resources. In *Encyclopedia of Climate and Weather*; Schneider, S.H., Ed.; Oxford University Press: New York, NY, USA, 1996; Volume 2, pp. 817–823.
2. Hussain, R.; Ali, L.; Hussain, I.; Khattak, S.A. Source identification and assessment of physico-chemical parameters and heavy metals in drinking water of Islampur area, Swat. *Pakistan. J. Himal. Earth Sci.* **2014**, *47*, 99–106.
3. Chaoua, S.; Boussaa, S.; El Gharmali, A.; Boumezzough, A. Impact of irrigation with wastewater on accumulation of heavy metals in soil and crops in the region of Marrakech in Morocco. *J. Saudi Soc. Agric. Sci.* **2019**, *18*, 429–436.
4. Guadie, A.; Yesigat, A.; Gatew, S.; Worku, A.; Liu, W.; Ajibade, F.O.; Wang, A. Evaluating the health risks of heavy metals from vegetables grown on soil irrigated with untreated and treated wastewater in Arba Minch, Ethiopia. *Sci. Total Environ.* **2021**, *761*, 143302.
5. Natasha Shahid, M.; Khalid, S.; Murtaza, B.; Anwar, H.; Shah, A.H.; Sardar, A.; Shabbir, Z.; Niazi, N.K. A critical analysis of wastewater use in agriculture and associated health risks in Pakistan. *Environ. Geochem. Health* **2020**, 1–20. [CrossRef]
6. UNW-DPC. *Safe Use of Wastewater in Agriculture*; United Nations-Water Decade Programme on Capacity Development: Bonn, Germany, 2013.
7. Mapanda, F.; Mangwayana, E.; Nyamangara, J.; Giller, K. Uptake of heavy metals by vegetables irrigated using wastewater and the subsequent risks in Harare, Zimbabwe. *Phys. Chem. Earth A/B/C* **2007**, *32*, 1399–1405.
8. Alghobar, M.A.; Suresha, S. Evaluation of metal accumulation in soil and tomatoes irrigated with sewage water from Mysore city, Karnataka, India. *J. Saudi Soc. Agric. Sci.* **2017**, *16*, 49–59.
9. Ali, M.; Ali, Z.; Ayyash, A. The effect of reuse of treated wastewater in irrigation on some chemical properties of soil and the growth of wheat yield. *J. Misurata Uni. Agric. Sci.* **2019**, *1*, 196–208.
10. Khalid, S.; Shahid, M.; Shah, A.H.; Saeed, F.; Ali, M.; Qaisrani, S.A.; Dumat, C. Heavy metal contamination and exposure risk assessment via drinking groundwater in Vehari, Pakistan. *Environ. Sci. Pollut. Control Ser.* **2020**, *27*, 39852–39864.
11. WHO. *Health Statistics and Health Information Systems: Global Burden of Disease*; World Health Organization: Geneva, Switzerland, 2011.
12. Qureshi, A.S.; Ismail, S. Evaluating health risks of using treated wastewater for vegetables under desert conditions. *J. Arid Land Stud.* **2016**, *26*, 111–119.
13. Ugulu, I.; Khan, Z.I.; Rehman, S.; Ahmad, K.; Munir, M.; Bashir, H. Effect of wastewater irrigation on trace metal accumulation in spinach (*Spinacia oleracea* L.) and human health risk. *Pak. J. Anal. Environ. Chem.* **2020**, *21*, 92–101. [CrossRef]
14. Nabulo, G.; Young, S.D.; Black, C.R. Assessing risk to human health from tropical leafy vegetables grownon contaminated urban soils. *Sci. Total Environ.* **2010**, *408*, 5338–5351. [CrossRef]
15. Nnaji, J.C.; Iweha, B.I.; Ogbuewu, I. Human health risk assessment of heavymetals in foodstuffs processed with diesel powered metallic disc grinders in Umuahia, Nigeria. *J. Chem Soc. Nigeria* **2020**, *45*, 458–468.
16. Amin, N.; Hussain, A.; Alamzeb, S.; Begum, S. Accumulation of heavy metals in edible parts of vegetables irrigated with wastewater and their daily intake to adults and children, District Mardan, Pakistan. *Food Chem.* **2013**, *136*, 1515–1523. [CrossRef] [PubMed]

17. Maleki, A.; Amini, H.; Nazmara, S.; Zandi, S.; Mahvi, A.H. Spatial distribution of heavy metals in soil, water, and vegetables of farms in Sanandaj, Kurdistan, Iran. *J. Environ. Health Sci. Eng.* **2014**, *12*, 136. [CrossRef]
18. Liu, J.; Ma, K.; Qu, L. Ecological risk assessments and context-dependence analysis of heavy metal contamination in the sediments of mangrove swamp in Leizhou Peninsula, China. *Mar. Pollut. Bull.* **2015**, *100*, 224–230. [CrossRef]
19. Singh, A.; Sharma, R.K.; Agrawal, M.; Marshall, F.M. Health risk assessment of heavy metals via dietary intake of foodstuffs from the wastewater irrigated site of a dry tropical area of India. *Food Chem. Toxicol.* **2010**, *48*, 611–619. [CrossRef]
20. Chary, N.S.; Kamala, C.T.; Raj, D.S.S. Assessing risk of heavy metals from consuming food grown on sewage irrigated soils and food chain transfer. *Ecotoxicol. Environ. Safety* **2008**, *69*, 513–524. [CrossRef]
21. Hassan, N.U.; Mahmood, Q.; Waseem, A.; Irshad, M.; Faridullah Pervez, A. Assement of heavy metals in wheat plants irrigated with contaminated waste water. *Pol. J. Environ. Stud.* **2013**, *22*, 115–123.
22. Gupta, N.; Yadav, K.K.; Kumar, V.; Prasad, S.; Cabral-Pinto, M.M.S.; Jeon, B.-H.; Kumar, S.; Abdellattif, M.H.; Alsukaibia, A.K.D. Investigation of Heavy Metal Accumulation in Vegetables and Health Risk to Humans From Their Consumption. *Front. Environ. Sci.* **2022**, *10*, 791052. [CrossRef]
23. Gelaye, Y.; Musie, S. Impacts of Heavy Metal Pollution on Ethiopian Agriculture: A Review on the Safety and Quality of Vegetable Crops. *Adv. Agric.* **2023**, *2023*, 1457498. [CrossRef]
24. Khan, M.N.; Aslam, M.A.; Muhsinah, A.B.; Uddin, J. Heavy Metals in Vegetables: Screening Health Risks of Irrigation with Wastewater in Peri-Urban Areas of Bhakkar, Pakistan. *Toxics* **2023**, *11*, 460. [CrossRef]
25. Aftab, K.; Iqbal, S.; Khan, M.R.; Busquets, R.; Noreen, R.; Ahmad, N.; Kazimi, S.G.T.; Karami, A.M.; Al Suliman, N.M.S.; Ouladsmane, M. Wastewater-Irrigated Vegetables Are a Significant Source of Heavy Metal Contaminants: Toxicity and Health Risks. *Molecules* **2023**, *28*, 1371. [CrossRef] [PubMed]
26. Atta, M.I.; Zehra, S.S.; Dai, D.Q.; Ali, H.; Naveed, K.; Ali, I.; Abdel-Hameed, U.K. Amassing of heavy metals in soils, vegetables and crop plants irrigated with wastewater: Health risk assessment of heavy metals in Dera Ghazi Khan, Punjab, Pakistan. *Front. Plant Sci.* **2023**, *13*, 1080635.
27. Fawad, A.; Hidayat, U.; Ikhtiar, K. Heavy metals accumulation in vegetables irrigated with industrial influents and possible impact of such vegetables on human health. *Sarhad J. Agric.* **2017**, *33*, 489–500.
28. Wang, Y.; Qiao, M.; Liu, Y.; Zhu, Y. Health risk assessment of heavy metals in soils and vegetables from wastewater irrigated area, Beijing-Tianjin city cluster, China. *J. Environ. Sci.* **2012**, *24*, 690–698. [CrossRef]
29. Jan, F.A.; Ishaq, M.; Khan, S.; Ihsanullah, I.; Ahmad, I.; Shakirullah, M. A comparative study of human health risks via consumption of food crops grown on wastewater irrigated soil (Peshawar) and relatively clean water irrigated soil (lower Dir). *J. Hazard. Mater.* **2010**, *179*, 612–621. [CrossRef]
30. Khaled, S.; Muhammad, B.; Ashraf, A. Field accumulation risks of heavy metals in soil and vegetable crop irrigated with sewage water in western region of Saudi Arabia. *Saudi J. Biol. Sci.* **2016**, *23*, 32–44.
31. Kumar, V.; Srivastava, S.; Chauhan, R.K.; Singh, J.; Kumar, P. Contamination, enrichment and translocation of heavy metals in certain leafy vegetables grown in composite effluent irrigated soil. *Arch. Agric. Environ. Sci.* **2018**, *3*, 252–260. [CrossRef]
32. Verger Eric, O.; Agnes, L.P.; Augustin, B.; Gabriel, B.; Mourad, M.; Mathilde, S.; François, M.; Martin-Prevel, Y. Dietary Diversity Indicators and Their Associations with Dietary Adequacy and Health Outcomes: A Systematic Scoping Review. *Adv. Nutr.* **2021**, *12*, 1659–1672. [CrossRef] [PubMed]
33. Butt, M.S.; Sharif, K.; Bajwa, B.E. Hazardous effects of sewage contaminated water on the environment focus on heavy metals and chemical composition of soil and vegetables. *Manag. Environ. Qual.* **2005**, *16*, 338–346. [CrossRef]
34. Wani, P.A.; Khan, M.S.; Zaidi, A. Toxic effects of heavy metals on germination and physiological processes of plants. In *Toxicity of Heavy Metals to Legumes and Bioremediation*; Springer: Vienna, Austria, 2012; pp. 45–66.
35. Dhanasekarapandian, M.; Chandran, S.; Kumar, V.; Surendran, U. Assessment of heavy metals in soil, paddy straw and SEM analysis of the soil for the impact of wastewater irrigation in Girudhumal sub basin of Tamil Nadu, India. *Glob. Nest J.* **2019**, *21*, 310–319.
36. Gupta, N.; Khan, D.; Santra, S. An Assessment of Heavy Metal Contamination in Vegetables Grown in Wastewater-Irrigated Areas of Titagarh, West Bengal, India. *Bull. Environ. Contam. Toxicol.* **2008**, *80*, 115–118. [CrossRef] [PubMed]
37. Rattan, R.K.; Datta, S.P.; Chhonkar, P.K.; Suribabu, K.; Singh, A.K. Long-term impact of irrigation with sewage effluents on heavy metal content in soils, crops and groundwater-a case study. *Agric. Ecosyst. Environ.* **2005**, *109*, 310–322.
38. Batool, F.; Hussain, M.I.; Nazar, S.; Bashir, H.; Khan, Z.I.; Ahmad, K.; Yang, H.H. Potential of sewage irrigation for heavy metal contamination in soil–wheat grain system: Ecological risk and environmental fate. *Agric. Water Manag.* **2023**, *278*, 108144.
39. Zakaria, Z.; Zulkafflee, N.S.; Mohd Redzuan, N.A.; Selamat, J.; Ismail, M.R.; Praveena, S.M.; Tóth, G.; Abdull Razis, A.F. Understanding Potential Heavy Metal Contamination, Absorption, Translocation and Accumulation in Rice and Human Health Risks. *Plants* **2021**, *10*, 1070. [CrossRef] [PubMed]
40. Liu, W.X.; Li, H.H.; Li, S.R.; Wang, Y.W. Heavy metal accumulation of edible vegetable cultivated in agricultural soil in the suburb of Zhengzhou city, peoples republic of China. *Bull. Environ. Contam. Toxicol.* **2006**, *76*, 163–170.
41. Hu, B.; Wang, J.; Jin, B.; Li, Y.; Shi, Z. Assessment of the potential health risks of heavy metals in soils in a coastal industrial region of the Yangtze River Delta. *Environ. Sci. Pollut. Res.* **2017**, *24*, 19816–19826.
42. Latif, A.; Bilal, M.; Asghar, W.; Azeem, M.; Ahmad, M.I.; Abbas, A. Heavy metal accumulation in vegetables and assessment of their potential health risk. *J. Environ. Anal. Chem.* **2018**, *5*, 1–7. [CrossRef]

43. Bi, X.; Feng, X.; Xeng, Y.; Qin, G.; Li, F.; Liu, T.; Fu, Z.; Jing, Z. Environmental contamination of heavy metals from zinc smelting area in Hezhang country, Western Guizhou. *China Environ. Int.* **2006**, *32*, 883–890.
44. Chen, F.; Ma, J.; Akhtar, S.; Khan, Z.I.; Ahmad, K.; Ashfaq, A.; Nawaz, H.; Nadeem, M. Assessment of chromium toxicity and potential health implications of agriculturally diversely irrigated food crops in the semi-arid regions of South Asia. *Agric. Water Manag.* **2022**, *272*, 107833.
45. Abbas, M.; Parveen, Z.; Iqbal, M.; Riazuddin, M.; Iqbal, S.; Ahmed, M.; Bhutto, R. Monitoring of toxic metals (cadmium, lead, arsenic and mercury) in vegetables of Sindh, Pakistan. *Kathmandu Univ. J. Sci. Eng. Technol.* **2010**, *6*, 60–65. [CrossRef]

Disclaimer/Publisher's Note: The statements, opinions and data contained in all publications are solely those of the individual author(s) and contributor(s) and not of MDPI and/or the editor(s). MDPI and/or the editor(s) disclaim responsibility for any injury to people or property resulting from any ideas, methods, instructions or products referred to in the content.

Article

Multivariate Statistical Methods and GIS-Based Evaluation of Potable Water in Urban Children's Parks Due to Potentially Toxic Elements Contamination: A Children's Health Risk Assessment Study in a Developing Country

Junaid Ghani [1], Javed Nawab [2,*], Zahid Ullah [3], Naseem Rafiq [4], Shah Zaib Hasan [5], Sardar Khan [6], Muddaser Shah [7,8] and Mikhlid H. Almutairi [9]

1. Department of Biological, Geological and Environmental Sciences, Alma Mater Studiorum University of Bologna, 40126 Bologna, Italy; junaid.ghani2@unibo.it
2. Department of Environmental Sciences, Kohat University of Science and Technology, Kohat 26000, Pakistan
3. School of Environmental Sciences, China University of Geosciences, Wuhan 430074, China; 2201890048@cug.edu.cn
4. Department of Zoology, Abdul Wali Khan University Mardan, Mardan 23200, Pakistan; naseem@awkum.edu.pk
5. Faculty of Environmental Sciences, Czech University of Life Sciences Prague, 16500 Prague, Czech Republic; hasans@fzp.czu.cz
6. Department of Environmental Sciences, University of Peshawar, Peshawar 25120, Pakistan; sardar@uop.edu.pk
7. Department of Botany, Abdul Wali Khan University, Mardan 23200, Pakistan; muddasershah@awkum.edu.pk
8. Natural and Medical Sciences Research Center, University of Nizwa, Nizwa 616, Oman
9. Zoology Department, College of Science, King Saud University, P.O. Box 2455, Riyadh 11451, Saudi Arabia; malmutari@ksu.edu.sa
* Correspondence: javednawab@kust.edu.pk

Abstract: Contamination of potentially toxic elements (PTEs) has received widespread attention in urban children's parks (UCPs) worldwide in the past few decades. However, the risk assessment of PTEs in drinking water sources of UCPs is still unknown particularly in developing countries. Hence, the present study investigated the spatial distribution, sources for PTEs (Cd, Cr, Pb, Ni, and Cu), and health risk assessment in drinking water sources of UCPs in Khyber Pakhtunkhwa, Pakistan. Among PTEs, Cd, Cr, and Pb had low to high concentrations and exceeded the safe limits of WHO and PAK-EPA in most UCPs. PCA results showed high anthropogenic and low natural sources, contributing to the release of PTEs in all UCPs. Heavy-metal pollution index (PTE-PI) results showed low to high pollution levels for all UCPs, with the highest values of 113 and 116 for Sardaryab Park Charsadda (SPC) and Zoo Park Peshawar (ZPP), respectively. Heavy-metal evaluation index (PTE-EI) results also showed low to high pollution levels for all UCPs. UCPs samples (50%) showed low pollution levels in PTE-PI results. To the contrary, UCPs samples (50%) exhibited high pollution levels in PTE-EI results. The non-carcinogenic risk of HQ and HI values of all PTEs were below the permissible limit (<1) for adults and children via ingestion and dermal contact. CR and TCR results showed that PTEs (Cr, Cd, Pb, and Ni) had the highest carcinogenic risk (>1.00×10^{-4}) for both adults and children in all UCPs, except Cd and Ni for adults via the ingestion route, while Cr values (>1.00×10^{-4}) were exceeded for children in some of the UCPs via the dermal route. Consequently, long-term exposure to toxic PTEs could pose a carcinogenic risk to the local population. Thus, the present study suggests that the government should implement enforcement with firm protocols and monitoring guidelines of environmental regulations to mitigate PTEs originating from anthropogenic sources in order to reduce health risks and improve public health safety in urban areas.

Keywords: toxic elements; pollution indices; source identification; water quality; risk assessment

Citation: Ghani, J.; Nawab, J.; Ullah, Z.; Rafiq, N.; Hasan, S.Z.; Khan, S.; Shah, M.; Almutairi, M.H. Multivariate Statistical Methods and GIS-Based Evaluation of Potable Water in Urban Children's Parks Due to Potentially Toxic Elements Contamination: A Children's Health Risk Assessment Study in a Developing Country. *Sustainability* **2023**, *15*, 13177. https://doi.org/10.3390/su151713177

Academic Editor: Jia Wen

Received: 3 August 2023
Revised: 27 August 2023
Accepted: 28 August 2023
Published: 1 September 2023

Copyright: © 2023 by the authors. Licensee MDPI, Basel, Switzerland. This article is an open access article distributed under the terms and conditions of the Creative Commons Attribution (CC BY) license (https://creativecommons.org/licenses/by/4.0/).

1. Introduction

According to the world urbanization prospects of the United Nations (UN), 55% of the world's population resides in urban areas [1]. Urban parks play a critical role in maintaining the urban ecosystem with the rapid development of urbanization [2,3]. As a result, high amounts of pollutants are accumulated in the urban environment due to rapid urbanization, high energy consumption, and migrations of people into big cities [4,5]. Pollutants like potentially toxic elements (PTEs) are released into the urban park soils and the surrounding environment as a consequence of high urbanization and industrialization [6–8]. The primary sources of PTEs are heavy traffic and intensive human activities (such as the dumping of municipal and industrial wastes, home heating, industrial emissions, and energy generation), leading to contamination of parks and roadside lawns in urban environments [9–11]. PTE contamination in urban children's parks (UCPs) is associated with traffic emissions, fuel combustion, tire and brake wear, particles of weathered street dust, waste discharge from industrial activities, weathering particles from buildings and pavement surfaces, urban infrastructure development, wastewater disposal systems, and dumping of municipal solid waste [12–17]. Moreover, urban soils are exposed to PTE input from natural sources, wastewater irrigation, and atmospheric deposition [18–21], resulting in being discharged into water bodies by natural weathering, the erosion process, and anthropogenic activities [22]. Typically, urban runoff contains a range of pollutants that could contaminate water quality [23,24], and such pollutants like PTEs degrade the drinking water sources and leach into groundwater system.

Residents can be exposed to PTEs by ingestion, inhalation, and skin contact, especially children and adults who often visit urban parks [3]. PTEs can pose substantial risks to the public's health once they have bioaccumulated, especially to the most susceptible groups, such as children and senior citizens [13]. Furthermore, visitors and locals commonly engage in recreational activities in public parks, and exposure to elevated PTEs can hinder a child's development, cause abnormal behavior, or lead to other chronic disorders, while others have the potential to cause cancer when exposed to high concentrations [25]. In general, prolonged exposure to PTEs may cause a variety of harmful health effects, including cancer, immune system disruption, neurological problems, cardiovascular and liver diseases, kidney and bone malfunction, and immune system disruption [26]. Due to the fact that children and other park visitors are more susceptible to PTE toxicity, it is crucial to identify the risks of exposure to drinking water sources in urban parks [27]. So far, multivariate statistical analyses like principle component analysis multiple linear regression (PCA-MLR) have been widely used for source identification of PTEs [28,29]. Therefore, it is crucial to identify the potential sources of PTEs and high risk areas in urban parks in order to prevent and control PTE pollution [30].

Several studies have shown contamination by PTEs such as Cr, Ni, Cd, As, Cu, Pb, and Zn in urban soils in parks, sports playgrounds, and schools [31–34], with most of the recent studies focused on PTE concentration, potential sources, and health risk assessment in UCP soils. PTE pollution was also reported in various research in some of the urban lakes [35,36]. The PTEs in urban environments have been a major concern for researchers and environmentalists worldwide. Research on PTE contamination in drinking water sources of urban parks and the urban environment is very limited. However, to the best of our knowledge, this will be the first study to assess PTE contamination, its potential source, and health risk evaluation in UCP drinking water sources of the Khyber Pakhtunkhwa (KPK) province in Pakistan. Parks in this province serve as popular recreation areas and tourism destinations for the local citizens. PTEs in the UCP drinking water could pose a serious threat to human health. As the proportion of people living in urban areas is predicted to expand from 50% today to 66% by 2050 worldwide [29], identifying and remediating PTEs-contaminated soils and drinking water may become more critical. Evidently, research into the health risks posed by PTEs in drinking water sources of UCPs is still unknown in developing countries like Pakistan. To fill this research gap, a scientific approach is needed to evaluate the PTEs pollution in UCPs and its origins to minimize the health risks via drinking water ingestion and dermal contact of exposure pathways for children and adults. This study could be important to provide scientific support for controlling and prevention of PTEs contamination in order to improve healthy urban environments worldwide, as proposed by the World Health Organization (WHO).

Therefore, we have selected a total of 20 UCPs in famous regions of the KPK province. The primary objectives of the present study were as follows: (1) to investigate the spatial distribution of PTEs (Cd, Cr, Pb, Ni, and Cu) in selected UCPs; (2) to assess the PTEs contamination and potential sources by PCA-MLR; and (3) to assess their possible risks (carcinogenic and non-carcinogenic) to human health, via ingestion and dermal contact, for the local population (adults and children). This research can be used as a reference basis in future studies for proper management of PTE contamination in urban parks. The findings of this study can help decision makers and can alert inhabitants around the world to manage and remediate PTEs contamination in soil and drinking water sources with the guidelines and remediation strategies in urban environments.

2. Materials and Methods

2.1. Study Area, Water Sampling, and Analysis

The water samples were collected from a total of 20 UCPs of 10 cities in the Khyber Pakhtunkhwa province of Pakistan in 2019, as shown in Figure 1. All the UCPs were selected based on high population and industrialized areas, and the parks were also used for different recreational activities by the local children and adults. The UCPs are categorized with specific acronyms, including Baghicha Park Charsadda (BPC), Sardaryab Park Charsadda (SPC), Zoo Park Peshawar (ZPP), Tatara Park Peshawar (TPP), Jannah Park Nowshera (JPN), Disney Water Park Nowshera (DWPN), Pak Wonder Land Park Haripur (PWLPH), Ayub Park Haripur (APH), Sher Khan International Park Swat (SKIPS), Wonder World Park Swat (WWPS), Albela Park Swabi (APS), Gohati Cricket Ground Park Swabi (GCGPS), Sports Complex Park Mardan (SCPM), Younus Stadium Park Mardan (YSPM), Gol National Park Chitral (GNPC), Pakistan Tour Park Chitral (PTPC), Haq Nawaz Park Dera Ismail Khan (HNPDK), Insaf Park Dera Ismail Khan (IPDK), Lady Garden Public Park Abbottabad (LGPPA), and Shimla Pahari Park Abbottabad (SPPA).

The research methodology (see Figure 2) was followed for drinking water sample collection with the proper guidelines, and all the samples were collected in the selected UCPs. The water samples were randomly collected from different drinking water sources including bore wells, hand pumps, and tap water in UCPs. The total geographical area of each park was varied, and water samples were collected from the abovementioned sources. Based on the total area of each UCP, 10 water samples were collected from drinking water sources. A total of (n = 200) drinking water samples were collected from all the 20 UCPs in all selected cities. The drinking water resources in these UCPs are regularly used for drinking and other domestic purposes. Before sample collection, water was allowed to run for a few minutes during the sampling time from the specific target in each park. The polyethylene bottles were first cleaned with distilled water, then water samples were taken. A few drops of acid (HNO_3, 0.5% v/v) were added to bottles prior to PTE analysis. All the water samples were filtered by Whatman filter paper (0.45 μm). All the samples were sealed, labeled, and well preserved using the method used by [37].

The water samples were directly used for the physicochemical parameters analysis (Figure 2). pH and total dissolved solids (TDS) were determined using a digital pH meter (Model C93, Turnhout, Belgium) and a TDS meter (Model S518877), respectively, while electrical conductivity (EC) was measured by a conductivity meter (Model HI98303). The water samples were analyzed for PTEs including cadmium (Cd), chromium (Cr), lead (Pb), nickel (Ni) and copper (Cu) by graphite furnace atomic absorption spectrophotometry (PerkinElmer, Waltham, MA, USA, ASS-PEA-700) using standard procedures adopted by [15].

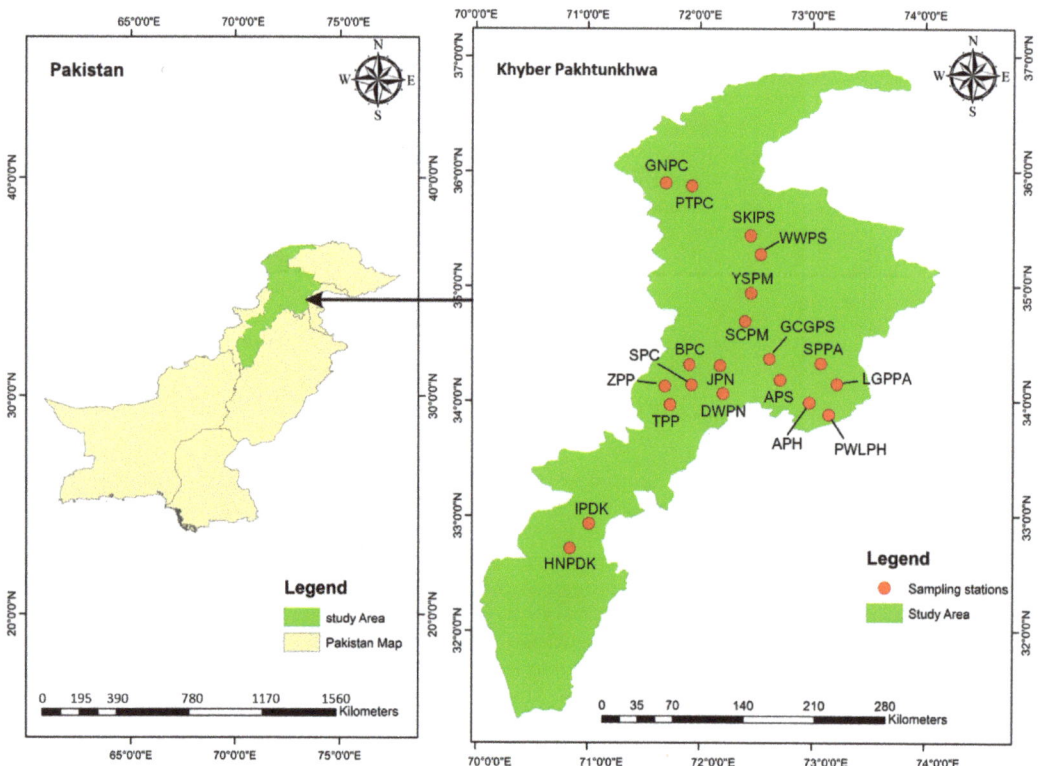

Figure 1. Study area map showing the sampling points of 20 urban children's parks in Khyber Pakhtunkhwa province, Pakistan. Baghicha Park Charsadda (BPC), Sardaryab Park Charsadda (SPC), Zoo Park Peshawar (ZPP), Tatara Park Peshawar (TPP), Jannah Park Nowshera (JPN), Disney Water Park Nowshera (DWPN), Pak Wonder Land Park Haripur (PWLPH), Ayub Park Haripur (APH), Sher Khan International Park Swat (SKIPS), Wonder World Park Swat (WWPS), Albela Park Swabi (APS), Gohati Cricket Ground Park Swabi (GCGPS), Sports Complex Park Mardan (SCPM), Younus Stadium Park Mardan (YSPM), Gol National Park Chitral (GNPC), Pakistan Tour Park Chitral (PTPC), Haq Nawaz Park Dera Ismail Khan (HNPDK), Insaf Park Dera Ismail Khan (IPDK), Lady Garden Public Park Abbottabad (LGPPA), and Shimla Pahari Park Abbottabad (SPPA).

2.2. PTEs Pollution Indices

2.2.1. Potentially Toxic Elements Pollution Index (PTE-PI)

The PTE-PI is a measurement that reflects the combined impact of several dissolved PTEs and indicates the overall quality of water with regard to PTE concentration [38]. PTE-PI is an effective technique used to assign a specific rating/weighting (Wi) for each selected parameter. The rating is based on a value between 0 and 1, indicates the relative importance of each parameter, and can be interpreted as inversely related to the suggested standard (Si) for individual parameter [39,40].

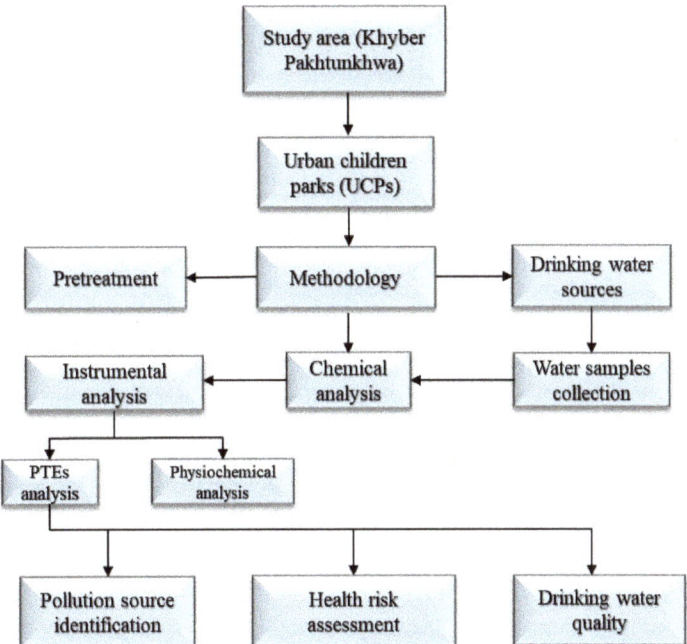

Figure 2. Research flow chart of the PTEs study in drinking water sources of UCPs.

Different PTE-PI ranges have been specified and used to classify water quality as "poor", "good", "very good", and "excellent". The critical level of the PTE-PI value is 100, and it has been reported that PTE-PI levels above 100 have a stronger negative impact on health [41]. PTE-PI can be calculated by using Equations (1) and (2):

$$\text{PTE-PI} = \frac{\sum_{i=1}^{n}(Qi \times Wi)}{\sum_{i=1}^{n} Wi} \quad (1)$$

$$Qi = \frac{Mi}{Si} \times 100 \quad (2)$$

where Qi and Wi represent the sub-index and unit weightage of the ith parameter, respectively, and n is the total number of the parameters. Mi is the concentration of the measured ith parameter, and Si is the concentration of standard maximum permissible values (µg/L) of WHO [42], respectively, as shown in Table S1, described by the European Directive on water quality [43]. The weightage unit (Wi) signifies an inverse proportional to the maximum concentration level.

$$Wi = \frac{1}{Si} \quad (3)$$

2.2.2. Potentially Toxic Elements Evaluation Index (PTE-EI)

The PTE-EI is also used to evaluate the combined impact of measured PTEs on the overall drinking water quality [39]. In the present study, five PTEs (Cd, Cr, Pb, Ni, and Cu) were selected and assessed for PTE-PI and PTE-EI calculation. HEI was computed by the given Equation (4):

$$\text{PTE-EI} = \sum_{i=1}^{n} \frac{Hc}{Hmac} \quad (4)$$

where Hc is the observed value of the ith parameter (mg/L), while $Hmac$ is the maximum permissible concentration of the ith parameter [44].

2.3. Human Health Risk Model

2.3.1. Non-Carcinogenic Risk Assessment

The health risk model is used to determine the extent of human exposure to toxic PTEs, developed by the United States Environmental Protection Agency (USEPA) [45], in the drinking water of UCPs. Risk assessment is described as the method of estimating the likelihood of any given likely amount of detrimental health impacts occurring over a specified time period [46]. Human health risk assessment was estimated in recent studies to demonstrate the potential health risks posed by PTEs in water, soil, and air [47,48]. The risk assessment is one of the most effective ways of providing crucial information, highlighting the most critical metals among heavy metals, which have a negative influence on human health and the environment due to their toxicity. The most useful aspect of health risk assessment is that it provides information and depicts an inhabitant's health status. Children and adults are usually exposed to toxic PTEs in UCPs drinking water by two main exposure pathways, i.e., ingestion and dermal contact. Furthermore, adults are regarded as the general population, while children are considered the sensitive group. The daily metal intake (*DMI*) was used to quantify the exposure to PTEs in UCPs water through ingestion and dermal contact. The daily metal intake via ingestion (*DMIing*) and the daily metal intake via dermal contact (*DMIder*) of PTEs in UCP water were calculated according to Equations (5) and (6), respectively.

$$DMIing = \frac{C \times IR \times EF \times ED}{BW \times AT} \quad (5)$$

$$DMIder = \frac{C \times SA \times Kp \times ET \times ED \times EF \times CF \times ABS}{BW \times AT} \quad (6)$$

The descriptions of the parameters for the health risk model are presented in Table S2. Non-carcinogenic toxic risk was estimated by calculating the hazard quotients (*HQ*) and hazard index (*HI*). The hazard quotient (*HQ*) was calculated as the proportion of the *DMI* and the reference dose (*RfD*) for a specific PTE. The total non-carcinogenic risks (*HI*) were identified by the sum of the *HQ* values of all selected PTEs in the UCPs water. The *HQing*, *HQder*, and *HI* of all PTEs in UCPs water was calculated according to Equations (7)–(9).

$$HQing = \frac{DMIing}{RfDing} \quad (7)$$

$$HQder = \frac{DMIder}{RfDder} \quad (8)$$

$$HI = \sum HQ = \sum \frac{DMI}{RfD} \quad (9)$$

where *DMI* is the average daily exposure dose through ingestion and dermal contact. *RfDs* are the reference doses of selected PTEs for ingestion and dermal contact. All *RfD* values of non-carcinogenic risk for both exposure pathways are presented by the United States Department of Energy (USDOE) [49], as shown in Table S2. The values of *HQ* and *HI* greater than 1 indicate non-carcinogenic risk to human health in the exposed population, while *HQ* and *HI* values less than 1 are considered safe.

2.3.2. Carcinogenic Risk Assessment

The carcinogenic risk of toxic PTEs was calculated by using the carcinogenic risk (*CR*), and the total cancer risk (*TCR*) represents the sum of the potential carcinogenic risks. The estimated *CR* value shows the probability of developing cancer risk for an individual during a lifetime exposure to carcinogenic toxic chemicals. The *CR* was calculated for PTEs (Cr, Cd, Pb, and Ni) in the present study, based on their observed corresponding *DMI* and available cancer slope factor (*CSF*) values. The non-carcinogenic and total carcinogenic risks can be identified by combining the overall risks for three exposure routes in the UCPs drinking water. The CR for two exposure pathways and TCR were calculated by using Equations (10)–(12) [50]:

$$CRing = DMIing \times CSF \quad (10)$$

$$CRder = DMIder \times CSF \quad (11)$$

$$TCR = \sum CR \quad (12)$$

where CSF is the cancer slope factors (per mg/kg-day) as shown in Table S2. In general, the carcinogenic risk is unacceptable if the CR values exceed (>10^{-4}), while CR values ($1 \times 10^{-6} < CR < 1 \times 10^{-4}$) are assumed to be acceptable with no carcinogenic risk [51], whereas the CR values (<10^{-6}) imply that the carcinogenic risk can be negligible [45].

2.4. Quality Assurance and Quality Control

The standard procedure was used to certify the PTEs quality in order to confirm the accuracy of the results. Double distilled water and certified standard solution (1000 mg/L) of Fluka Kamica (Buchs, Switzerland) were used for quality control of the standard solution and analysis of PTEs. The analytical chemical spectroscopic purity of 99.9% (Merck Darmstadt, Germany) was also used for the sample formation and PTEs analysis. The reagent blanks, duplicate samples, and standards were used with different concentrations. Analytical estimated error was less than or equal to 10%, and the reproducibility of the analytical results was within 5%.

2.5. Statistical Analysis

Descriptive statistics were performed using Microsoft Excel (2016) and SPSS (version 21). PCA-MLR was used for PTEs contamination sources using XLSTAT (2022). Origin Lab (2018) and SigmaPlot (14.0) were used to make all the figures. Arc Geographic Information System (Arc-GIS 10) software was used for the study area map and distribution maps of PTEs.

3. Results

3.1. Physiochemical Characteristics of UCPs Water

The physiochemical properties of all UCPs water samples are presented in Table S3 and Figure 3. There is a substantial variation in the concentration of physiochemical parameters in drinking water of all UCPs. The overall pH varied from 6.10 to 8.46 for drinking water of all the UCPs, suggesting that it is neutral to alkaline in nature. The lowest mean value of pH (6.95) was recorded for BPC and PTPC, while a high pH value (8.03) was observed for LGPPA. The EC values were varied for drinking water of all UCPs, ranged from 192 to 1394 µS/cm. The lowest EC mean value was recorded for LGPPA (331 µS/cm), and the highest EC mean value was detected for APH (959 µS/cm). Furthermore, the concentration of total dissolved solids (TDS) of all UCPs samples was observed between 220–1245 µg/L. The highest TDS value was recorded for DWPN, with a mean value of 986 µg/L, while the lowest TDS was observed for BPC with a mean value of 313 µg/L. These physiochemical parameters possibly influence and control the occurrence and bioavailability of PTEs in groundwater, which is further discussed in Section 4 below.

3.2. Spatial Distribution of PTEs in UCPs Water

The basic descriptive statistics and spatial distribution of PTEs concentrations in UCPs drinking water are presented in Table S3 and Figure 3. The distribution levels of PTEs in the drinking water of all UCPs are greatly varied. Among PTEs, Cu had the highest concentration for all UCPs, followed by Cr, Ni, Pb, and Cd. Cu was observed with elevated concentration, ranging from 38.8–222 µg/L with a mean value of 130 ug/L for PWLPH. The lowest concentration was recorded for Cd, ranged from 0.09–0.19 µg/L with a mean value of 0.14 ug/L in the UCP water of PTPC. Cr concentrations exceeded the permissible limit (50 µg/L) set by WHO [52] in UCPs water of SPC, PWLPH, WWPS, SCPM, and YSPM, while its concentrations were found to be within acceptable limits of WHO [52] and PAK-EPA [53].

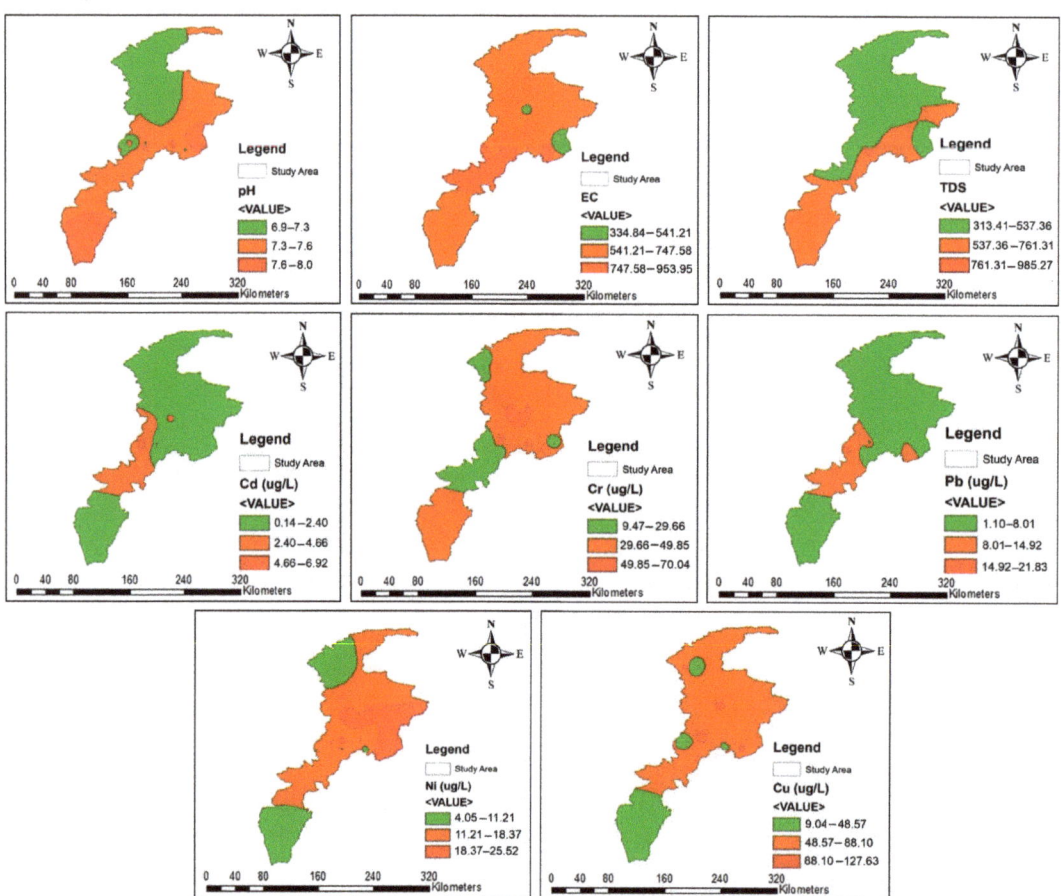

Figure 3. Spatial distribution of pH, EC, TDS, and PTEs (Cd, Cr, Pb, Ni, and Cu) in UCPs.

For BPC and SPC, Cu exhibited high mean concentrations (31.1 and 19.0 µg/L), followed by Ni (19.5 and 14.6 µg/L) and Cr (11.6 and 12.5 µg/L), respectively. A low mean concentration was recorded for Pb (8.49 and 9.37 µg/L) and followed by Cd (4.35 and 6.95 µg/L), respectively. For ZPP, Pb showed a high mean concentration of 21.9 µg/L, followed by Ni (19.2 µg/L) and Cu (14.0 µg/L), respectively. The concentrations of all selected PTEs were found to be within allowable limits of WHO in all the UCPs water, indicating low contamination, except Cd. Noticeably, Cd concentration was above the permissible limit (3 µg/L) of WHO, with mean values of 4.35 and 4.05 µg/L in UCPs water of BPC and ZPP, respectively. Similarly, Cu showed high abundance with mean values of 106 and 75.5 µg/L, while Cd and Pb showed lowest concentrations with mean values (3.60 and 1.12 µg/L) and (9.54 and 2.45 µg/L) in TPP and JPN, respectively. Likewise, Cu exhibited high mean concentration (80.6 µg/L), while the lowest mean concentration (0.42 µg/L) was recorded for Cd in DWPN. Furthermore, the PTEs such as Cr and Ni concentrations were relatively comparable for PWLPH, APH, and SKIPS, with mean values

(51.4, 49.2, and 46.1 µg/L) and (25.8, 18.2 and 16.5 µg/L), respectively. To the contrary, the concentrations of Cu were comparatively higher than the other PTEs, with mean values of 130, 54.2, and 98.8 µg/L, respectively, in the above-mentioned UCPs. Cr and Pb mean concentrations (51.4 and 15.2 µg/L) in PWLPH were found to be higher than the threshold values of WHO [52] and PAK-EPA [53], respectively. The mean concentrations of PTEs in drinking water showed high variation between different UCPs of KPK, indicating low to high contamination of Cd, Cr, and Pb, which exceeded the permissible limits of WHO [52] and PAK-EPA [53]. The PTEs concentrations were noticeably low to moderate level for WWPS and APS. Similarly, Cr was remarkably observed with high mean values of 68.3, 71.1, 61.5, and 61.3 µg/L for WWPS, GCGPS, SCPM, and YSPM, respectively and exceeded the threshold limits of WHO [52] and PAK-EPA [53]. Cd exhibited low concentration levels (0.67 and 0.99 µg/L), followed by Ni (2.60 and 19.2 µg/L), for APS and YSPM, respectively. Among UCPs, the PTEs concentrations were comparatively lower, especially for Cd, Pb, and Ni in UCPs drinking water of GNPC, PTPC, HNPDK, and IPDK. Cu showed high concentration for GNPC with mean value of 104 µg/L, while to the contrary, PTPC, HNPDK, and IPDK showed low Cu concentration with mean values of 8.5, 10.0, and 19.1 µg/L, respectively, and Cr had moderate mean concentration in the aforementioned UCPs.

Similarly, LGPPA and SPPA showed relatively similar mean values of PTEs such as Cd (0.55 and 0.58 µg/L) and Cu (60.1 and 51.1 µg/L), respectively. The spatial distribution of high abundance of PTEs (especially Cu, Cr, Pb, and Cd) in all UCPs water confirms high contamination levels that could be subjected to the input of potential sources in the study areas. Overall, in the present study, the mean concentrations of PTEs occurred in descending order of Cu > Cr > Ni > Pb > Cd for all UCPs.

3.3. Source Apportionment of PTEs

Principle component analysis (PCA) is an effective technique used for source apportionment of PTEs [54]. In this study, we used PCA multilinear regression (PCA-MLR) to extract three major component factors in terms of eigenvalues (eigenvalue > 1) and estimated total variance. The positive and negative loadings for different datasets of PTEs in UCPs are presented in Table 1. Overall, four significant factors (F1, F2, F3, and F4) were obtained, and the first two significant loading factors were observed for PTEs in UCPs, as presented in Figure 4. The positive loading factors imply that the presence of the water variables could influence the groundwater or surface water samples. Contrarily, the negative loading factors show that the groundwater and surface water quality are not affected by the water variables. The four factors of the UCPs water sources described 80.8% of the total variation with an eigenvalue of 6.56, as shown in Table 1.

Table 1. Principal components analysis of selected parameters of UCPs.

Parameters	F1	F2	F3	F4
pH	−0.69	0.09	0.14	**0.62**
EC	0.08	0.00	**0.90**	−0.33
TDS	−0.68	0.02	**0.66**	0.15
Cd mg/L	**0.78**	−0.31	0.15	0.26
Cr mg/L	−0.09	**0.75**	−0.22	−0.05
Pb mg/L	**0.75**	−0.05	0.34	0.21
Ni mg/L	**0.50**	**0.63**	0.10	0.45
Cu mg/L	0.13	**0.78**	0.18	−0.28
Eigenvalue	2.40	1.67	1.49	1.00
Variability (%)	29.9	20.9	18.6	11.3
Cumulative %	29.9	50.9	69.5	80.8

Note: Bold values are the main contributors to PCA.

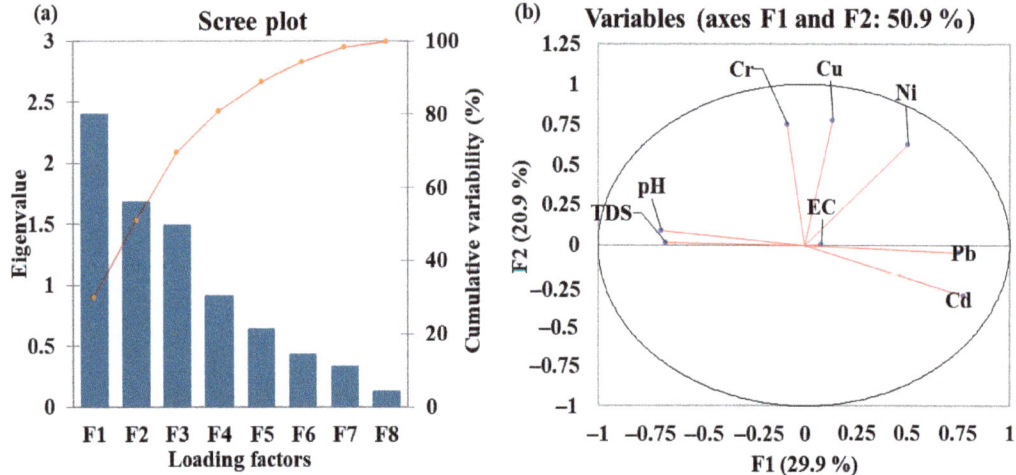

Figure 4. (**a**) Significant loading factors and (**b**) overall loading factors of results of UCPs.

Factor 1 (F1) described 29.9% of the total variance with an eigenvalue of 2.40. F1 is predominantly loaded with PTEs such as Cd, Pb, and Ni, showing moderate positive loadings (0.78), (0.75), and (0.50), respectively, except pH and TDS have moderate negative loadings (−0.69) and (−0.68), respectively. F1 contributed to the moderate loadings of PTEs for the UCPs drinking groundwater, suggesting anthropogenic and natural sources. The negative loadings of pH and TDS could be associated with the features and mobility of PTEs [55]. Factor 2 (F2) described 20.9% of the total variance with an eigenvalue of 1.67. F2 had strong positive loadings for Cr (0.75), Ni (0.63), and Cu (0.78), respectively. The contribution of the strong loadings of PTEs in F2 results showed anthropogenic origin in drinking water sources of UCPs.

Thus, F1 and F2 results showed moderate to high loadings of physicochemical and PTEs contribution, suggesting anthropogenic and natural sources in the drinking water sources of UCPs. Factor 3 (F3) accounted for 18.6% of the total variance with an eigenvalue of 1.49 and was strongly characterized with EC (0.90) and moderate loading with TDS (0.66). The high loading of EC and moderate TDS reflect anthropogenic sources in drinking water sources of UCPs. Factor 4 (F4) explained 11.3% of the total variance with an eigenvalue of 1.00. F4 showed a low contribution in the PCA results, and only pH had a positive moderate loading of 0.62 in F4. Moderate loadings of the physiochemical parameters might be attributed to their moderate to high concentration in the study area.

3.4. Pollution Assessment of PTEs

Pollution indices (PTE-PI and PTE-EI) of PTEs were carried out for the drinking water quality of UCPs. PTE-PI and PTE-EI represent different classes with low to high risk levels. The PTE-PI values were calculated based on all PTEs for each UCP (Table 2). The mean concentration value of selected PTEs (Cd, Cr, Pb, Ni, and Cu) were considered while calculating the PTE-PI and PTE-EI of the UCPs water.

Results showed that PTE-PI values of PTEs were observed from low to high values for all the drinking water samples of UCPs, ranging from 11.4 to 116 with the mean PTE-PI value (48.2), which is below the critical threshold pollution index value of 100. The lowest PTE-PI value was recorded for GNPC, while the highest PTE-PI value was observed for ZPP. Overall, the PTE-PI values for all the UCPs water samples were found to be below the critical limit of PTE-PI (100), except SPC (113) and ZPP (116), which exceeded the threshold critical value. These UCPs (SPC and ZPP) showed high PTE-PI values, indicating contamination in drinking water in comparison with other UCPs. In most of the cases (16

UCPs), the mean PTE-PI was much lower than the allowable index value of 100 suggested for drinking water [56]. The scales were slightly altered utilizing multiples of the median as a criterion in order to apply these PTE-PI indices. To distinguish between different levels of contamination, the data are divided into three classes: low (<40), medium (40–80), and high (>80). According to the PTE-PI results, 50% of 10 UCPs samples showed low pollution levels (low risk) with "Excellent" water quality, 40% of eight UCPs had medium pollution levels (medium risk) with "Good" and "Very Good" water quality, and 10% of two UCPs indicated high pollution levels (high risk) with "Poor" water quality.

Table 2. Overall Potentially Toxic Elements Pollution Index (PTE-PI), Potentially Toxic Elements Evaluation Index (PTE-EI), and quality classification (as per PTE-PI scale) of all PTEs of UCPs water.

UCPs	PTE-PI	PTE-EI	Quality as per PTE-PI Scale
BPC	79.8	2.25	Good
SPC	113	2.80	Poor
ZPP	116	3.48	Poor
TPP	73.7	2.15	Good
JPN	23.9	1.16	Excellent
DWPN	22.9	1.19	Excellent
PWLPH	73.4	3.35	Good
APH	57.4	2.55	Very good
SKIPS	33.1	1.95	Excellent
WWPS	55.5	2.74	Very good
APS	24.1	1.11	Excellent
GCGPS	44.2	2.55	Very good
SCPM	74.3	3.24	Good
YSPM	43.2	2.55	Very good
GNPC	11.4	0.810	Excellent
PTPC	17.1	1.16	Excellent
HNPDK	28.7	1.59	Excellent
IPDK	20.9	1.43	Excellent
LGPPA	31.5	1.50	Excellent
SPPA	20.0	1.30	Excellent
Average	48.2	2.04	

Furthermore, PTE-EI was also used for the brief interpretation of the pollution index [57] of PTEs in UCPs water. The PTE-EI values of PTEs ranged from 0.810 to 3.48 for all UCPs water, with a mean value of 2.04 (Table 2). The respective mean values of the samples were used to construct different PTE-EI values, and the various levels of contamination are characterized by the mean values. Moreover, the PTE-EI values were classified on the basis of pollution levels as low (PTE-EI < 1), medium (PTE-EI = 1–2), and high (>2) as described previously by [50]. Therefore, the results were observed according to the proposed PTE-EI criteria as follows: 5% of samples (1 UCP) showed low pollution level (low risk), 45% of (9 UCPs) samples had medium pollution levels (medium risk), and 50% of (10 UCPs) samples exhibit high pollution levels (high risk), as shown in Table 2. Similar observations and results of PTE-PI and PTE-EI were found in agreement with the study reported by [58].

3.5. Health Risk Assessment

The health risks of selected PTEs in UCPs drinking water sources were evaluated for the adults and children via two exposure pathways, i.e., ingestion and dermal contact. The statistical results of DMI for both adults and children via the two exposure pathways are shown in Tables S4 and S5. The results showed that DMI values were higher for children than adults in both exposure routes. Cd had the lowest DMI values for both exposure routes as compared to other PTEs. The non-carcinogenic risks of PTEs also showed low HQ values for adults in comparison with children. Similarly, the ingestion route is the dominant exposure pathway for non-carcinogenic risks of PTEs, followed by the dermal route. The highest HQ mean value (2.67×10^{-3}) was observed for Cr in GCGPS, while the lowest HQ mean value (3.77×10^{-6}) was recorded for Ni in APS via the ingestion route. For the dermal route, Cr had the highest HQ mean value (6.18×10^{-3}) for GCGPS, while Ni showed the lowest HQ mean value (9.13×10^{-8}) for SKIPS. Based on non-carcinogenic risk results, the HQ values of all PTEs for both exposure routes were less than the standard permissible limit (HQ < 1), as shown in Tables S6 and S7, suggesting no non-carcinogenic risk to the local population. The decreasing trend of the HQ mean values of PTEs was Cr > Pb > Cd > Ni > Cu, via ingestion for both children and adults, while for the dermal route, it was as follows: Cr > Cd > Pb > Cu > Ni, for both children and adults.

The HI values of all PTEs in UCPs were less than 1 for both the adults and children, as shown in Figure 5. HI values of PTEs ranged from 2.03×10^{-3} to 5.98×10^{-4} for children, and for adults they were 4.83×10^{-4} to 1.44×10^{-4}, via both exposure pathways, respectively. This indicates that exposure of PTEs in the UCPs could not pose non-carcinogenic risks to adults and children through the two exposure routes. Among UCPs, HQ and HI values were relatively lower in JPN, DWPN, APS, GNPC, and PTPC than the others. Generally, the HQ and HI values of children were relatively higher than those of adults in all UCPs, signifying that the non-carcinogenic risk of children is higher than that of adults via both exposure pathways (Tables S6 and S7 and Figure 5).

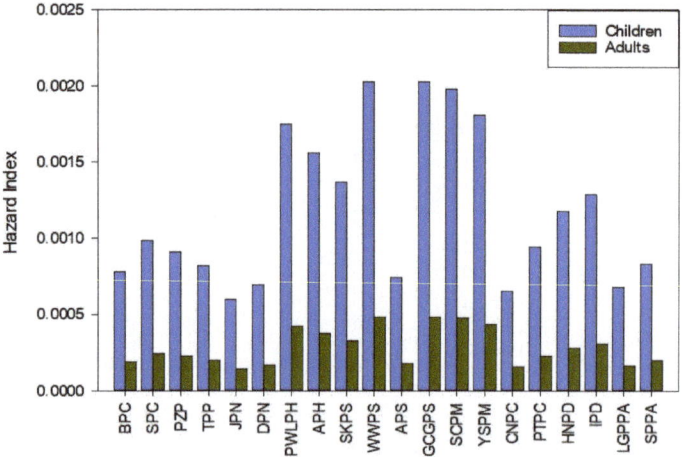

Figure 5. Hazard index (HI) of non-carcinogenic effects of all PTEs in drinking water of UCPs.

Furthermore, CR results showed that PTEs (Cr, Cd, Pb, and Ni) had the highest potential carcinogenic risk for both adults and children and surpassed the threshold value (1.00×10^{-4}) in all UCPs, except Cd and Ni for adults (<1.00×10^{-4}) via the ingestion route (Table 3). The highest value was recorded for Cr (3.28×10^{-1}) in GCGPS via the ingestion route, while the lowest CR value was observed for Ni (6.34×10^{-5}) in APS. Moreover, CR values of all PTEs were lower than the standard limit, except Cr had a high CR value (>1.00×10^{-4}) for children via the dermal route. High CR values of Cr (>1.00×10^{-4}) were recorded for all the UCPs, except BPC, SPC, ZPP, TPP, and JPN via the dermal route. The carcinogenic risk levels of PTEs in UCPs are found to be acceptable, except Cd and Ni for adults (<1.00×10^{-4}) via the ingestion route and Cr (>1.00×10^{-4}) for children via the dermal route are prone to cause potential carcinogenic risk. The TCR mean values were identified for selected PTEs (Cr, Cd, Pb, and Cu) and were found to be varied in the present study. The TCR values for Cr, Cd, Pb, and Cu exceeded the acceptable limit (1.00×10^{-4}) except Cd and Ni for adults (Table 3). High TCR values of Cr, Cd, Pb, and Cu indicate high total carcinogenic risk for children and adults exposed to drinking water in all UCPs, and were found to be in agreement with the results of high cancer risk observed for adults and children [59].

Table 3. Carcinogenic risk of selected PTEs (Cr, Cd, Pb, and Ni) via ingestion and dermal route in drinking water of UCPs.

UCPs	Children								Adults							
	Cr		Cd		Pb		Ni		Cr		Cd		Pb		Ni	
	CRing	CRder	CRing	CRder	CRing	CRder	CRing	CRder	CRing	CRder	CRing	CRder	CRing	CRder	CRing	CRder
BPC	5.36×10^{-2}	6.22×10^{-4}	2.99×10^{-3}	1.73×10^{-5}	8.12×10^{-3}	4.71×10^{-5}	1.84×10^{-3}	2.13×10^{-6}	1.36×10^{-2}	1.42×10^{-4}	7.58×10^{-4}	3.96×10^{-6}	2.06×10^{-3}	1.08×10^{-5}	4.67×10^{-4}	4.88×10^{-6}
SPC	5.76×10^{-2}	6.68×10^{-4}	4.77×10^{-3}	2.77×10^{-5}	8.96×10^{-3}	5.20×10^{-5}	1.38×10^{-3}	1.60×10^{-6}	1.46×10^{-2}	1.53×10^{-4}	1.21×10^{-3}	6.32×10^{-6}	2.28×10^{-3}	1.19×10^{-5}	3.50×10^{-4}	3.66×10^{-6}
ZPP	4.34×10^{-2}	5.03×10^{-4}	2.73×10^{-3}	1.61×10^{-5}	2.10×10^{-2}	1.22×10^{-4}	1.81×10^{-3}	2.10×10^{-6}	1.12×10^{-2}	1.15×10^{-4}	7.06×10^{-4}	3.68×10^{-6}	5.34×10^{-3}	2.79×10^{-5}	4.61×10^{-4}	4.81×10^{-6}
TPP	6.25×10^{-2}	7.24×10^{-4}	2.47×10^{-3}	1.43×10^{-5}	9.12×10^{-3}	5.29×10^{-5}	9.15×10^{-4}	1.06×10^{-6}	1.59×10^{-2}	1.66×10^{-4}	6.27×10^{-4}	3.28×10^{-6}	2.32×10^{-3}	1.21×10^{-5}	2.32×10^{-4}	2.43×10^{-6}
JPN	7.60×10^{-2}	8.82×10^{-4}	7.69×10^{-3}	4.46×10^{-5}	2.34×10^{-2}	1.36×10^{-4}	2.03×10^{-3}	2.35×10^{-6}	1.93×10^{-2}	2.02×10^{-4}	1.95×10^{-3}	1.02×10^{-5}	5.95×10^{-3}	3.11×10^{-5}	5.15×10^{-4}	5.37×10^{-6}
DWPN	9.72×10^{-2}	1.13×10^{-3}	2.88×10^{-3}	1.67×10^{-5}	4.75×10^{-3}	2.76×10^{-5}	8.61×10^{-4}	9.99×10^{-7}	2.47×10^{-2}	2.58×10^{-4}	7.32×10^{-4}	3.82×10^{-6}	1.21×10^{-3}	6.30×10^{-6}	2.19×10^{-4}	2.28×10^{-6}
PWLPH	2.37×10^{-1}	2.75×10^{-3}	1.15×10^{-2}	6.69×10^{-5}	1.46×10^{-2}	8.48×10^{-5}	2.44×10^{-3}	2.82×10^{-6}	6.02×10^{-2}	6.29×10^{-4}	2.93×10^{-3}	1.53×10^{-5}	3.71×10^{-3}	1.94×10^{-5}	6.18×10^{-4}	6.46×10^{-6}
APH	2.27×10^{-1}	2.63×10^{-3}	1.45×10^{-2}	8.40×10^{-5}	8.08×10^{-3}	4.69×10^{-5}	1.72×10^{-3}	1.99×10^{-6}	5.76×10^{-2}	6.01×10^{-4}	3.68×10^{-3}	1.92×10^{-5}	2.05×10^{-3}	1.07×10^{-5}	4.36×10^{-4}	4.55×10^{-6}
SKIPS	2.12×10^{-1}	2.46×10^{-3}	5.56×10^{-3}	3.22×10^{-5}	5.44×10^{-3}	3.16×10^{-5}	1.56×10^{-3}	1.81×10^{-6}	5.40×10^{-2}	5.63×10^{-4}	1.41×10^{-3}	7.37×10^{-6}	1.38×10^{-3}	7.21×10^{-6}	3.96×10^{-4}	4.14×10^{-6}
WWPS	3.15×10^{-1}	3.65×10^{-3}	1.89×10^{-2}	1.09×10^{-4}	4.16×10^{-3}	2.41×10^{-5}	2.33×10^{-3}	2.70×10^{-6}	8.00×10^{-2}	8.35×10^{-4}	4.79×10^{-3}	2.50×10^{-5}	1.06×10^{-3}	5.51×10^{-6}	5.92×10^{-4}	6.18×10^{-6}
APS	1.12×10^{-1}	1.30×10^{-3}	4.60×10^{-3}	2.67×10^{-5}	4.20×10^{-3}	2.43×10^{-5}	2.49×10^{-3}	2.89×10^{-6}	2.85×10^{-2}	2.98×10^{-4}	1.17×10^{-3}	6.10×10^{-6}	1.07×10^{-3}	5.57×10^{-6}	6.34×10^{-4}	6.61×10^{-6}
GCGPS	3.28×10^{-1}	3.80×10^{-3}	1.20×10^{-2}	6.97×10^{-5}	4.36×10^{-3}	2.53×10^{-5}	1.85×10^{-3}	2.14×10^{-6}	8.33×10^{-2}	8.69×10^{-4}	3.05×10^{-3}	1.59×10^{-5}	1.11×10^{-3}	5.78×10^{-6}	4.69×10^{-4}	4.90×10^{-6}
SCPM	2.84×10^{-1}	3.29×10^{-3}	1.83×10^{-2}	1.06×10^{-4}	1.09×10^{-2}	6.32×10^{-5}	1.84×10^{-3}	2.14×10^{-6}	7.20×10^{-2}	7.52×10^{-4}	4.64×10^{-3}	2.42×10^{-5}	2.77×10^{-3}	1.44×10^{-5}	4.68×10^{-4}	4.89×10^{-6}
YSPM	2.83×10^{-1}	3.28×10^{-3}	6.79×10^{-3}	3.94×10^{-5}	7.35×10^{-3}	4.27×10^{-5}	1.82×10^{-3}	2.11×10^{-6}	7.18×10^{-2}	7.49×10^{-4}	1.73×10^{-3}	9.01×10^{-6}	1.87×10^{-3}	9.75×10^{-6}	4.62×10^{-4}	4.82×10^{-6}
GNPC	9.87×10^{-2}	1.14×10^{-3}	2.95×10^{-3}	1.71×10^{-5}	1.03×10^{-3}	5.99×10^{-6}	8.07×10^{-4}	9.36×10^{-7}	2.51×10^{-2}	2.62×10^{-4}	7.49×10^{-4}	3.91×10^{-6}	2.62×10^{-4}	1.37×10^{-6}	2.05×10^{-4}	2.14×10^{-6}
PTPC	1.60×10^{-1}	1.86×10^{-3}	9.61×10^{-3}	5.57×10^{-5}	3.53×10^{-3}	2.05×10^{-5}	4.42×10^{-4}	5.13×10^{-7}	4.06×10^{-2}	4.24×10^{-4}	2.44×10^{-3}	1.27×10^{-5}	8.96×10^{-4}	4.68×10^{-6}	1.12×10^{-4}	1.17×10^{-6}
HNPDK	1.92×10^{-1}	2.23×10^{-3}	3.98×10^{-3}	2.31×10^{-5}	5.27×10^{-3}	3.06×10^{-5}	5.26×10^{-4}	6.11×10^{-7}	4.88×10^{-2}	5.09×10^{-4}	1.01×10^{-3}	5.28×10^{-6}	1.34×10^{-3}	6.99×10^{-6}	1.34×10^{-4}	1.40×10^{-6}
IPDK	2.20×10^{-1}	2.53×10^{-3}	4.60×10^{-3}	2.67×10^{-5}	2.15×10^{-3}	1.25×10^{-5}	7.13×10^{-4}	8.28×10^{-7}	5.59×10^{-2}	5.83×10^{-4}	1.17×10^{-3}	6.10×10^{-6}	5.46×10^{-4}	2.85×10^{-6}	1.81×10^{-4}	1.89×10^{-6}
LGPPA	8.62×10^{-2}	1.00×10^{-3}	3.77×10^{-3}	2.19×10^{-5}	6.93×10^{-3}	4.02×10^{-5}	1.63×10^{-3}	1.92×10^{-6}	2.19×10^{-2}	2.29×10^{-4}	9.59×10^{-4}	5.00×10^{-6}	1.76×10^{-3}	9.19×10^{-6}	4.21×10^{-4}	4.40×10^{-6}
SPPA	1.28×10^{-1}	1.48×10^{-3}	3.98×10^{-3}	2.31×10^{-5}	2.73×10^{-3}	1.58×10^{-5}	2.05×10^{-3}	2.38×10^{-6}	3.24×10^{-2}	3.39×10^{-4}	1.01×10^{-3}	5.28×10^{-6}	6.92×10^{-4}	3.61×10^{-6}	5.21×10^{-4}	5.43×10^{-6}
TCR	1.66×10^{-1}		1.27×10^{-3}		6.79×10^{-3}		1.44×10^{-3}		4.20×10^{-2}		3.23×10^{-4}		1.72×10^{-3}		3.67×10^{-4}	

CRing: Carcinogenic risk via ingestion, CRder: Carcinogenic risk via dermal contact.

4. Discussion

The physiochemical properties play a critical role in influencing the bioavailability of PTEs in groundwater. All the drinking water samples had relatively low to high pH levels in the present study of UCPs. By comparison, the mean pH values in UCPs water were slightly varied, reflecting the existence of alkalizing products such as calcium-magnesium carbonate or calcium carbonate in flagstones, gravel, concrete, cement, and mortar as well as the atmospheric particulate deposition [60,61], resulting in release into the groundwater system. As a result, the drinking water sources could be affected by low to high pH and influence the PTEs concentration. TDS mean values were high for most of the UCPs water, and the presence of high TDS could be attributed to the leaching of ions into the groundwater system [62].

The comparison of PTEs concentrations in all samples with the permissible limits of WHO revealed varying degrees of contamination in drinking water sources of UCPs. Our results indicated that PTEs (Cu, Cr, and Cd) concentrations were generally higher in the UCPs drinking water than in the previous studies. For instance, the PTEs (Cu, Cd, Ni) concentrations (Figure 3) in the present study were relatively higher in most of the UCPs than in a previous study of lakes along urban gradient lakes in Wuhan, China [63]. Similarly, high Cu, Cr, Pb, Ni, and Cd were found to be higher than the previous study of the urban Houguan Lake in Wuhan, China [64], and also higher than the reported values of PTEs (Cu, Pb, and Cd) in urban groundwater of Nnewi, Southeast Nigeria [65]. In contrast, the mean values of PTEs concentrations were relatively lower than the results of the urban Mariout Lake, Egypt, reported by [66], and were lower than the mean values of Cu, Pb, Ni, and Cr (except Cd) in the groundwater of an industrial park situated at the southeast of Zhejiang Province, China [67]. In another study, PTEs such as Cr, Cu, and Cd were relatively higher, while Pb concentrations were lower than in the previous study of drinking water sources in the industrial city of Sialkot, Punjab Pakistan [26], and were higher than in our recent studies of [68–70]. The variation in PTEs concentration level demonstrates numerous factors depending upon various potential sources of anthropogenic activities and natural geological backgrounds, resulting in UCPs water contamination. This variation in PTEs concentration also reflects the impact of physiochemical properties of the soil, which might affect the mobilization and bioavailability of PTEs in the groundwater system. For instance, the bioavailability and mobility of PTEs can potentially be influenced by soil pH, organic matter, and cation-exchange mechanism. In addition, due to the precipitation mechanism between PTEs and anions, PTEs can be more readily accessible and mobilized in comparison to high pH [71]. Another reason is that urbanization in UCPs areas could also affect the soil texture, soil pH, cation-exchange capacity, bulk density, and thus PTEs discharge and deposit in the soils [72] and release into the groundwater system. Moreover, long-term disposal of solid waste and wastewater irrigation in parks result in UCPs soil contamination with high level of PTEs. Additional factors that contribute to the contamination of the soil with PTEs include inadequate sludge and septic tank treatment plans, sanitary system leakage, and raw sewage water from installations [73].

Some other factors like pH may have an impact on bioavailability of PTEs in the drinking water of UCPs. Evidently, Li et al. also reported that the physiochemical properties of the soil ecosystem (soil pH, cation exchange mechanism, and organic matter) could influence bioavailability of PTEs. The dissolution of PTEs occur in soil due to redox potential under acidic to alkaline conditions (pH 5.0 to 8.0) and reach into the groundwater through leaching [74]. In the present study, some of the UCPs (BPC, SKIPS, and PTPC) had low pH level with moderate to high PTEs concentration as compared to the other UCPs. PTEs are bioavailable and easier to mobilize at low pH levels compared to high pH levels because of soil precipitations between various anions and metals [75]. Some of the anthropogenic sources such as domestic waste disposal contribute to release toxic pollutants [76] in the UCPs. According to a previous study of PTEs in Peshawar urban parks in Pakistan, Cd originates from industrial facilities (power stations, coal combustion, and the metallurgical sector), whereas Ni is derived from agricultural practices including wastewater irrigation and solid waste disposal [73], and Pb is primarily found in urban soil due to vehicle exhaust emissions [77]. Furthermore, ref. [5] reported that the soil's characteristics, pollutant emissions, and the geochemical background had a significant impact on the spatial distribution of PTEs.

In the present study, moderate and high loadings of selected PTEs in PCA results demonstrate numerous factors as shown in Figure 4, showing their anthropogenic and natural sources. The loading factors of pH in F1 and F4 factors reflect the presence of alkaline salts such as calcium-magnesium carbonate $CaMg(CO_3)_2$ and calcium carbonate ($CaCO_3$) in flagstones, concrete, gravel stones, and cement in soils [71], resulting in influencing the PTEs availability in UCPs. Despite the anthropogenic sources, previous research demonstrated that soil parent materials had an impact on the distribution of PTEs like Cd [3]. The PTEs such as Ni could originate from waste disposal and agricultural activity of wastewater irrigation [73], which could be released to the drinking water sources of UCPs. Other PTEs (Cu and Pb) in urban parks are mainly derived from vehicular exhaust emissions [77].

The results of health risk assessment showed that children are more exposed to average daily exposure doses due to smaller skin surface area and lower body weight. Children are also more susceptible to external environment during their developmental stages and growth. This could be a reason that children are more exposed to drinking water sources in UCPs via ingestion and dermal contact than adults. Previous studies also reported that children are more vulnerable to average daily exposure dose and high non-carcinogenic risks than adults [78]. We concluded that long-term exposure to these toxic PTEs (Cd, Ni, and Cr) through drinking water consumption of UCPs presents a carcinogenic risk to the local population. Our results of HQ, HI, and CR are consistent with the recent study conducted by [79] and [34]. TCR results also showed high values of Cr, Cd, Pb, and Cu, indicating high total carcinogenic risk for children and adults exposed to drinking water in all UCPs. In the present study, children present high non-carcinogenic risk and carcinogenic risk for both exposure routes. In general, children are typically more susceptible than adults to exposure to toxic contaminants due to their continuous outdoor activity [80] and their propensity to play on the ground, put objects in their mouth, and use their hands [81,82], confirming the risk of drinking water contaminated with PTEs in UCPs.

Additionally, heavy traffic and industrialized regions can influence and prominently increase these contaminants in the soil [83], resulting in contaminating the drinking water sources as well in local UCPs. The public urban parks are more likely accessed and visited by children and adults, and contaminated soil can be introduced in UCPs by their footsteps. The soil in these locations also has a significant impact on the population's health, posing a concern if pollutant concentrations are high [84]. As a result, PTEs can possibly reach the groundwater system though leaching due to contaminated soil and continuous wastewater irrigation. A high content of PTEs is released as a result of long-term wastewater discharge and solid waste disposal in parks. Additionally, improper treatment and leakage in sludge and septic tank schemes, sanitary system seepage from installations, and discharge of raw sewage water [73] also contribute to contaminating the drinking water sources with PTEs in UCPs. Therefore, with the primary attention on prevention, and gaining control of potential health concerns needs to be undertaken related to exposure to PTEs, particularly for children in UCPs of Khyber Pakhtunkhwa, Pakistan.

5. Conclusions

In the present study, the mean concentrations of PTEs (Cd, Cr, Pb, Ni, and Cu) showed high variation in drinking water sources of UCPs. PTEs such as Cd, Cr, and Pb had high contamination and exceeded the permissible limits of WHO and PAK-EPA. The PCA results of PTEs revealed high anthropogenic sources of traffic and industrial emissions, solid waste disposal, wastewater irrigation, and soil parent materials. The PTE-PI and PTE-EI results showed low to high pollution levels for all the UCPs, with high values for Sardaryab Park Charsadda (SPC) and Zoo Park Peshawar (ZPP), indicating unhealthy drinking water quality in comparison with other UCPs. The HQ and HI values of all PTEs were less than the permissible limit (<1), while CR values surpassed the threshold value (1.00×10^{-4}) of PTEs for both adults and children in all UCPs, except Cd and Ni for adults via the ingestion route. For the dermal route, CR values of Cr were higher than the standard limit among PTEs in most of the UCPs. The TCR values of PTEs were higher for children and adults via drinking water consumption in all UCPs. Our results revealed high carcinogenic risk of PTEs (Cd, Ni, and Cr) exposed to the local population in all UCPs. Therefore, the present study proposed that the government should control and mitigate the PTEs pollution in drinking water sources by enforcing environmental regulations as well as implementing the proper management strategies to remediate its pollution in urban children's parks and reduce potential health risk in urban areas. The assessment of PTEs contamination and the health risk evaluation used in this work should be used in future risk assessments, as it can aid in the implementation of more appropriate metal risk management in urban environments worldwide.

Supplementary Materials: The following supporting information can be downloaded at https://www.mdpi.com/article/10.3390/su151713177/s1: Table S1. The mean Wi and Qi calculation for groundwater samples of the study area. Table S2. Parameter values used for daily metal intake in health risk models. Table S3. Physiochemical and PTEs mean concentration (µg/L) in drinking water sources of UCPs in KPK, Pakistan. Table S4. The DMI values of selected PTEs via ingestion for drinking water in UCPs. Table S5. The DMI values of selected PTEs via dermal contact for drinking water in UCPs. Table S6. Non-carcinogenic risks posed by each PTE via ingestion of UCPs water in KPK. Table S7. Non-carcinogenic risks posed by each PTE via dermal exposure pathway of UCPs water in KPK.

Author Contributions: Conceptualization, J.G. and J.N.; Methodology, J.G.; writing—original draft preparation, J.G.; formal analysis, J.N. and Z.U.; validation, J.N.; investigation, Z.U. and S.Z.H.; reviewing and editing, N.R., S.K., and M.S.; funding acquisition, M.H.A. All authors have read and agreed to the published version of the manuscript.

Funding: This research was funded by King Saud University through a Research group project under grant number (RSP2023R191).

Institutional Review Board Statement: Not applicable.

Informed Consent Statement: Not applicable.

Data Availability Statement: Data will be made available on request.

Acknowledgments: The authors extend their appreciation to the Researchers Supporting Project number (RSP2023R191), King Saud University, Riyadh, Saudi Arabia.

Conflicts of Interest: The authors declare no conflict of interest.

References

1. UN. *World Urbanization Prospects*; The Population Division of the Department of Economic and Social Affairs: New York, NY, USA, 2018.
2. Coulon, F.; Jones, K.; Li, H.; Hu, Q.; Gao, J.; Li, F.; Chen, M.; Zhu, Y.-G.; Liu, R.; Liu, M. China's soil and groundwater management challenges: Lessons from the UK's experience and opportunities for China. *Environ. Int.* **2016**, *91*, 196–200. [CrossRef] [PubMed]
3. Liu, L.; Liu, Q.; Ma, J.; Wu, H.; Qu, Y.; Gong, Y.; Yang, S.; An, Y.; Zhou, Y. Heavy metal (loid) s in the topsoil of urban parks in Beijing, China: Concentrations, potential sources, and risk assessment. *Environ. Pollut.* **2020**, *260*, 114083. [CrossRef]
4. Gu, Y.-G.; Gao, Y.-P. Bioaccessibilities and health implications of heavy metals in exposed-lawn soils from 28 urban parks in the megacity Guangzhou inferred from an in vitro physiologically-based extraction test. *Ecotoxicol. Environ. Saf.* **2018**, *148*, 747–753. [CrossRef] [PubMed]
5. Wu, S.; Zhou, S.; Bao, H.; Chen, D.; Wang, C.; Li, B.; Tong, G.; Yuan, Y.; Xu, B.J. Improving risk management by using the spatial interaction relationship of heavy metals and PAHs in urban soil. *J. Hazard. Mater.* **2019**, *364*, 108–116. [CrossRef] [PubMed]
6. Liu, X.; Zhong, L.; Meng, J.; Wang, F.; Zhang, J.; Zhi, Y.; Zeng, L.; Tang, X.; Xu, J. A multi-medium chain modeling approach to estimate the cumulative effects of cadmium pollution on human health. *Environ. Pollut.* **2018**, *239*, 308–317. [CrossRef] [PubMed]
7. Yadav, I.C.; Devi, N.L.; Singh, V.K.; Li, J.; Zhang, G. Spatial distribution, source analysis, and health risk assessment of heavy metals contamination in house dust and surface soil from four major cities of Nepal. *Chemosphere* **2019**, *218*, 1100–1113. [CrossRef] [PubMed]

8. Adewumi, A. Heavy metals in soils and road dust in Akure City, Southwest Nigeria: Pollution, sources, and ecological and health risks. *Expos. Health* **2022**, *14*, 375–392. [CrossRef]
9. Santorufo, L.; Van Gestel, C.A.; Maisto, G. Ecotoxicological assessment of metal-polluted urban soils using bioassays with three soil invertebrates. *Chemosphere* **2012**, *88*, 418–425. [CrossRef]
10. Penteado, P.B.; Nogarotto, D.C.; Baltazar, J.P.; Pozza, S.A.; Canteras, F.B. Inorganic pollution in urban topsoils of Latin American cities: A systematic review and future research direction. *Catena* **2022**, *210*, 105946. [CrossRef]
11. Singh, K.K.; Tewari, G.; Kumar, S.; Busa, R.; Chaturvedi, A.; Rathore, S.S.; Singh, R.K.; Gangwar, A. Understanding urban groundwater pollution in the Upper Gangetic Alluvial Plains of northern India with multiple industries and their impact on drinking water quality and associated health risks. *Groundw. Sustain. Dev.* **2023**, *21*, 100902. [CrossRef]
12. Antoniadis, V.; Golia, E.E.; Liu, Y.-T.; Wang, S.-L.; Shaheen, S.M.; Rinklebe, J. Soil and maize contamination by trace elements and associated health risk assessment in the industrial area of Volos, Greece. *Environ. Int.* **2019**, *124*, 79–88. [CrossRef] [PubMed]
13. Huang, J.; Wu, Y.; Li, Y.; Sun, J.; Xie, Y.; Fan, Z. Do trace metal (loid) s in road soils pose health risks to tourists? A case of a highly-visited national park in China. *J. Environ. Sci.* **2022**, *111*, 61–74. [CrossRef] [PubMed]
14. Khan, K.; Lu, Y.; Saeed, M.A.; Bilal, H.; Sher, H.; Khan, H.; Ali, J.; Wang, P.; Uwizeyimana, H.; Baninla, Y. Prevalent fecal contamination in drinking water resources and potential health risks in Swat, Pakistan. *J. Environ. Sci.* **2018**, *72*, 1–12. [CrossRef] [PubMed]
15. Nawab, J.; Khan, S.; Ali, S.; Sher, H.; Rahman, Z.; Khan, K.; Tang, J.; Ahmad, A. Health risk assessment of heavy metals and bacterial contamination in drinking water sources: A case study of Malakand Agency, Pakistan. *Environ. Monit. Assess.* **2016**, *188*, 286. [CrossRef]
16. Wu, L.; Liang, Y.; Fu, S.; Huang, Y.; Chen, Z.; Chang, X. Biomonitoring trace metal contamination in Guangzhou urban parks using Asian tramp snails (*Bradybaena similaris*). *Chemosphere* **2023**, *334*, 138960. [CrossRef]
17. Boum-Nkot, S.N.; Nlend, B.; Komba, D.; Ndondo, G.N.; Bello, M.; Fongoh, E.; Ntamak-Nida, M.-J.; Etame, J. Hydrochemistry and assessment of heavy metals groundwater contamination in an industrialized city of sub-Saharan Africa (Douala, Cameroon). Implication on human health. *HydroResearch* **2023**, *6*, 52–64. [CrossRef]
18. Pinto, M.M.C.; Silva, M.M.; da Silva, E.A.F.; Dinis, P.A.; Rocha, F. Transfer processes of potentially toxic elements (PTE) from rocks to soils and the origin of PTE in soils: A case study on the island of Santiago (Cape Verde). *Environ. Monit. Assess.* **2017**, *183*, 140–151.
19. Wong, C.S.; Li, X.; Thornton, I. Urban environmental geochemistry of trace metals. *Environ. Pollut.* **2006**, *142*, 1–16. [CrossRef]
20. Derakhshan-Babaei, F.; Mirchooli, F.; Mohammadi, M.; Nosrati, K.; Egli, M. Tracking the origin of trace metals in a watershed by identifying fingerprints of soils, landscape and river sediments. *Sci. Total Environ.* **2022**, *835*, 155583. [CrossRef]
21. Pan, Y.; Peng, H.; Hou, Q.; Peng, K.; Shi, H.; Wang, S.; Zhang, W.; Zeng, M.; Huang, C.; Xu, L. Priority control factors for heavy metal groundwater contamination in peninsula regions based on source-oriented health risk assessment. *Sci. Total Environ.* **2023**, *894*, 165062. [CrossRef]
22. Guo, C.; Chen, Y.; Xia, W.; Qu, X.; Yuan, H.; Xie, S.; Lin, L.-S. Eutrophication and heavy metal pollution patterns in the water supplying lakes of China's south-to-north water diversion project. *Sci. Total Environ.* **2020**, *711*, 134543. [CrossRef] [PubMed]
23. Wang, L.; Lyons, J.; Kanehl, P.; Bannerman, R. Impacts of urbanization on stream habitat and fish across multiple spatial scales. *Environ. Manag.* **2001**, *28*, 255–266. [CrossRef] [PubMed]
24. Valtanen, M.; Sillanpää, N.; Setälä, H. The effects of urbanization on runoff pollutant concentrations, loadings and their seasonal patterns under cold climate. *Water Air Soil Pollut.* **2014**, *225*, 1977. [CrossRef]
25. Oginawati, K.; Susetyo, S.H.; Rosalyn, F.A.; Kurniawan, S.B.; Abdullah, S.R.S. Risk analysis of inhaled hexavalent chromium (Cr 6+) exposure on blacksmiths from industrial area. *Environ. Sci. Pollut. Res.* **2021**, *28*, 14000–14008. [CrossRef] [PubMed]
26. Ahmad, W.; Alharthy, R.D.; Zubair, M.; Ahmed, M.; Hameed, A.; Rafique, S. Toxic and heavy metals contamination assessment in soil and water to evaluate human health risk. *Sci. Rep.* **2021**, *11*, 17006. [CrossRef]
27. Peña-Fernández, A.; González-Muñoz, M.; Lobo-Bedmar, M. Establishing the importance of human health risk assessment for metals and metalloids in urban environments. *Environ. Int.* **2014**, *72*, 176–185. [CrossRef]
28. Dash, S.; Borah, S.S.; Kalamdhad, A.S. Application of positive matrix factorization receptor model and elemental analysis for the assessment of sediment contamination and their source apportionment of Deepor Beel, Assam, India. *Ecol. Indic.* **2020**, *114*, 106291. [CrossRef]
29. Gholizadeh, M.H.; Melesse, A.M.; Reddi, L. Water quality assessment and apportionment of pollution sources using APCS-MLR and PMF receptor modeling techniques in three major rivers of South Florida. *Sci. Total Environ.* **2016**, *566*, 1552–1567. [CrossRef]
30. Bisone, S.; Chatain, V.; Blanc, D.; Gautier, M.; Bayard, R.; Sanchez, F.; Gourdon, R. Geochemical characterization and modeling of arsenic behavior in a highly contaminated mining soil. *Environ. Earth Sci.* **2016**, *75*, 306. [CrossRef]
31. Han, Q.; Wang, M.; Cao, J.; Gui, C.; Liu, Y.; He, X.; He, Y.; Liu, Y. Health risk assessment and bioaccessibilities of heavy metals for children in soil and dust from urban parks and schools of Jiaozuo, China. *Ecotoxicol. Environ. Saf.* **2020**, *191*, 110157. [CrossRef]
32. Rodríguez-Oroz, D.; Vidal, R.; Fernandoy, F.; Lambert, F.; Quiero, F. Metal concentrations and source identification in Chilean public children's playgrounds. *Environ. Monit. Assess.* **2018**, *190*, 703. [CrossRef] [PubMed]
33. Vega, A.S.; Arce, G.; Rivera, J.I.; Acevedo, S.E.; Reyes-Paecke, S.; Bonilla, C.A.; Pastén, P. A comparative study of soil metal concentrations in Chilean urban parks using four pollution indexes. *Appl. Geochem.* **2022**, *141*, 105230. [CrossRef]

34. Ghani, J.; Nawab, J.; Faiq, M.E.; Ullah, S.; Alam, A.; Ahmad, I.; Ali, S.W.; Khan, S.; Ahmad, I.; Muhammad, A. Multi-geostatistical analyses of the spatial distribution and source apportionment of potentially toxic elements in urban children's park soils in Pakistan: A risk assessment study. *Environ. Pollut.* **2022**, *311*, 119961. [CrossRef]
35. Yang, Y.; Wei, L.; Cui, L.; Zhang, M.; Wang, J. Profiles and risk assessment of heavy metals in Great Rift Lakes, Kenya. *CLEAN–Soil Air Water* **2017**, *45*, 1600825.
36. Li, H.-B.; Yu, S.; Li, G.-L.; Deng, H.; Xu, B.; Ding, J.; Gao, J.-B.; Hong, Y.-W.; Wong, M.-H. Spatial distribution and historical records of mercury sedimentation in urban lakes under urbanization impacts. *Sci. Total Environ.* **2013**, *445*, 117–125. [CrossRef] [PubMed]
37. Muhammad, N.; Nafees, M.; Ge, L.; Khan, M.H.; Bilal, M.; Chan, W.P.; Lisak, G. Assessment of industrial wastewater for potentially toxic elements, human health (dermal) risks, and pollution sources: A case study of Gadoon Amazai industrial estate, Swabi, Pakistan. *J. Hazard. Mater.* **2021**, *419*, 126450. [CrossRef] [PubMed]
38. Sirajudeen, J.; Arulmanikandan, S.; Manivel, V. Heavy metal pollution index of groundwater of Fathima Nagar area near Uyyakondan channel Tiruchirappalli district, Tamil Nadu, India. *World J. Pharm. Pharm. Sci.* **2015**, *4*, 967–975.
39. Prasanna, M.; Praveena, S.; Chidambaram, S.; Nagarajan, R.; Elayaraja, A. Evaluation of water quality pollution indices for heavy metal contamination monitoring: A case study from Curtin Lake, Miri City, East Malaysia. *Environ. Earth Sci.* **2012**, *67*, 1987–2001. [CrossRef]
40. Prasad, B.; Mondal, K.K. The impact of filling an abandoned open cast mine with fly ash on ground water quality: A case study. *Mine Water Environ.* **2008**, *27*, 40–45. [CrossRef]
41. Tokatlı, C.; Varol, M.J.E.R. Impact of the COVID-19 lockdown period on surface water quality in the Meriç-Ergene River Basin, Northwest Turkey. *Environ. Res.* **2021**, *197*, 111051. [CrossRef]
42. WHO. *Guidelines for Drinking-Water Quality*; World Health Organization: Geneva, Switzerland, 2002.
43. Schiller, G.J.Z. Directive 2009/12/EC of the European Parliament and of the Council of 11 March 2009 on Airport Charges/Neue Gemeinschaftsrechtliche Vorgaben zur Festsetzung von Flughafenentgelten: Die Richtlinie 2009/12/EG uber Flughafenentgelte/La Directive 2009/12/CE du Parlement Europeen et du Conseil sur les Redevances Aeroportuaires. *ZLW* **2009**, *58*, 356.
44. Varol, M.; Tokatlı, C. Impact of paddy fields on water quality of Gala Lake (Turkey): An important migratory bird stopover habitat. *Environ. Pollut.* **2021**, *287*, 117640. [CrossRef] [PubMed]
45. USEPA. *Regional Screening Levels for Chemical Contaminants at Superfund Sites*; United States Environmental Protection Agency: Washington, DC, USA, 2010.
46. Wongsasuluk, P.; Chotpantarat, S.; Siriwong, W.; Robson, M. Heavy metal contamination and human health risk assessment in drinking water from shallow groundwater wells in an agricultural area in Ubon Ratchathani province, Thailand. *Environ. Geochem. Health* **2014**, *36*, 169–182. [CrossRef] [PubMed]
47. Behrooz, R.D.; Kaskaoutis, D.; Grivas, G.; Mihalopoulos, N. Human health risk assessment for toxic elements in the extreme ambient dust conditions observed in Sistan, Iran. *Chemosphere* **2021**, *262*, 127835. [CrossRef] [PubMed]
48. Tepanosyan, G.; Sahakyan, L.; Maghakyan, N.; Saghatelyan, A. Identification of spatial patterns, geochemical associations and assessment of origin-specific health risk of potentially toxic elements in soils of Armavir region, Armenia. *Chemosphere* **2021**, *262*, 128365. [CrossRef]
49. US Department of Energy (USDOE). *The Risk Assessment Information System (RAIS)*; US Department of Energy's Oak Ridge Operations Office (ORO): Argonne, IL, USA, 2011.
50. Jiang, C.; Zhao, Q.; Zheng, L.; Chen, X.; Li, C.; Ren, M. Distribution, source and health risk assessment based on the Monte Carlo method of heavy metals in shallow groundwater in an area affected by mining activities, China. *Ecotoxicol. Environ. Saf.* **2021**, *224*, 112679. [CrossRef]
51. Brtnický, M.; Pecina, V.; Hladký, J.; Radziemska, M.; Koudelková, Z.; Klimánek, M.; Richtera, L.; Adamcová, D.; Elbl, J.; Galiová, M.V. Assessment of phytotoxicity, environmental and health risks of historical urban park soils. *Chemosphere* **2019**, *220*, 678–686. [CrossRef]
52. WHO. *Guidelines for Drinking-Water Quality*; World Health Organization: Geneva, Switzerland, 2011; Volume 38, pp. 104–108.
53. Pak-EPA, Government of Pakistan. *National Standards for Drinking Water Quality*; Pakistan Environmental Protection Agency: Islamabad, Pakistan, 2008.
54. Gu, Y.G.; Li, Q.S.; Fang, J.H.; He, B.Y.; Fu, H.B.; Tong, Z. Identification of heavy metal sources in the reclaimed farmland soils of the pearl river estuary in China using a multivariate geostatistical approach. *Ecotoxicol. Environ. Saf.* **2014**, *105*, 7–12. [CrossRef]
55. Song, Z.; Dong, L.; Shan, B.; Tang, W. Assessment of potential bioavailability of heavy metals in the sediments of land-freshwater interfaces by diffusive gradients in thin films. *Chemosphere* **2018**, *191*, 218–225. [CrossRef]
56. Mohan, S.V.; Nithila, P.; Reddy, S. Estimation of heavy metals in drinking water and development of heavy metal pollution index. *J. Environ. Sci. Health A J.* **1996**, *31*, 283–289. [CrossRef]
57. Edet, A.; Offiong, O. Evaluation of water quality pollution indices for heavy metal contamination monitoring. A study case from Akpabuyo-Odukpani area, Lower Cross River Basin (southeastern Nigeria). *GeoJournal* **2002**, *57*, 295–304. [CrossRef]
58. Rezaei, A.; Hassani, H.; Hassani, S.; Jabbari, N.; Mousavi, S.B.F.; Rezaei, S. Evaluation of groundwater quality and heavy metal pollution indices in Bazman basin, southeastern Iran. *Groundw. Sustain. Dev.* **2019**, *9*, 100245. [CrossRef]

59. Rahman, M.S.; Khan, M.; Jolly, Y.; Kabir, J.; Akter, S.; Salam, A. Assessing risk to human health for heavy metal contamination through street dust in the Southeast Asian Megacity: Dhaka, Bangladesh. *Sci. Total Environ.* **2019**, *660*, 1610–1622. [CrossRef] [PubMed]
60. Scharenbroch, B.C.; Lloyd, J.E.; Johnson-Maynard, J.L. Distinguishing urban soils with physical, chemical, and biological properties. *Pedobiologia* **2005**, *49*, 283–296. [CrossRef]
61. Li, Z.-G.; Zhang, G.-S.; Liu, Y.; Wan, K.-Y.; Zhang, R.-H.; Chen, F. Soil nutrient assessment for urban ecosystems in Hubei, China. *PLoS ONE* **2013**, *8*, e75856. [CrossRef] [PubMed]
62. Loh, Y.S.A.; Akurugu, B.A.; Manu, E.; Aliou, A.-S. Assessment of groundwater quality and the main controls on its hydrochemistry in some Voltaian and basement aquifers, northern Ghana. *Groundw. Sustain. Dev.* **2020**, *10*, 100296. [CrossRef]
63. Xia, W.; Wang, R.; Zhu, B.; Rudstam, L.G.; Liu, Y.; Xu, Y.; Xin, W.; Chen, Y. Heavy metal gradients from rural to urban lakes in central China. *Ecol. Process* **2020**, *9*, 47. [CrossRef]
64. Dou, Y.; Yu, X.; Liu, L.; Ning, Y.; Bi, X.; Liu, J. Effects of hydrological connectivity project on heavy metals in Wuhan urban lakes on the time scale. *Sci. Total Environ.* **2022**, *853*, 158654. [CrossRef] [PubMed]
65. Ayejoto, D.A.; Egbueri, J.C. Human health risk assessment of nitrate and heavy metals in urban groundwater in Southeast Nigeria. *Acta Ecol. Sin.* **2023**, in press. [CrossRef]
66. El-Magd, S.A.A.; Taha, T.; Pienaar, H.H.; Breil, P.; Amer, R.; Namour, P. Assessing heavy metal pollution hazard in sediments of Lake Mariout, Egypt. *J. Afr. Earth Sci.* **2021**, *176*, 104116. [CrossRef]
67. Xiang, Z.; Wu, S.; Zhu, L.; Yang, K.; Lin, D. Pollution characteristics and source apportionment of heavy metal (loid)s in soil and groundwater of a retired industrial park. *J. Environ. Sci.* 2023, in press. [CrossRef]
68. Nawab, J.; Rahman, A.; Khan, S.; Ghani, J.; Ullah, Z.; Khan, H.; Waqas, M. Drinking water quality assessment of government, non-government and self-based schemes in the disaster affected areas of khyber pakhtunkhwa, Pakistan. *Expos. Health* **2022**, 1–17. [CrossRef]
69. Bhatti, Z.I.; Ishtiaq, M.; Khan, S.A.; Nawab, J.; Ghani, J.; Ullah, Z.; Khan, S.; Baig, S.A.; Muhammad, I.; Din, Z.U. Contamination level, source identification and health risk assessment of potentially toxic elements in drinking water sources of mining and non-mining areas of Khyber Pakhtunkhwa, Pakistan. *J. Water Health* **2022**, *20*, 1343–1363. [CrossRef] [PubMed]
70. Nawab, J.; Khan, S.; Xiaoping, W. Ecological and health risk assessment of potentially toxic elements in the major rivers of Pakistan: General population vs. Fishermen. *Chemosphere* **2018**, *202*, 154–164. [CrossRef] [PubMed]
71. Li, H.; Qian, X.; Hu, W.; Wang, Y.; Gao, H. Chemical speciation and human health risk of trace metals in urban street dusts from a metropolitan city, Nanjing, SE China. *Sci. Total Environ.* **2013**, *456*, 212–221. [CrossRef]
72. Amjadian, K.; Sacchi, E.; Rastegari Mehr, M. Heavy metals (HMs) and polycyclic aromatic hydrocarbons (PAHs) in soils of different land uses in Erbil metropolis, Kurdistan Region, Iraq. *Environ. Monit. Assess.* **2016**, *188*, 605. [CrossRef] [PubMed]
73. Khan, S.; Munir, S.; Sajjad, M.; Li, G. Urban park soil contamination by potentially harmful elements and human health risk in Peshawar City, Khyber Pakhtunkhwa, Pakistan. *J. Geochem. Explor.* **2016**, *165*, 102–110. [CrossRef]
74. Chuan, M.; Shu, G.; Liu, J. Solubility of heavy metals in a contaminated soil: Effects of redox potential and pH. *Water Air Soil Pollut.* **1996**, *90*, 543–556. [CrossRef]
75. Li, J.; Rate, A.; Gilkes, R. Fractionation of trace elements in some non-agricultural Australian soils. *J. Soil Res.* **2003**, *41*, 1389–1402. [CrossRef]
76. Grigg, N.; Ahmad, S.; Podger, G.; Kirby, M.; Colloff, M. *Water Quality in the Ravi and Sutlej Rivers, Pakistan: A System View*; South Asia Sustainable Development Investment Portfolio CSIRO: Canberra, Australia, 2018.
77. Chen, T.-B.; Zheng, Y.-M.; Lei, M.; Huang, Z.-C.; Wu, H.-T.; Chen, H.; Fan, K.-K.; Yu, K.; Wu, X.; Tian, Q.-Z. Assessment of heavy metal pollution in surface soils of urban parks in Beijing, China. *Chemosphere* **2005**, *60*, 542–551. [CrossRef]
78. Varol, M.; Tokatlı, C. Seasonal variations of toxic metal (loid)s in groundwater collected from an intensive agricultural area in northwestern Turkey and associated health risk assessment. *Environ. Res.* **2022**, *204*, 111922. [CrossRef]
79. Shi, H.; Zeng, M.; Peng, H.; Huang, C.; Sun, H.; Hou, Q.; Pi, P. Health Risk Assessment of Heavy Metals in Groundwater of Hainan Island Using the Monte Carlo Simulation Coupled with the APCS/MLR Model. *Int. J. Environ. Res. Public Health* **2022**, *19*, 7827. [CrossRef] [PubMed]
80. Ahmad, I.; Khan, B.; Asad, N.; Mian, I.; Jamil, M. Traffic-related lead pollution in roadside soils and plants in Khyber Pakhtunkhwa, Pakistan: Implications for human health. *Int. J. Environ. Sci. Technol.* **2019**, *16*, 8015–8022. [CrossRef]
81. Wang, J.; Li, S.; Cui, X.; Li, H.; Qian, X.; Wang, C.; Sun, Y. Bioaccessibility, sources and health risk assessment of trace metals in urban park dust in Nanjing, Southeast China. *Ecotoxicol. Environ. Saf.* **2016**, *128*, 161–170. [CrossRef] [PubMed]
82. Kravchenko, J.; Lyerly, H.K. The impact of coal-powered electrical plants and coal ash impoundments on the health of residential communities. *N. C. Med. J.* **2018**, *79*, 289–300. [CrossRef] [PubMed]

83. Mitchell, R.G.; Spliethoff, H.M.; Ribaudo, L.N.; Lopp, D.M.; Shayler, H.A.; Marquez-Bravo, L.G.; Lambert, V.T.; Ferenz, G.S.; Russell-Anelli, J.M.; Stone, E.B. Lead (Pb) and other metals in New York City community garden soils: Factors influencing contaminant distributions. *Environ. Pollut.* **2014**, *187*, 162–169. [CrossRef]
84. Figueiredo, A.M.G.; Tocchini, M.; dos Santos, T.F. Metals in playground soils of Sao Paulo city, Brazil. *Procedia Environ. Sci.* **2011**, *4*, 303–309. [CrossRef]

Disclaimer/Publisher's Note: The statements, opinions and data contained in all publications are solely those of the individual author(s) and contributor(s) and not of MDPI and/or the editor(s). MDPI and/or the editor(s) disclaim responsibility for any injury to people or property resulting from any ideas, methods, instructions or products referred to in the content.

Article

Trace Metals in Rice Grains and Their Associated Health Risks from Conventional and Non-Conventional Rice Growing Areas in Punjab-Pakistan

Nukshab Zeeshan [1], Zia Ur Rahman Farooqi [1], Iftikhar Ahmad [2,*], Ghulam Murtaza [1], Aftab Jamal [3,*], Saifullah [1], Ayesha Abdul Qadir [1] and Emanuele Radicetti [4]

[1] Institute of Soil and Environmental Sciences, University of Agriculture, Faisalabad 38040, Pakistan
[2] Department of Environmental Sciences, COMSATS University Islamabad, Vehari Campus, Vehari 61100, Pakistan
[3] Department of Soil and Environmental Sciences, Faculty of Crop Production Sciences, The University of Agriculture, Peshawar 25130, Pakistan
[4] Department of Chemical, Pharmaceutical and Agricultural Sciences, University of Ferrara, Via Luigi Borsari n. 46, 44121 Ferrara, Italy
* Correspondence: iftikharahmad@cuivehari.edu.pk (I.A.); aftabses98@gmail.com (A.J.)

Abstract: Rice (*Oryza sativa* L.) is cultivated and consumed worldwide, but the contamination of rice grains with trace metals (TMs) could cause adverse impacts on human health. The aims of this study were to determine the concentrations of TMs in different rice varieties available for sale in local markets and to determine whether consumers are likely to be at risk via the consumption of these rice cultivars. For this purpose, samples of rice grains were collected from 12 rice growing districts (administrative units) in Punjab, Pakistan. These districts were further classified based on rice growing methods due to specific soil type. In conventional districts, the puddling method was used, while direct seeding was used for rice cultivation in non-conventional districts. The samples were collected and analyzed for the determination of essential (Cu, Fe, Zn, and Mn) and non-essential (Cd, Ni, and Pb) TMs using an atomic absorption spectrophotometer (AAS). The results showed that the maximum respective concentrations of Cd, Ni, and Pb (0.54, 0.05, 1.10 mg kg^{-1}) were found in rice grains in conventional areas, whereas values of 0.47, 0.20, and 1.20 mg kg^{-1} were found in non-conventional rice growing areas. The maximum concentrations of essential TMs (Cu, Fe, Mn, and Zn) were 4.54, 66.01, 4.82, and 21.51 mg kg^{-1} in conventional areas and 3.76, 74.11, 5.66, 19.63 mg kg^{-1} in non-conventional areas. In the conventional rice growing areas, Fe and Zn concentrations exceeded the permissible limits in the 27 and 7% samples, respectively. In the non-conventional rice areas, the concentrations of Cu, Fe, and Mn exceeded the permissible limits in the 15, 26, and 3% samples, respectively, while its Zn concentration was found within the permissible limits. The estimated weekly intake (EWI) and maximum tolerable dietary intake (MTDI) values for all studied metals were found within the permissible values set by WHO, except for Fe, in both sampled areas. It was concluded that no health risks were associated by utilizing the rice grains. However, the mean values of TMs were found considerably higher in collected rice samples from non-conventional areas than the conventional areas. Therefore, the concentrations of TMs should be monitored properly.

Keywords: fine and coarse rice; food chain; health risks; metals contamination

1. Introduction

Trace metals (TMs) pollution has become a serious problem worldwide. There is a continuous buildup of TMs in agricultural soils through the application of solid waste and low quality irrigation water [1,2]. The continued use of raw or partially treated wastewater for growing fodder and cereals has resulted in the contamination of food crops, and thus their entry into the food chain [3,4]. The TMs in soils could contaminate the environment and damage human health through various exposure pathways, including direct ingestion, dermal contact, and inhalation [5]. TMs concentration in edible parts of several crops such as spinach, clover, grape vines, shrubs, barley, and wheat have been reported by various reports [6–8]; thus, the risk to humans of excessive TMs in edible parts of food crops should be a matter of concern. The crops absorb these metals from contaminated soils through their roots or from surface-deposited particles [9]. As TMs in crops enter the human body through the food chain and present health risks to humans, food consumption is also an important exposure pathway [10]. Vitamin C, Fe, and other nutrients stored in the body are significantly decreased if humans consume food contaminated by these TMs, leading to a decline in immunity, a deterioration of human function, and disabilities associated with malnutrition [9,11].

As rice is fast growing crop and produces large biomasses, several studies have indicated that rice grown in TM-contaminated soils have higher concentrations of metals than those grown in un-contaminated soils [12,13]. In un-contaminated soils, a major pathway of soil contamination is through the atmospheric deposition of TMs from point sources such as mining, smelting, and industrial activities [14]. Other sources of agricultural soil contamination include agricultural inputs such as fertilizers, pesticides, sewage sludge, organic manures, and composts [15,16]. Therefore, some eco-friendly alternatives could be applied for clean production and sustainable cultivation. The sustainable production of quality food for mankind is a big challenge of the era. Biofertilizers are being used to not only improve the yield but also improve the nutrition status of crop plants. Scientists are using nitrogen-fixing bacteria along with phosphorus-fixing bacteria and getting good results [17]. The use of biofertilizers and compost instead of chemical fertilizers improves soil health and decreases TM contamination in soil, improving plant quality [18]. The balanced use of fertilizer is very effective technique in producing crops and fulfilling their deficiencies [19]. Rice is a major staple food in many countries, particularly in Asian countries such as Bangladesh, India, Thailand, China, Pakistan, and Vietnam, where soil and groundwater pollution with high levels of TMs have been reported [20–23]. Increased levels of TMs in agricultural soils and their uptake in rice, vegetables, and other food crops have become a real health issue in this region [24]. A significant number of studies have focused on studying TM concentrations in Pakistani, Bangladeshi, Chinese, and Indian food items including vegetables [25–28]. The work of Wasim et al. [29] assessed the TM concentrations in rice, but sampling was performed only from one site, i.e., Karachi, Pakistan. Additionally, this study did not involve different rice growing districts and comparisons between conventional and non-conventional rice growing areas from Punjab, Pakistan. The conventional rice growing areas of Punjab are Sheikhupura, Gujranwala, Hafizabad, Sialkot, Narowal, and Gujranwala; rice is grown using the puddling method because of the specific soil type (clay loam), as clay loam soil puddling is easy due to the water retention capacity of soil. However, soil puddling enhances the possibility of TM availability to the plants, as the metals could be solubilized in the water and converted into the more available forms to the plants [30], i.e., zinc sulphide (ZnS) can be converted to cadmium sulphide (CdS) with the exchange of their ions during this practice. The non-conventional rice growing areas are Bahawalnagar, Chiniot, Jhang, Toba Tek Singh, Shorkot, and Mandi Bahauddin; rice is usually grown by using the direct seeded method because the soil is sandy loam. However, soil puddling is difficult in sandy loam because water retention and anaerobic conditions are not possible, moreover, TMs are leached down in these types of soils, thereby reducing the risks of TM contamination. Thus, there is a possibility that conventional rice growing areas could have more TM concentrations than the non-conventional growing areas. This study is based on the same hypothesis, i.e., TMs could behave differently due to the use of conventional and non-conventional rice growing techniques. Previous studies have not reported on this aspect along with the potential health risks to humans. Therefore, the objectives of the present study were to: (a) determine the concentrations of TMs in different rice varieties available for sale in local markets, produced from different conventional and non-conventional rice growing areas of

Punjab, Pakistan; and (b) whether consumers are likely to be at risk of TM exposure via the consumption of these rice cultivars.

2. Materials and Methods

2.1. Sampling

This experiment was conducted to find out the concentrations of different essential (Cu, Fe, Mn, and Zn) and non-essential (Cd, Ni, and Pb) TMs in different types of crop (rice) varieties available in markets from conventional and non-conventional rice growing areas of Punjab, Pakistan. Fifty-five (55) rice samples (in triplicate) from conventional and non-conventional rice growing districts (6 conventional: Sheikhupura, Gujranwala, Hafizabad, Sialkot, Narowal, and Gujranwala; 6 non-conventional: Bahawalnagar, Chiniot, Jhang, Toba Tek Singh, Shorkot, and Mandi Bahauddin) of Punjab, Pakistan were collected during the month of January 2015 (Figure 1, Table 1).

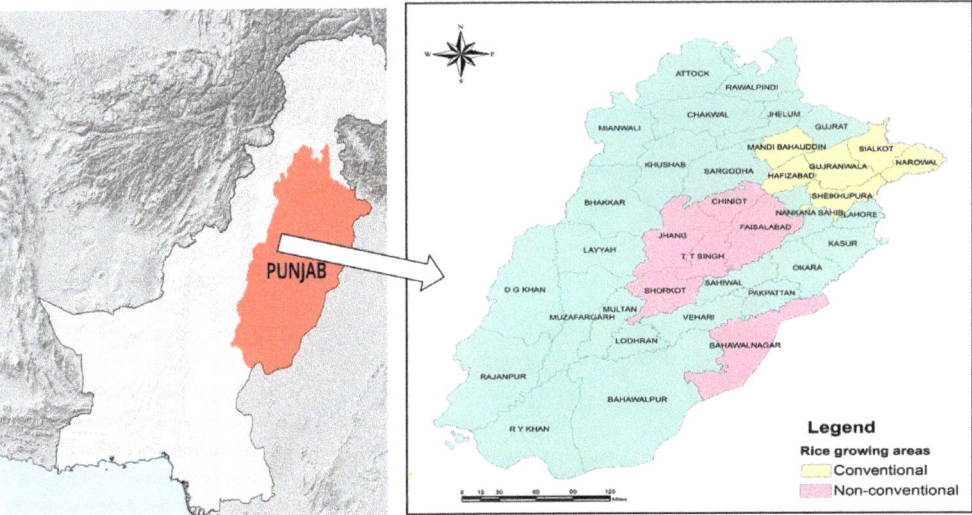

Figure 1. Base map showing sampling districts of Punjab, Pakistan.

Table 1. Coordinates of sampling sites.

Sr. No.	Sampling Location	Longitude	Latitude
1	Chiniot	31.710762	72.992123
2	Bahawalnagar	29.613738	73.137526
3	Jhang	30.843993	71.850886
4	Shorkot	30.844500	72.085357
5	Toba Tek Singh	30.970966	72.479624
6	Faisalabad	31.413924	73.074134
7	Narowal	32.096795	74.862740
8	Gujranwala,	31.982594	74.130777
9	Mandi Bahauddin	32.597835	73.476281
10	Hafizabad	32.052014	73.659907
11	Sialkot	32.386376	74.408410
12	Sheikhupura	31.715192	73.982211

The collected rice varieties were preferred to be cultivated by the farmers of the respective sampling districts. The collected samples details (sampling district, rice variety, and sample ID) are given in Supplementary Table S1. The collected samples were stored at 4 °C prior to further processing. All samples were washed with tap water (three times) followed by deionized (DI) water twice. The samples were then dried in open air for 24 h and in an oven at 65 ± 5 °C for 48 h. Then, the samples were homogenized by grinding with a ceramic mortar and pestle and processed for metals analysis.

2.2. Sample Preparation and Analysis

One gram of ground grain sample was taken in a 25 mL conical flask and kept overnight by adding 10 mL of digestion mixture (2:1 ratio of HNO_3 and $HClO_4$, respectively). The next day, samples were digested by placing on a hot plate at a temperature of 250 ± 5 °C until a clear solution was obtained. Following digestion, tubes were removed, and the samples were diluted to 20 mL with DI water after cooling. Prior to analysis, all samples were filtered through 0.45 μm filter paper (Whatman No. 41). A flame AAS (Solaar S-100, CiSA) was used for the determination of Cd, Cu, Fe, Mn, Ni, Pb, and Zn in rice samples (AOAC, 1990). Elemental concentrations in rice were determined on a dry weight basis.

2.3. Quality Control

The blank reagents and standard reference materials such as Batch 1701-3, BCR no. 150 and Fluka Kamica, Busch Switzerland were used to verify the accuracy and precision of the digestion procedure and subsequent analyses. Each sample batch was analyzed in triplicate under the standard operating conditions within the confidence limit of 95%. The validity of the method was further ascertained by cross method checks, spiked recovery, and replicate analysis.

2.4. Health Risk Assessment

The potential human health risk assessment was conducted by considering the following parameters according to Onsanit et al. [31]. The estimated weekly intake (EWI) and provisional tolerance weekly intake (PTWI) were jointly established by FAO/WHO [32]. The EWI (mg kg^{-1} body weight/week) was calculated using the following equation:

$$EWI = Crice \times \frac{WCrice}{BW}$$

where C rice = average trace metal concentrations in rice (mg kg^{-1} dry weight), WCrice = weekly rice consumption (g week^{-1}) per capita (18 kg/capita/year or ≈50 g per capita per day × 7) (PACRA, 2020), and BW = average body weight (kg) of the Pakistani population (72 kg). The values are expressed in mg week^{-1} person^{-1}. The calculated EWI values were compared with levels for typical daily exposure and provisional tolerable weekly intake (PTWI) set by the Codex committee on food additives and contaminants of the joint FAO/WHO food standards program for TM.

2.5. Statistical Analysis

XLSTAT 2018 and Origin 2018 were used for the descriptive statistics and further analysis. Data presented as means ± SD. The Pearson correlation heatmap with dendrograms was developed using Origin v2018 (Origin Lab Corp., Northampton, MA, USA).

3. Results

The mean and range of TM concentrations in rice grains from different conventional and non-conventional growing areas are presented in Supplementary Information (SI) Table S1. All the permissible limits set by WHO/FAO are presented in Table 2 for reference and comparison to our data presented in this paper. The weekly intake rates of essential and non-essential metals are presented in Table 3.

Table 2. Permissible limits of trace metals in rice grains.

Metals	Minimum Value		Maximum Value		Limits (mg kg−1)	References
	CA *	NCA **	CA *	NCA **		
Cd	0	0.10	0.54	0.47	0.2	WHO/FAO [29]
Pb	0.12	0.30	1.10	1.20	0.2	
Ni	0	0.01	0.05	0.20	1.0	
Fe	25.43	31.11	66.01	74.11	15–50	
Cu	1.13	1.01	4.54	3.76	4–15	
Zn	12.61	12.74	21.51	19.63	20	
Mn	1.05	1.13	4.82	5.66	05	

* Conventional Rice Growing Areas; ** Non-Convention Rice Growing Areas.

Table 3. EWI rate of essential and non-essential metals.

Rice Growing Area	Estimated Weekly Intakes (EWI)						
	Essential Metals				Non-Essential Metals		
	Cu	Fe	Mn	Zn	Cd	Ni	Pb
Non-conventional	18.27	360.25	27.51	95.42	2.28	0.97	5.83
Conventional	22	321	23.43	105	2.65	0.24	5.34
EWI by WHO	70 *	315 *	77 *	280 *	7	35	25

EWI: estimated weekly intake for toxic elements, based on WHO guidelines. * There is no PTWI set for Cu, Fe, Mn, and Zn, but it is used as MTDI in mg kg^{-1} body weight in a week).

3.1. Trace Metal Concentration in Rice Grains from Non-Conventional Growing Areas

3.1.1. Essential Metals

There were six non-conventional rice growing areas (Bahawalnagar, Chiniot, Jhang, Shorkot, Toba Tek Singh and Faisalabad), from which we collected rice samples. From non-conventional rice-producing areas, the highest Cu, Fe, Mn, and Zn concentrations in the rice samples were recorded having values of 3.76, 74.11, 5.66, and 19.63 mg kg^{-1}, respectively. The average Cu concentrations were 2.16, 1.57, 1.79, 2.16, 2.93, and 3.25 mg kg^{-1} in Bahawalnagar, Chiniot, Jhang, Shorkot, Toba Tek Singh, and Faisalabad, respectively. Iron (56.38, 40.45, 38.58, 33.90, 36.65, and 50.10 mg kg^{-1}), Mn (2.00, 2.46, 3.32, 2.41, 3.54, and 3.84 mg kg^{-1}), and Zn (16.75, 15.34, 16.72, 16.66, 17.54, and 16.35 mg kg^{-1}) concentrations were recorded, respectively (Figure 2).

3.1.2. Non-Essential Metals

From non-conventional rice-producing areas, the highest Cd, Ni, and Pb concentrations in collected rice samples were recorded having values of 0.47, 0.20, and 1.20 mg kg^{-1}, respectively. The average Cd concentrations were recorded as 0.23, 0.20, 0.31, 0.28, 0.31, and 0.24 mg kg^{-1} in Bahawalnagar, Chiniot, Jhang, Shorkot, Toba Tek Singh, and Faisalabad, respectively. Average Ni (0.08, 0.06, 0.02, 0.04, 0.01, and 0.01 mg kg^{-1}) and Pb (0.59, 0.88, 0.84, 0.98, 0.92, and 0.87 mg kg^{-1}) concentrations were found in the same order of the rice growing areas, respectively (Figure 3).

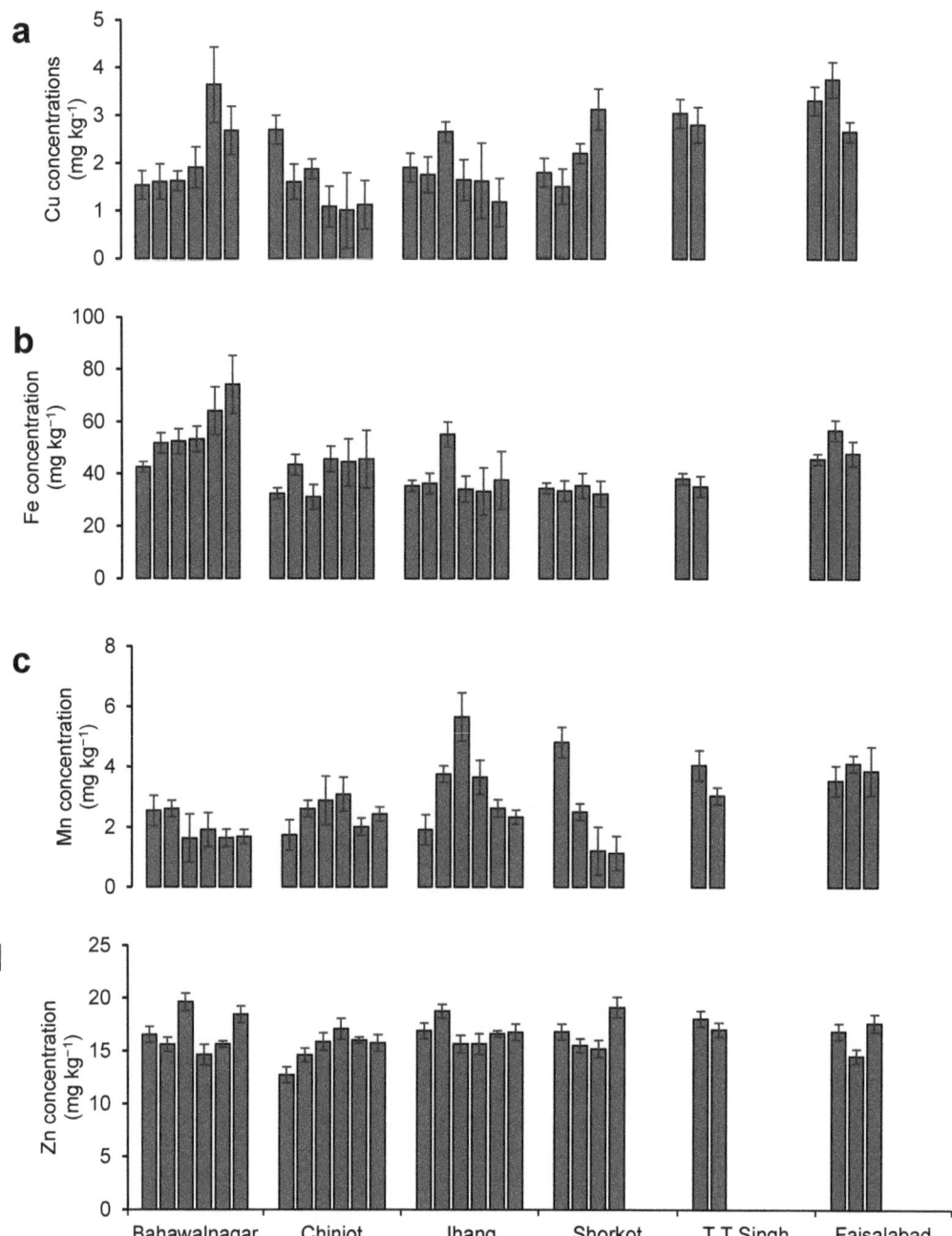

Figure 2. Concentrations of (**a**) Cu, (**b**) Fe, (**c**) Mn, and (**d**) Zn in rice grain samples from non-conventional rice growing districts; Bahawalnagar, Chiniot, Jhang, Shorkot, Toba Tek Singh (T.T. Singh), and Faisalabad, respectively; No. of bars show no. of samples taken from each district, and height shows metals' concentrations; error bars show standard deviation.

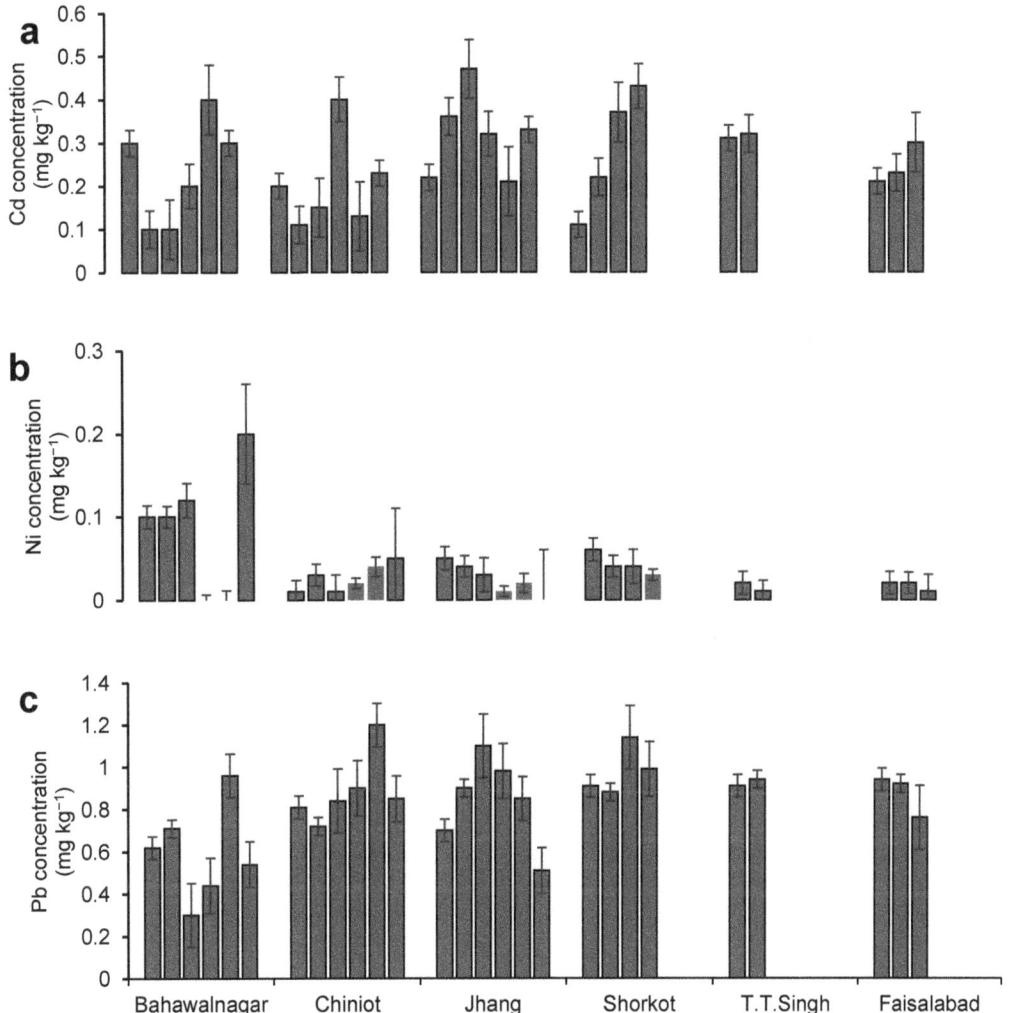

Figure 3. Concentrations of (**a**) Cd, (**b**) Ni, and (**c**) Pb in rice grain samples from non-conventional rice growing districts; Bahawalnagar, Chiniot, Jhang, Shorkot, Toba Tek Singh (T.T. Singh), and Faisalabad, respectively; No. of bars show no. of samples taken from each district, and height shows metals' concentrations; error bars show standard deviation.

3.2. Trace Metal Concentration in Rice Grains from Conventional Rice Growing Areas

3.2.1. Essential Metals

From conventional rice-producing areas, the highest Cu, Fe, Mn, and Zn concentrations in collected rice samples were recorded as 4.54, 66.01, 4.82, and 21.51 mg kg^{-1}, respectively. The average concentrations of Cu were found as 2.12, 2.89, 2.74, 1.97, 3.62, and 2.21 mg kg^{-1} in Narowal, Gujranwala, Hafizabad, Mandi Bahauddin, Sialkot, and Sheikhupura, respectively. Iron (42.75, 48.31, 34.00, 44.70, 45.19 and 45.81 mg kg^{-1}), Mn (2.53, 2.89, 1.99, 2.76, 4.37, and 3.89 mg kg^{-1}), and Zn (18.69, 14.64, 14.74, 16.51, 15.87 and 16.59 mg kg^{-1}) concentrations were found in the same order of the conventional rice growing areas, respectively (Figure 4).

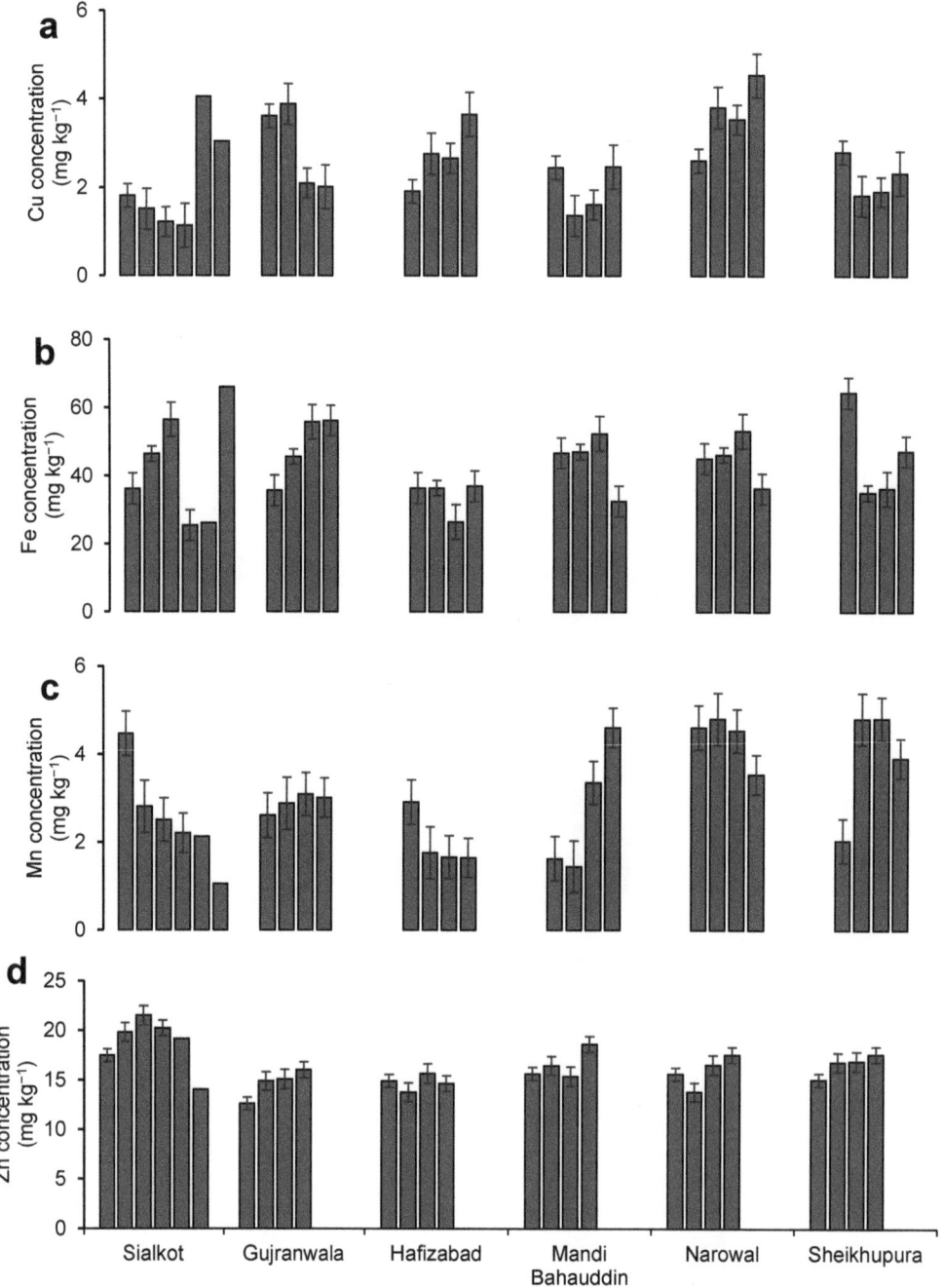

Figure 4. Concentrations of (**a**) Cu, (**b**) Fe, (**c**) Mn, and (**d**) Zn, in rice grain samples from conventional rice growing districts; Sialkot, Gujranwala, Hafizabad, Mandi Bahauddin, Narowal, and Sheikhupura, respectively; No. of bars show no. of samples taken from each district, and height shows metals' concentrations; error bars show standard deviation.

3.2.2. Non-Essential Metals

From conventional rice-producing areas, the highest Cd, Ni and Pb concentrations in the collected rice samples were recorded as 0.54, 0.05, and 1.10 mg kg^{-1}, respectively. The average concentrations of Cd were found as 0.37, 0.31, 0.09, 0.29, 0.30, and 0.36 mg kg^{-1} in Narowal, Gujranwala, Hafizabad, Mandi Bahauddin, Sialkot, and Sheikhupura, respectively. Nickel (0.02, 0.01, 0.01, 0.02, 0.01, and 0.02 mg kg^{-1}) and Pb (0.85, 0.82, 0.80, 0.66, 0.78 and 0.81 mg kg^{-1}) concentrations were found in the same order of the conventional rice growing areas, respectively (Figure 5).

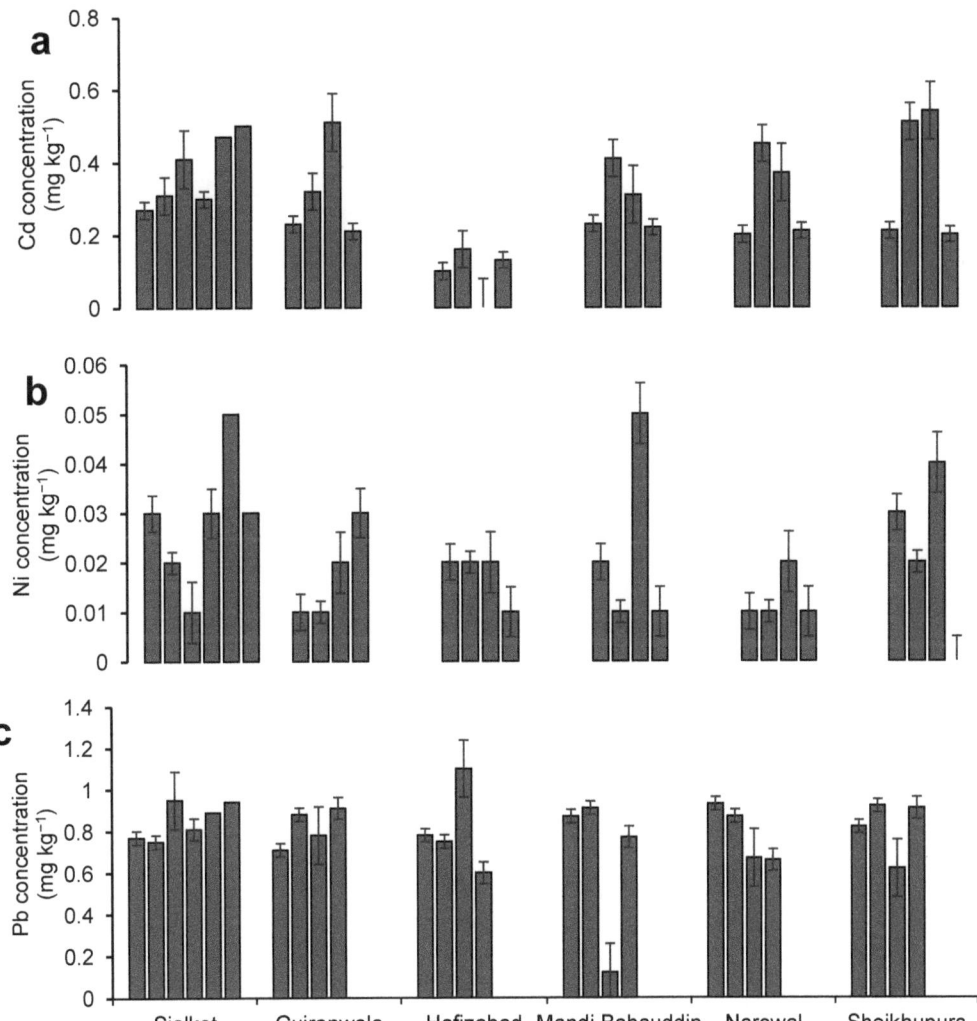

Figure 5. Concentrations of (**a**) Cd, (**b**) Ni, and (**c**) Pb in rice grain samples from conventional rice growing districts; Sialkot, Gujranwala, Hafizabad, Mandi Bahauddin, Narowal, and Sheikhupura, respectively; No. of bars show no. of samples taken from each district, and height shows metals' concentrations; error bars show standard deviation.

3.3. Relation between the Toxic Metals

Table 4 shows the Pearson correlation (R) between essential and non-essential toxic metals found in rice grains from conventional and non-conventional rice growing areas. At both places, the TMs displayed varied responses. In non-conventional areas, Cd had a significant positive effect with Zn ($R^2 = 0.22$, $p < 0.01$) and Cu ($R^2 = 0.34$, $p < 0.01$), whereas Ni had positive relation with Fe ($R^2 = 0.47$, $p < 0.05$), as well as Cd with Pb ($R^2 = 0.35$, $p < 0.05$). In the conventional areas, although the R^2 showed slight effect, the Zn had a significant positive effect on Mn ($R^2 = 0.14$, $p < 0.01$), Cd had positive relationship with Fe ($R^2 = 0.27$, $p < 0.01$), Mn ($R^2 = 0.23$, $p < 0.01$) and Zn ($R^2 = 0.18$, $p < 0.01$), while Ni also had positive relation with Cd ($R^2 = 0.30$, $p < 0.01$).

Table 4. Correlation matrix of metals in rice grains from conventional and non-conventional growing areas.

Metals	Essential Metals				Non-Essential Metals		
	Cu	Fe	Mn	Zn	Cd	Ni	Pb
Metal concentrations in rice grains from non-conventional growing areas							
Cu	1.00						
Fe	0.30	1.00					
Mn	0.14	−0.02	1.00				
Zn	0.00	0.05	0.00	1.00			
Cd	0.34 *	0.08	0.11	0.22 **	1.00		
Ni	−0.15	0.47 *	−0.22	0.40 *	−0.21	1.00	
Pb	0.23	−0.28	0.30	−0.21	0.35 *	−0.42	1.00
Metal concentrations in rice grains from conventional growing areas							
Cu	1.00						
Fe	−0.10	1.00					
Mn	−0.01	−0.11	1.00				
Zn	−0.42	−0.20	0.14 **	1.00			
Cd	−0.11	0.27 *	0.23 *	0.18 *	1.00		
Ni	−0.18	−0.02	−0.10	0.09	0.30 *	1.00	
Pb	0.01	−0.03	−0.19	0.12	−0.03	−0.38	1.00

* Values with single asterisk are statistically significant at $p < 0.00$; ** values with double asterisk are statistically significant at $p < 0.005$.

3.4. Principal Component Analysis and Heatmap

The first three components of the PCA accumulated 61.7% of the total variance (Figure 6). The first principal component explained 24.4% of the variance and reflected the negative coordination with Pb and positive with Ni. The second principal component explained 19.0% of the variance and reflected a covariation of Zn and Cd with loading factors of 0.75 and 0.67, respectively. The third principal component explained 18.3% of the variance and was linked to Fe and Cu (loading factor 0.78 and 0.56). The biplot shows the overlapping of the confidence eclipse for conventional and non-conventional rice growing areas that means the differences in metal concentrations, with respect to areas, were non-significant. The points near the lines originating from the center depicted higher values as compared to distant points. The heat map (Figure 7) gives the clearer grouping of metals on the basis of the Pearson's correlation. It can be seen from Figure 5 that Pb and Cu; Mn, Cd, and Zn; and Ni and Fe were grouped together in three clusters.

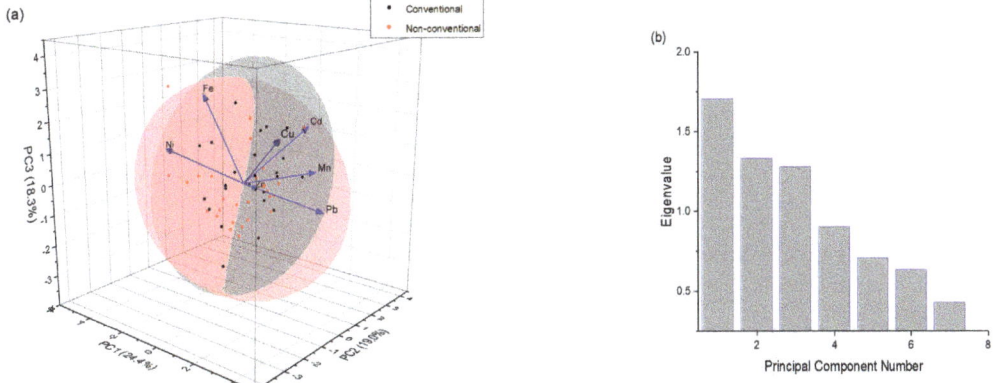

Figure 6. Principal component analysis of studied metals analyzed in rice varieties grown at conventional and non-conventional rice growing areas of Punjab Pakistan. (**a**) Biplot between PC1, PC2, and PC3. The blue lines originating from central point of biplot indicate positive or negative correlations of different metals. The black and red dots represent the tested collected rice samples from conventional and non-conventional areas, respectively. Gray and pink circles represent the 95% confidence eclipse for conventional and non-conventional areas. (**b**) Variance decomposition of principal components.

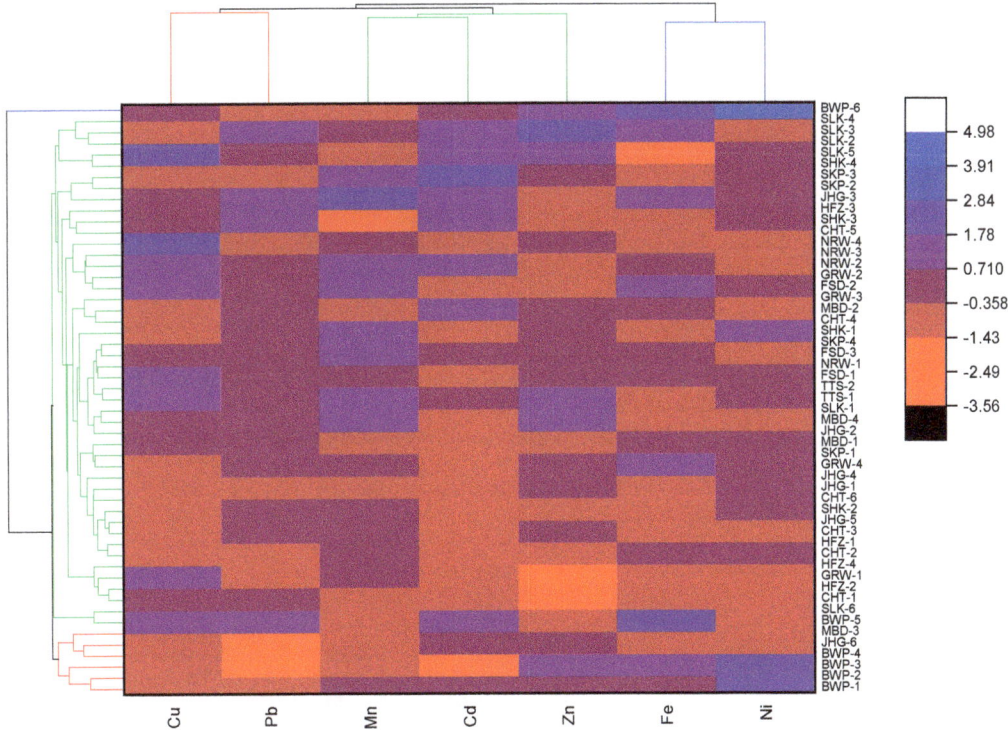

Figure 7. The Heatmap based on Pearson's correlation.

4. Discussion

4.1. Sources of Metals in Rice Growing Areas of Pakistan

Agricultural soil contaminations with TMs have gradually increased in the last decade due to urban sprawl, population increase, and industrialization [33,34]. It is considered a major global environmental issue because of their persistence and associated risks to both the biotic and abiotic factors of the environment. Trace metals such as Cd, Cu, Fe, Mn, Ni, Pb, and Zn are naturally present in soil through rocks weathering and volcanic eruptions, but their concentrations may be locally elevated as a result of nearby anthropogenic activities such as mining, resource extraction, industrial and vehicular emissions, etc. [35,36]. Additionally, soil preparation, crop growing, and management practices also greatly influence the TM concentration in rice growing areas. In conventional rice growing areas, puddling is done before the rice nursery transplanting. Puddling could increase the bioavailability of TMs to rice crop by enhancing their solubilization [30]. Higher concentrations of TMs in agricultural soils are responsible for their transfer from soil to food crops and grains, leading into long-term risks to the ecosystem. Vegetables, grains, and other essential foods may have a range of toxic and essential trace metals [37,38].

Additionally, TM pollution in rice growing areas is attributed to adjacent industrial sites and their effluent's use to irrigate agricultural soils in Pakistan, as indicated by Nawaz et al. [39], Azam et al. [40], and Qi et al. [41]. In addition to the use of industrial wastewater for irrigation, some farmers are over-using fertilizers and pesticides recommended for crop growth [42,43]. Phosphatic fertilizers are a rich source of Cd, and their repetitive application results in the magnification of Cd in soil and the edible parts of plants [44,45]. Moreover, submerged conditions during rice growth and development stages result in more mobility of TMs in soil solutions and their translocation to plants' vegetative and edible parts [46,47].

4.2. Non-Essential Metals in Rice

Cadmium is a metallic element that occurs naturally at low levels in the environment. Food, rather than air or water, represents the major source of Cd exposure [48,49]. The highest Cd concentration in the rice grains was found as 0.54 mg kg^{-1} (Supplementary Table S1). Cd concentrations in all the rice samples in the present study ranged from 0 to 0.47 mg kg^{-1}. The Cd concentration ranged from 0 to 0.54 in conventional areas, while it was 0.10–0.47 mg kg^{-1} in non-conventional sites. Thus, the mean Cd concentrations were considerably higher in non-conventional areas as compared to conventional areas. The Cd concentrations in rice samples from non-conventional areas were found higher than the WHO limits in Bahawalnagar and Chiniot (33%), Jhang (100%), 75% of the samples from Shorkot, and 100% samples from Toba Tek Singh and Faisalabad, respectively. These results are supported by the previous work of Dehghani and Mosaferi [50], who found higher Cd and Pb concentrations in imported Pakistani rice grains available in Irani markets and Cd, Cr, Ni, Fe, and Pb in Shakargarh Tehsil, District Sialkot, Pakistan [51]. This is probably due to fact that non-conventional areas receive relatively more contaminated irrigational water as compared to conventional areas. The conventional areas fall in upper Punjab, Pakistan, which has more slopes compared to the non-conventional areas (higher than sea level, due to which the water flow is high, minimizing the chances of water contamination) [52]. Secondly, conventional areas are not densely populated and are mainly agricultural lands [53], thus the risks of water contamination from sewage effluents are rare. In contradiction, the rice in non-conventional areas is mostly grown in the peri-urban areas and regularly receives sewage water for use in irrigation. The maximum concentrations of Cd in rice from conventional areas were found to be higher in 75% samples from Narowal, 100% from Gujranwala, 25% from Hafizabad, and 100% in Mandi Bahauddin, Sialkot, and Sheikhupura. Khan et al. [33] reported similar results of TM (Cd, Pb and Ni) contaminations in the Khyber Pakhtunkhwa Province of Pakistan. Nickel concentrations in rice grains from conventional areas of Punjab, Pakistan were 0–0.05 and 0.01–0.05 mg kg^{-1} in non-conventional areas. All the samples collected from conventional and non-conventional areas remained within the permissible limits (1.0 mg kg^{-1}) of the WHO. Lead ranged from 0.12 to 1.10 and 0.30 to 1.20 mg kg^{-1} in conventional and non-conventional areas, respectively. All the samples from non-conventional areas exceeded Pb concentration limits in rice prescribed by WHO (except one sample from Mandi Bahauddin), indicating a significantly higher Pb pollution level in the Punjab. The higher Pb levels could be due to the aerial deposition of Pb from vehicular emissions and their subsequent entry to rice grains through plant tissues and mixing in soil solution and uptake [54].

4.3. Essential Metals in Rice

Copper concentration in rice grains was found highest (4.54 mg kg^{-1}) in conventional areas (Figure 4). A previous study by Xu et al. [55] reported a mean Cu concentration in Chinese rice of up to 18 mg kg^{-1}, which is considerably higher than that found in the present study. Mean Mn concentrations in the conventional and non-conventional areas were 43.57 and 44.03 mg kg^{-1}, respectively. It was notable that Mn concentrations in the rice samples exceeded permissible limits. Roychowdhury [56] reported a mean Mn concentration of 9.9 mg kg^{-1} in rice in West Bengal, India. The mean Zn concentrations in rice in the present study did not differ significantly for different areas. The highest mean Zn concentration (16.42 mg kg^{-1}) in rice grains was found in the conventional area, which was comparable to that in the non-conventional areas' rice grains 16.41 mg kg^{-1}.

4.4. Health Risk Assessment

As peer reviewed data on all the parameters used for the health risk assessment of rice in this study are not available, we could not estimate the health risks associated with the consumption of essential metals (Cu, Fe, Mn, and Zn), but we calculated the health risks associated with consumption of non-essential metals (Cd, Pb, and Ni) through comparing PTWI values devised by the FAO/WHO. The same procedures were used by Naseri et al. [57] and Ghoreishy et al. [58] in Iran. Based on the results, it was recorded that the EWI of the samples from both areas fell within the permissible limits, except for Fe, whose EWI limits exceeded the MTDI in the conventional as well as non-conventional areas (Table 3).

5. Conclusions

In this study, most of the mean metal concentrations (Cd, Pb, Fe, Zn, and Mn) were found to be substantially higher in conventional rice growing areas, followed by non-conventional areas. Specifically, the mean Cd concentrations were considerably higher in rice grains collected from non-conventional areas compared to conventional rice growing areas. The Cd, Pb, and Ni concentrations in rice samples from non-conventional areas were found higher than the WHO limits in the 70, 100, and 0% collected samples, respectively. In conventional areas, Cd, Pb, and Ni concentrations were found higher than WHO limits in the 77, 99, and 0% samples, respectively. A health risk assessment indicated that all the EWI and/or MTDI values were found within the permissible values set by the WHO due to less rice consumption by Pakistani people, except for Fe in both areas.

Supplementary Materials: The following supporting information can be downloaded at: https://www.mdpi.com/article/10.3390/su15097259/s1, Table S1: title: Details of the sampling districts, IDs along with concentrations of essential and non-essential metals.

Author Contributions: Conceptualization, N.Z. and Z.U.R.F.; methodology, N.Z. and G.M.; software, S.; validation, A.A.Q. and S.; formal analysis, Z.U.R.F.; investigation, N.Z.; resources, I.A.; data curation, A.J.; writing—original draft preparation, N.Z.; writing—review and editing, A.J., I.A. and E.R.; visualization, A.J.; supervision, S.; project administration and funding acquisition, A.J. All authors have read and agreed to the published version of the manuscript.

Funding: This research received no external funding.

Institutional Review Board Statement: Not applicable.

Informed Consent Statement: Not applicable.

Data Availability Statement: The data presented in this study are available on request from the first corresponding author.

Acknowledgments: We are thankful to the laboratory staff of Institute of Soil and Environmental Sciences, University of Agriculture, Faisalabad (Pakistan) for assistance and technical support in this work.

Conflicts of Interest: The authors declare no conflict of interest.

References

1. Mehmood, A.; Mirza, M.A.; Choudhary, M.A.; Kim, K.-H.; Raza, W.; Raza, N.; Lee, S.S.; Zhang, M.; Lee, J.-H.; Sarfraz, M. Spatial distribution of heavy metals in crops in a wastewater irrigated zone and health risk assessment. *Environ. Res.* **2019**, *168*, 382–388. [CrossRef]
2. Nzediegwu, C.; Prasher, S.; Elsayed, E.; Dhiman, J.; Mawof, A.; Patel, R. Effect of biochar on heavy metal accumulation in potatoes from wastewater irrigation. *J. Environ. Manag.* **2019**, *232*, 153–164. [CrossRef] [PubMed]
3. Onakpa, M.M.; Njan, A.A.; Kalu, O.C. A review of heavy metal contamination of food crops in Nigeria. *Ann. Glob. Health* **2018**, *84*, 488. [CrossRef] [PubMed]
4. Sharma, S.; Nagpal, A.K.; Kaur, I. Heavy metal contamination in soil, food crops and associated health risks for residents of Ropar wetland, Punjab, India and its environs. *Food Chem.* **2018**, *255*, 15–22. [CrossRef] [PubMed]
5. Sarim, M.; Jan, T.; Khattak, S.A.; Mihoub, A.; Jamal, A.; Saeed, M.F.; Soltani-Gerdefaramarzi, S.; Tariq, S.R.; Fernández, M.P.; Mancinelli, R. Assessment of the Ecological and Health Risks of Potentially Toxic Metals in Agricultural Soils from the Drosh-Shishi Valley, Pakistan. *Land* **2022**, *11*, 1663. [CrossRef]
6. Ikram, M.; Ali, N.; Jan, G.; Jan, F.G.; Rahman, I.U.; Iqbal, A.; Hamayun, M. IAA producing fungal endophyte *Penicillium roqueforti* Thom., enhances stress tolerance and nutrients uptake in wheat plants grown on heavy metal contaminated soils. *PLoS ONE* **2018**, *13*, e0208150. [CrossRef]
7. Zhu, H.; Teng, Y.; Wang, X.; Zhao, L.; Ren, W.; Luo, Y.; Christie, P. Changes in clover rhizosphere microbial community and diazotrophs in mercury-contaminated soils. *Sci. Total Environ.* **2021**, *767*, 145473. [CrossRef] [PubMed]
8. Shilev, S.; Babrikova, I.; Babrikov, T. Consortium of plant growth—Promoting bacteria improves spinach (*Spinacea oleracea* L.) growth under heavy metal stress conditions. *J. Chem. Technol. Biotechnol.* **2020**, *95*, 932–939. [CrossRef]
9. Xiao, R.; Wang, S.; Li, R.; Wang, J.J.; Zhang, Z. Soil heavy metal contamination and health risks associated with artisanal gold mining in Tongguan, Shaanxi, China. *Ecotoxicol. Environ. Saf.* **2017**, *141*, 17–24. [CrossRef]
10. Adimalla, N.; Chen, J.; Qian, H. Spatial characteristics of heavy metal contamination and potential human health risk assessment of urban soils: A case study from an urban region of South India. *Ecotoxicol. Environ. Saf.* **2020**, *194*, 110406. [CrossRef]
11. Mishra, S.; Bharagava, R.N.; More, N.; Yadav, A.; Zainith, S.; Mani, S.; Chowdhary, P. Heavy metal contamination: An alarming threat to environment and human health. In *Environmental Biotechnology: For Sustainable Future*; Sobti, R., Arora, N., Kothari, R., Eds.; Springer: Singapore, 2019; pp. 103–125.
12. Du, F.; Yang, Z.; Liu, P.; Wang, L. Accumulation, translocation, and assessment of heavy metals in the soil-rice systems near a mine-impacted region. *Environ. Sci. Pollut. Res.* **2018**, *25*, 32221–32230. [CrossRef] [PubMed]
13. Wang, Z.; Qin, H.; Liu, X. Health risk assessment of heavy metals in the soil-water-rice system around the Xiazhuang uranium mine, China. *Environ. Sci. Pollut. Res.* **2019**, *26*, 5904–5912. [CrossRef]
14. Fan, Y.; Zhu, T.; Li, M.; He, J.; Huang, R. Heavy metal contamination in soil and brown rice and human health risk assessment near three mining areas in central China. *J. Health Eng.* **2017**, *2017*, 4124302. [CrossRef] [PubMed]
15. Barau, B.; Abdulhameed, A.; Ezra, A.; Muhammad, M.; Kyari, E.; Bawa, U. Heavy metal contamination of some vegetables from pesticides and the potential health risk in Bauchi, northern Nigeria. *Int. J. Sci. Technol.* **2018**, *7*, 1–11. [CrossRef]
16. Tian, D.; Li, Z.; O'Connor, D.; Shen, Z. The need to prioritize sustainable phosphate—Based fertilizers. *Soil Use Manag.* **2020**, *36*, 351–354. [CrossRef]
17. Kahil, A.; Hassan, F.; Ali, E. Influence of bio-fertilizers on growth, yield and anthocyanin content of *Hibiscus sabdariffa* L. plant under Taif region conditions. *Annu. Res. Rev. Biol.* **2017**, *17*, 1–15. [CrossRef]
18. Hassan, F.; Ali, E.; Mahfouz, S. Comparison between different fertilization sources, irrigation frequency and their combinations on the growth and yield of coriander plant. *Aust. J. Basic. Appl. Sci.* **2012**, *6*, 600–615.
19. Saboor, A.; Ali, M.A.; Ahmed, N.; Skalicky, M.; Danish, S.; Fahad, S.; Hassan, F.; Hassan, M.M.; Brestic, M.; El Sabagh, A. Biofertilizer-based zinc application enhances maize growth, gas exchange attributes, and yield in zinc-deficient soil. *Agriculture* **2021**, *11*, 310. [CrossRef]
20. Wagh, V.M.; Panaskar, D.B.; Mukate, S.V.; Gaikwad, S.K.; Muley, A.A.; Varade, A.M. Health risk assessment of heavy metal contamination in groundwater of Kadava River Basin, Nashik, India. *Model Earth Syst. Environ.* **2018**, *4*, 969–980. [CrossRef]
21. Chiamsathit, C.; Auttamana, S.; Thammarakcharoen, S. Heavy metal pollution index for assessment of seasonal groundwater supply quality in hillside area, Kalasin, Thailand. *Appl. Water Sci.* **2020**, *10*, 1–8. [CrossRef]

22. Jalees, M.I.; Farooq, M.U.; Anis, M.; Hussain, G.; Iqbal, A.; Saleem, S. Hydrochemistry modelling: Evaluation of groundwater quality deterioration due to anthropogenic activities in Lahore, Pakistan. *Environ. Dev. Sustain.* **2021**, *23*, 3062–3076. [CrossRef]
23. Qiao, J.; Zhu, Y.; Jia, X.; Niu, X.; Liu, J. Distributions of arsenic and other heavy metals, and health risk assessments for groundwater in the Guanzhong Plain region of China. *Environ. Res.* **2020**, *181*, 108957. [CrossRef] [PubMed]
24. Pateriya, A.; Verma, R.K.; Sankhla, M.S.; Kumar, R. Heavy metal toxicity in rice and its effects on human health. *Lett. Appl. NanoBio. Sci.* **2020**, *10*, 1833–1845.
25. Sultana, M.S.; Rana, S.; Yamazaki, S.; Aono, T.; Yoshida, S. Health risk assessment for carcinogenic and non-carcinogenic heavy metal exposures from vegetables and fruits of Bangladesh. *Cogent Environ. Sci.* **2017**, *3*, 1291107. [CrossRef]
26. Bi, C.; Zhou, Y.; Chen, Z.; Jia, J.; Bao, X. Heavy metals and lead isotopes in soils, road dust and leafy vegetables and health risks via vegetable consumption in the industrial areas of Shanghai, China. *Sci. Total Environ.* **2018**, *619*, 1349–1357. [CrossRef]
27. Rehman, Z.U.; Sardar, K.; Shah, M.T.; Brusseau, M.L.; Khan, S.A.; Mainhagu, J. Transfer of heavy metals from soils to vegetables and associated human health risks at selected sites in Pakistan. *Pedosphere* **2018**, *28*, 666–679. [CrossRef]
28. Gupta, N.; Yadav, K.K.; Kumar, V.; Krishnan, S.; Kumar, S.; Nejad, Z.D.; Khan, M.M.; Alam, J. Evaluating heavy metals contamination in soil and vegetables in the region of North India: Levels, transfer and potential human health risk analysis. *Environ. Toxicol. Pharm.* **2021**, *82*, 103563. [CrossRef]
29. Wasim, A.A.; Naz, S.; Khan, M.N.; Fazalurrehman, S. Assessment of heavy metals in rice using atomic absorption spectrophotometry–A study of different rice varieties in Pakistan. *Pak. J. Anal. Environ.* **2019**, *20*, 67–74. [CrossRef]
30. Huang, B.; Li, Z.; Li, D.; Yuan, Z.; Chen, Z.; Huang, J. Distribution characteristics of heavy metal (loid) s in aggregates of different size fractions along contaminated paddy soil profile. *Environ. Sci. Pollut. Res.* **2017**, *24*, 23939–23952. [CrossRef]
31. Onsanit, S.; Ke, C.; Wang, X.; Wang, K.-J.; Wang, W.-X. Trace elements in two marine fish cultured in fish cages in Fujian province, China. *Environ. Pollut.* **2010**, *158*, 1334–1342. [CrossRef]
32. Food and Agriculture Organization of the United Nations (FAO); World Health Organization (WHO). *Principles and Methods for the Risk Assessment of Chemicals in Food; Chapter 6 Dietary Exposure Assessment of Chemicals in Food*; FAO: Rome, Italy; WHO: Geneva, Switzerland, 2009; Volume 68.
33. Khan, A.Z.; Khan, S.; Khan, M.A.; Alam, M.; Ayaz, T. Biochar reduced the uptake of toxic heavy metals and their associated health risk via rice (*Oryza sativa* L.) grown in Cr-Mn mine contaminated soils. *Environ. Technol. Innov.* **2020**, *17*, 100590. [CrossRef]
34. Tran, L.D.; Phung, L.D.; Pham, D.V.; Pham, D.D.; Nishiyama, M.; Sasaki, A.; Watanabe, T. High yield and nutritional quality of rice for animal feed achieved by continuous irrigation with treated municipal wastewater. *Paddy Water Environ.* **2019**, *17*, 507–513. [CrossRef]
35. Hussain, A.; Alamzeb, S.; Begum, S. Accumulation of heavy metals in edible parts of vegetables irrigated with waste water and their daily intake to adults and children, District Mardan, Pakistan. *Food Chem.* **2013**, *136*, 1515–1523.
36. Karim, Z.; Qureshi, B.A. Health risk assessment of heavy metals in urban soil of Karachi, Pakistan. *Hum. Ecol. Risk Assess* **2014**, *20*, 658–667. [CrossRef]
37. Mahmood, A.; Malik, R.N. Human health risk assessment of heavy metals via consumption of contaminated vegetables collected from different irrigation sources in Lahore, Pakistan. *Arab. J. Chem.* **2014**, *7*, 91–99. [CrossRef]
38. Ullah, H.; Noreen, S.; Rehman, A.; Waseem, A.; Zubair, S.; Adnan, M.; Ahmad, I. Comparative study of heavy metals content in cosmetic products of different countries marketed in Khyber Pakhtunkhwa, Pakistan. *Arab. J. Chem.* **2017**, *10*, 10–18. [CrossRef]
39. Nawaz, A.; Khurshid, K.; Arif, M.S.; Ranjha, A. Accumulation of heavy metals in soil and rice plant (*Oryza sativa* L.) irrigated with industrial effluents. *Int. J. Agric. Biol.* **2006**, *8*, 391–393.
40. Azam, I.; Afsheen, S.; Zia, A.; Javed, M.; Saeed, R.; Sarwar, M.K.; Munir, B. Evaluating insects as bioindicators of heavy metal contamination and accumulation near industrial area of Gujrat, Pakistan. *Biomed. Res. Int.* **2015**, *2015*, 942751. [CrossRef]
41. Qi, D.; Yan, J.; Zhu, J. Effect of a reduced fertilizer rate on the water quality of paddy fields and rice yields under fishpond effluent irrigation. *Agric. Water Manag.* **2020**, *231*, 105999. [CrossRef]
42. Khan, K.; Lu, Y.; Khan, H.; Ishtiaq, M.; Khan, S.; Waqas, M.; Wei, L.; Wang, T. Heavy metals in agricultural soils and crops and their health risks in Swat District, northern Pakistan. *Food Chem. Toxicol.* **2013**, *58*, 449–458. [CrossRef] [PubMed]
43. Tariq, S.R.; Shafiq, M.; Chotana, G.A. Distribution of heavy metals in the soils associated with the commonly used pesticides in cotton fields. *Scientifica* **2016**, *2016*, 7575239. [CrossRef] [PubMed]
44. Park, H.J.; Kim, S.U.; Jung, K.Y.; Lee, S.; Choi, Y.D.; Owens, V.N.; Kumar, S.; Yun, S.W.; Hong, C.O. Cadmium phytoavailability from 1976 through 2016: Changes in soil amended with phosphate fertilizer and compost. *Sci. Total Environ.* **2021**, *762*, 143132. [CrossRef] [PubMed]
45. Murtaza, G.; Javed, W.; Hussain, A.; Wahid, A.; Murtaza, B.; Owens, G. Metal uptake via phosphate fertilizer and city sewage in cereal and legume crops in Pakistan. *Environ. Sci. Pollut. Res.* **2015**, *22*, 9136–9147. [CrossRef]
46. Jho, E.H.; Youn, Y.; Yun, S.H. Effect of CO2 exposure on the mobility of heavy metals in submerged soils. *Appl. Biol. Chem.* **2018**, *61*, 617–623. [CrossRef]
47. Wan, Y.; Huang, Q.; Camara, A.Y.; Wang, Q.; Li, H. Water management impacts on the solubility of Cd, Pb, As, and Cr and their uptake by rice in two contaminated paddy soils. *Chemosphere* **2019**, *228*, 360–369. [CrossRef] [PubMed]
48. Dennis, K.K.; Judd, S.E.; Alvarez, J.A.; Kahe, K.; Jones, D.P.; Hartman, T.J. Plant food intake is associated with lower cadmium body burden in middle-aged adults. *Eur. J. Nutr.* **2021**, *60*, 3365–3374. [CrossRef]

49. Li, N.; Feng, A.; Liu, N.; Jiang, Z.; Wei, S. Silicon application improved the yield and nutritional quality while reduced cadmium concentration in rice. *Environ. Sci. Pollut. Res.* **2020**, *27*, 20370–20379. [CrossRef]
50. Dehghani, M.; Mosaferi, F. Determination of heavy metals (cadmium, arsenic and lead) in Iranian, Pakistani and Indian rice consumed in Hormozgan Province, Iran. *J. Maz. Univ. Med.* **2016**, *25*, 363–367.
51. Tariq, S.R.; Rashid, N. Multivariate analysis of metal levels in paddy soil, rice plants, and rice grains: A case study from Shakargarh, Pakistan. *J. Chem.* **2013**, *2013*, 539251. [CrossRef]
52. Flood Map Pro. Pakistan Flood Map: Elevation Map, Sea Level Rise Map. 2020. Available online: https://www.floodmap.net/?ct=PK (accessed on 18 August 2021).
53. World Bank. Population, Total-Pakistan. 2021. Data. Available online: https://data.worldbank.org/indicator/SP.POP.TOTL?end=2020&locations=PK&start=1960&view=chart (accessed on 5 August 2022).
54. Zeng, F.; Wei, W.; Li, M.; Huang, R.; Yang, F.; Duan, Y. Heavy metal contamination in rice-producing soils of Hunan province, China and potential health risks. *Int. J. Environ. Res.* **2015**, *12*, 15584–15593. [CrossRef]
55. Xu, J.; Yang, L.; Wang, Z.; Dong, G.; Huang, J.; Wang, Y. Toxicity of copper on rice growth and accumulation of copper in rice grain in copper contaminated soil. *Chemosphere* **2006**, *62*, 602–607. [CrossRef] [PubMed]
56. Roychowdhury, T. Impact of sedimentary arsenic through irrigated groundwater on soil, plant, crops and human continuum from Bengal delta: Special reference to raw and cooked rice. *Food Chem. Toxicol.* **2008**, *46*, 2856–2864. [CrossRef] [PubMed]
57. Naseri, M.; Vazirzadeh, A.; Kazemi, R.; Zaheri, F. Concentration of some heavy metals in rice types available in Shiraz market and human health risk assessment. *Food Chem.* **2015**, *175*, 243–248. [CrossRef] [PubMed]
58. Ghoreishy, F.; Salehi, M.; Fallahzade, J. Cadmium and lead in rice grains and wheat breads in Isfahan (Iran) and human health risk assessment. *Hum. Ecol. Risk Assess* **2019**, *25*, 924–934. [CrossRef]

Disclaimer/Publisher's Note: The statements, opinions and data contained in all publications are solely those of the individual author(s) and contributor(s) and not of MDPI and/or the editor(s). MDPI and/or the editor(s) disclaim responsibility for any injury to people or property resulting from any ideas, methods, instructions or products referred to in the content.

Review

Potentially Toxic Elements in Pharmaceutical Industrial Effluents: A Review on Risk Assessment, Treatment, and Management for Human Health

Hussein K. Okoro [1,*], Muyiwa M. Orosun [2,*], Faith A. Oriade [1], Tawakalit M. Momoh-Salami [3], Clement O. Ogunkunle [4], Adewale G. Adeniyi [5], Caliphs Zvinowanda [6] and Jane C. Ngila [6]

1. Environmental-Analytical & Material Research Group, Department of Industrial Chemistry, Faculty of Physical Sciences, University of Ilorin, P.M.B. 1515, Ilorin 240003, Nigeria
2. Department of Physics, Faculty of Physical Sciences, University of Ilorin, P.M.B. 1515, Ilorin 240003, Nigeria
3. Department of Microbiology, Biological Sciences Unit, College of Basic Science, Lagos State University of Science and Technology, Ikorodu P.O. Box 249, Nigeria
4. Department of Plant Biology, Environmental Botany Unit, University of Ilorin, P.M.B. 1515, Ilorin 240003, Nigeria
5. Department of Chemical Engineering, Faculty of Engineering, University of Ilorin, P.M.B. 1515, Ilorin 240003, Nigeria
6. Analytical-Environmental and Membrane Nanotechnology Research Group, Department of Chemical Sciences, University of Johannesburg, P.O. Box 17011, Doornfontein, Johannesburg 2028, South Africa
* Correspondence: okoro.hk@unilorin.edu.ng (H.K.O.); orosun.mm@unilorin.edu.ng (M.M.O.)

Abstract: Potentially toxic elements (PTEs) are metallic chemicals with densities that are higher than that of water. Water pollution by PTEs due to the discharge of untreated pharmaceutical industrial effluents is a risk to human health and ecological integrity. The present review paper provides an overview of the threats to human health due to water contamination by PTEs such as lead, cobalt, cadmium, nickel, and arsenic originating from pharmaceutical industrial wastewater. This review reveals the associated advantages and shortcomings of the outmoded and the modern methods and the challenges involved in addressing the shortcomings. Additionally, due to the increasing amount of uncontrollable pharmaceutical effluents entering the ecosystem, this paper reviewed the management approach supported by the World Health Organization and the Environmental Protection Agency. Studies revealed that PTEs find their way into human bodies through different pathways, which include drinking water, edibles, and dermal vulnerability at intervals. This paper focuses on how pharmaceutical effluents can be handled and how regulations and strategies can be reinforced step by step. To preserve public health and the environment, a comprehensive study on the environmental evaluation of carcinogenic substances, particularly toxic elements and metalloids, should be supported and advocated. To protect living organisms and the welfare of consumers, efforts should be made to reduce the presence of potentially hazardous elements on land and water.

Keywords: potential toxic elements; pharmaceutical effluents; pathways; health risk assessment; advance treatment

Citation: Okoro, H.K.; Orosun, M.M.; Oriade, F.A.; Momoh-Salami, T.M.; Ogunkunle, C.O.; Adeniyi, A.G.; Zvinowanda, C.; Ngila, J.C. Potentially Toxic Elements in Pharmaceutical Industrial Effluents: A Review on Risk Assessment, Treatment, and Management for Human Health. *Sustainability* **2023**, *15*, 6974. https://doi.org/10.3390/su15086974

Academic Editor: Said Muhammad

Received: 16 February 2023
Revised: 16 April 2023
Accepted: 17 April 2023
Published: 21 April 2023

Copyright: © 2023 by the authors. Licensee MDPI, Basel, Switzerland. This article is an open access article distributed under the terms and conditions of the Creative Commons Attribution (CC BY) license (https://creativecommons.org/licenses/by/4.0/).

1. Introduction

There has been increasing concern in recent years regarding water treatment and reuse which necessitates the toughest standards due to the increasing demand for high-quality water worldwide, be it for drinking, sanitation, agriculture, or industrial usage [1]. Increased industrial activities have caused a number of water bodies to become contaminated with pollutants that go above what is permitted for effluent [2]. Among the different contaminants, potential toxic elements have posed the major concerns for environmental and human health due to their non-biodegradability and great accumulation capability [3]. Potentially toxic elements (PTEs), otherwise known as heavy metals, are metallic chemicals with a density that is higher than that of water. These metals are very toxic to people and can hurt them if they go beyond what is considered normal [4]. For instance, some of these metals have atomic masses ranging from 63.546 to 200.590 u. On the periodic table, these metals are positioned in a vertical column between groups III and V with related properties.

Priority metals are naturally present in groundwater, and while some of them are necessary for life, higher concentrations of these metals can actually be quite detrimental. These vital resources of life, such as water and land, are becoming pressured as a result of industrialization, poor water and land use practices, farming techniques, and population growth [5]. Waterbody contamination has become a global issue that has grown in both developed and developing countries, impeding economic growth and negatively harming public health [6]. Both the condition and the features of our finite freshwater resources are being negatively impacted by the discharge of toxic metal ions into the environment. The uncontrolled release of contaminants is threatening the sustainability and availability of drinking water, and this is becoming a worldwide public concern [7]. Due to their bioaccumulative nature, potential hazardous metals have raised health concerns even in minute quantities [8]. Many of these elements, including arsenic (As), cadmium (Cd), chrome (Cr), nickel (Ni), lead (Pb), and zinc (Zn), are non-biodegradable, and the Environmental Protection Agency has designated them as priority contaminants to be controlled [9].

Several industrial processes are discharging PTEs, and pharmaceutical industrial effluents have proven to be a major discussion point. International organizations such as the World Health Organization (WHO) and the United Nations Environmental Programme (UNEP) have comprehensively reviewed the health consequences related to several toxic elements. Over the years, PTEs have been discovered, and their intensely detrimental impacts on human health are well documented, but currently the use of PTEs has greatly increased, giving rise to their discharge into the environment and thus provoking sudden symptoms in the body [10].

The objective of this review paper was to evaluate any potential health risks linked to pharmaceutical wastewater discharge into the environment. A discussion on how to improve water quality by lowering water contamination through monitoring techniques, management choices, and treatment preference selection of toxic metals to safeguard ecosystems reliant on water was reviewed. Analysis of the existing methods for treatment and disposal practices of pharmaceutical wastes has been undertaken on the classification of the treatment methods under well-identified broad specific groups. The overall goal of this paper is to educate and influence the attitudes and actions of pharmaceutical industries, consumers, and farmers against the unauthorized discharge of effluents into the environment.

2. Sources and Entry Pathways of Potential Toxic Elements

Toxicity can be explained as the tendency of a substance to cause harm to a living organism. It is also the degree to which a chemical substance can exert a harmful influence on the life span, development, and replication of living things. Several studies have demonstrated that certain toxic metals have been identified as substances that alter normal body functions, leading to developmental malformations commonly known as cancer and genetic mutations in organisms based on the exposure rate to those metals and the amount taken at a particular time, which affects both humans and animals. The following sections describe the primary sources of pharmaceutical effluents and their end products in the environment [11–13].

- Pharmaceutical manufacturing generates effluents containing human drugs, suspension drugs, and solid wastes containing animal drugs in the form of boluses, and storm water runoff carries finely ground drugs.
- Consumers and clients also dispose used and untreated drugs into water bodies, which pollute the ecosystem and affect human health. Furthermore, numerous households also release wastes through excretion into the wastewater without any precautionary methods.

- Health centers and clinics discharge wastewater and pharmaceutical wastes down the drain.
- Agricultural residues and drugs administered to animals such as fowl, sheep, and cold-blooded animals; discharge of dissolved fertilizers and antibiotics mixed with farm animal feed and water. Figure 1 summarizes the environmental entrance pathways for the pharmaceuticals.

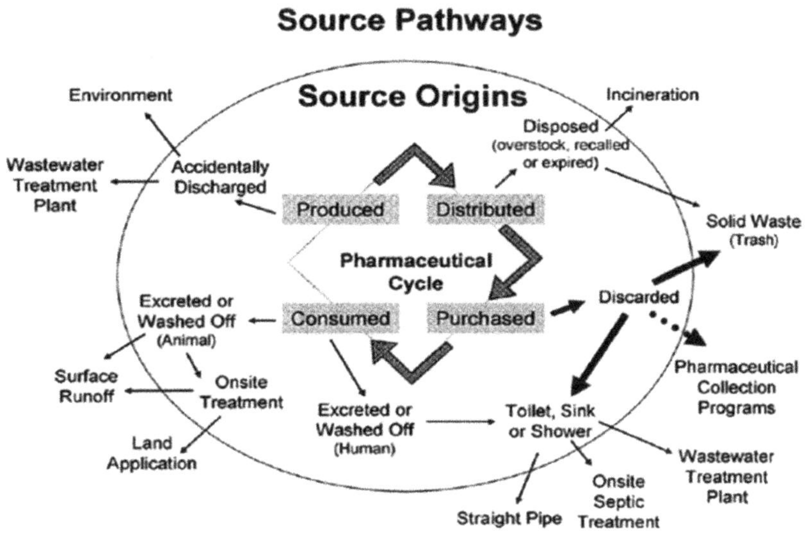

Figure 1. Environmental entrance pathways for pharmaceuticals. Reprinted from [14] with permission from Elsevier.

The sources through which these pharmaceuticals enter the environment may differ from one location to another, but it is widely known that the major pathway of pharmaceutical effluents into water bodies is through the release of unrefined wastewater from consumers and veterinary medicines. Other sources are domestic wastes, farming, and some industries that empty partially untreated effluent into the river. Potentially toxic elements from anthropogenic sources enter the rivers as inorganic complexes [5].

3. Routes of Uptake

Living things can be exposed to PTEs through the respiratory organs such as lungs, skin, and the process of taking in food [15,16]. There are three (3) ways in which these priority metals are taken up by humans and animals, as discussed below:

1. Ingestion: This is the process of swallowing foodstuffs or other substances such as coffee, water, and juice that are polluted by the PTEs (see Figure 2).
2. Skin or dermal adsorption: Skin adsorption is a process by which the harmful substance meets the body through the skin or gills.
3. Inspiration (inhalation): Living things breathe in poisonous gases or vapor as dust fumes in this process.

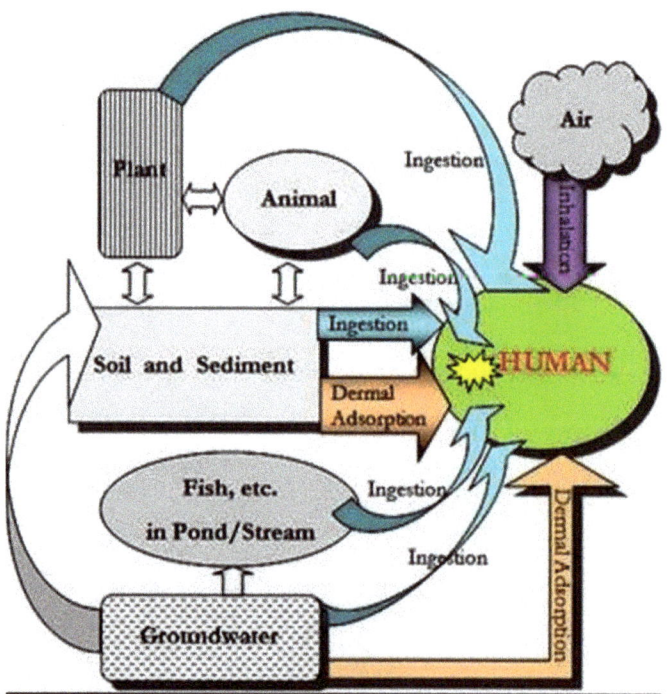

Figure 2. Reprinted with permission from [17].

4. Potentially Toxic Metals in Ground and Surface Water

Potentially toxic elements such as Cr, Pb, and Cd can harm living organisms, in addition to long-lasting organic pollutants such as pesticides and polychlorinated biphenyls (PCBs), which are some of the pollutants found in pharmaceutical effluent discharges. These elements are highly harmful to marine animals, even at trace amounts, because they change the morphology of the organisms' cells [18].

The evaluation of the influence of point-source contaminants on the sustainability of water bodies is becoming particularly crucial. The study of the risk of chemicals in rivers and the effects of adulteration on the freshness of streams is important because these sources of water play significant functions. The society ends up paying a heavy cost for inappropriate pharmaceutical sewage discharge because contaminated water is primarily used for both household and specific purposes by residents living near groundwater and surface water [19].

An examination of the literature revealed that dangerous metal concentrations in fresh water, drinkable water, and aquifers around the world frequently exceed the ranges permitted by regulatory bodies, thereby posing a risk to the ecosystem. The range of arsenic concentrations in groundwater sources was 0.0005–1.15 mg/L [20].

In 2015, according to the World Health Organization, 8.8 million cancer deaths occurred globally. The accumulation of harmful metals above legal limits in bodies of water is among the main reasons people die because these metals affect the central nervous system. The physiology of living tissues in the body is severely affected when foodstuffs or potable water comprising potentially harmful metals surpass their optimum threshold level [21]. The immunological and blood systems in humans and livestock are also harmed when they are exposed to metal combinations.

The existence of toxic metals beyond their maximum limit in the aquatic system is the major source of human endocrine disorders, according to a report by [22]. Potentially dangerous metals, including Pb, Hg, Cd, As, and Cr, put pressure on biological systems and produce cardiotoxicity, which harms tissues and causes age-related diseases such as Alzheimer's disease and mental retardation.

5. Classifications of Potentially Toxic Metals

Several scholars (e.g., [23,24]) have discovered that pollutants in groundwater are harmful metals. These harmful components are divided into four classes due to their health implications.

1. Essential: These are metals that play a biological role in living organisms but become hazardous when they exceed their permitted limits; examples are manganese, copper, iron, and zinc.
2. Less toxic: These are metals that are less toxic to human health; examples include tin and bismuth.
3. Highly toxic: These are known as metalloids such as cadmium, lead and arsenic.
4. Non-essential: These are metals that do not have a biological role in human beings; examples include aluminum and lithium.

Due to the obvious severity, permanence, and eventual biomagnifications of hazardous metals in the body, contamination of both land and marine environments by various harmful metals has become among the foremost problematic pollution concerns [25,26].

6. Pharmaceutical Industrial Effluents

Pharmaceutical effluents are liquid waste produced by drug companies during the development of new drugs; pharmaceutical effluent mainly contains the largest proportion of organic and inorganic compounds, animal drugs, and antibacterial drugs. During their synthesis, administration, and disposal, active pharmaceutical ingredients (APIs) are discharged into the surroundings, and there is an indication that API exposure in the environment has adverse implications on living organisms [27,28]. Potentially harmful metals have heavily contaminated the surroundings and their components. The biosphere is the most significant region of the ecosystem since it houses living things. Interactions between plants and animals and their environment also take place in the biosphere. Potentially, metals are detrimental to health, and their consumption has increased as a result of the activities carried out by drug manufacturing companies as well as global industrialization. In this regard, the potency of a contaminant in freshwater, the atmosphere, and food is examined [29,30]. Table 1 provides the distribution of pharmaceutical discharge in the environment.

Table 1. Distribution of pharmaceutical discharge in the environment.

Source	Pharmaceutical manufacturing plants	Waste water treatment plant (WWTP)	Agriculture
Pathways	Point source, e.g., refineries for oil, paper mills and chemicals	Diffuse source, e.g., fertilizer and pesticides in agriculture	
Concentration patterns	Continuous (e.g., WWTPs)	Seasonal (linked with farming practices)	Intermittent (linked with rainfall event)
Pharmaceutical attribute	Persistence (e.g., Half-life and solubility)	Biomagnification	Toxicity (individual effects and mixture effects)
Forms of receiving environment	Streams, lagoons	Aquifers, sea	Soil, sand
Context-dependent factors	Pharmaceutical properties	Illegal medication usage	Waste control and disposal methods

7. Selected Potentially Toxic Metals

7.1. Lead (Pb)

Pb is among the most common elements in nature all over the world. Due to its elemental physical and chemical properties, it has numerous industrial applications. It ranks as the fifth most widely used metal in the world as a result of increased industrial applications; it is used in over 900 industries, including oil refineries, drug manufacturing, and quarrying. There is a rise in the metal ion concentration in industrial effluent due to an increase in outflow from firms operating close to rivers. According to a study by [31], prolonged exposure of children to low concentrations of lead has been linked to reduced intelligent quotient (IQ), and acute lead exposure is known to cause sudden kidney injury. According to [32], acute Pb exposure can cause anorexia, fatigue, sleeplessness, disorganized thinking, dizziness, liver failure, high blood pressure, and joint pain, while chronic Pb exposure can cause congenital defects, intellectual disabilities, weight loss, learning disabilities, behavioral problems, muscle cramps, kidney diseases, neurological damage, and unconsciousness, and may eventually result in death.

7.2. Arsenic (As)

Arsenic is a substance that is found in nature and is broadly distributed across the atmosphere. It is widely distributed in nature because it can exist as a synthetic or natural compound and has variable oxidation states of -3, -1, 0, $+3$, and $+5$. It can be found in environmental matrices such as soil, air, water, and natural food sources. Arsenic is sometimes referred to as the "King of Poisons" due to its potent toxicity [33]. Arsenic is exceedingly hazardous in nature; studies have shown that its potential adverse health effects are linked to its exposure, which has gotten a lot of attention in the last two decades [34,35]. Prolonged exposure to low levels of arsenic affects the liver in humans and other mammals. Arsenic also causes swollen kidneys, structural abnormalities, and kidney failure in humans. Arsenic is highly hazardous and carcinogenic, and it is widely available as oxides or sulfides or as an iron, sodium, calcium, or copper salt. Arsenic causes oxidative processes and induces detrimental consequences in individuals, including cell apoptosis and mental retardation, memory deficits, hearing disorders, high blood pressure disorientation in old age, and kidney cancer [16,36].

7.3. Cadmium (Cd)

Cadmium exposure has been reported to be associated with fertility problems because it impairs sperm and lowers pregnancy outcomes. It is a cancer-causing agent and appears to play a role in heart disease and high blood pressure [37]. Cadmium destroys the gastrointestinal system, reduces renal antioxidant activities, and alters chemical fluid in the body, resulting in reproductive organ malfunction. The cognitive function of the brain is affected when Cd is present [38]. Cadmium is hazardous to tissues due to its ability to remove vitamins C and E from their metabolically active locations. Hence, Cd is a hormonal disrupter and is detrimental to child development. Immunological suppression and immune disorders are two symptoms of cadmium toxicity in fish. The accumulation of Cd in wastewater may disrupt body systems, causing short- or long-term problems [39,40].

7.4. Nickel (Ni)

Nickel is among the most common metals on earth and is present in a wide range of metal alloys used in the steel industry, as well as in colorants, taps, and dry cells. Nickel can end up in water bodies due to its presence in wastewater streams [41]. Respiratory illness, lung fever, hypersensitivities, cardiovascular disease, skin irritation, and mishaps are all caused by the excessive absorption of Ni by humans. There is a high risk of developing prostate cancer, nasal cancer, and laryngeal cancer due to human exposure. Ni is only found in trace amounts in the environment, but food plants grown in contaminated soils can accumulate large quantities. Numerous cytotoxic activities carried out by Ni include the production of free radicals, genetic regulation, and the control of transcription factors. It has been discovered that Ni contributes to the control of the expression of a few lengthy non-coding chromosomes. Nickel has also been shown to produce reactive species, which contribute to neurotoxic development [32].

7.5. Chromium (Cr)

Chromium is essential for glucose and fat breakdown, as well as lipoprotein consumption. Its biological function is intertwined with that of insulin, and most Cr-enhanced processes are insulin dependent. An excessive amount, on the other hand, could be harmful [42]. In soils treated with wastewater, toxic levels are typical. Over 7.5 million tons of chromium are thought to be produced annually on an international scale. In the presence of chromium ions, the discharge of collagen-type I, which aids in the repair of broken bones, is reduced. Nose ulcers, pneumonia, and genetic mutation in humans, as well as the destruction of red blood cells and urinary damage, are caused by chromium. While Cr (VI) is associated with a variety of anomalies and disorders, Cr (III) acts as an endocrine activator and is required in trace amounts for appropriate fat and amino acid metabolism [43–45]. Table 2 provides the summary of some studies highlighting the health risks of selected toxic metals.

Table 2. Summary of some studies highlighting the health risks of selected toxic metals.

s/n	PTE	Health Risk	Reference
1	Lead (Pb)	Long-term lead exposure may also give rise to kidney damage; acute exposure of Pb can cause loss of appetite, fatigue, sleeplessness, hallucinations, coma and even cause death	[32]
2	Arsenic (As)	Low concentrations of arsenic for long period damage liver in human and other animals; enlargement of kidney, skin and hair changes	[30] [34]
3	Cadmium (Cd)	Presence of Cd in contaminated water could disturb the necessary mechanisms in the body, possibly resulting in short-term or long-term disorders; it disrupts endocrine function and also affects reproduction rate in humans	[38–40]
4	Nickel (Ni)	Excess uptake of Ni by humans causes asthma, pneumonia, allergies, heart disorder, skin rashes, miscarriage and increases the chances of developing carcinoma and nose cancer	[32]
5	Chromium (Cr)	Cr in humans causes nose ulcers, asthma, DNA damage, hemolysis, damage to liver and kidney, and carcinoma	[42–44]

8. Human Health Risk Assessment of Pharmaceutical Industrial Effluents in the Environment

8.1. Health Risk Assessment

Human health risk assessment is a method of calculating the potential health consequences associated with the exposure of people to chemical hazards.

In the risk assessment process, there are four basic steps:
1. Hazard identification,
2. Assessment of exposure,
3. Toxicity/dose–response assessment, and
4. Characterization of risk.

8.1.1. Hazard Identification

The purpose of hazard identification is to look at the contaminants that are present in a given area, their level of accumulation, and their diffusion rate.

8.1.2. Assessment of Exposure

The purpose of the assessment of exposure is to determine the magnitude, reoccurrence, duration, or timing of individual exposure to pollutants (i.e., PTEs). The average daily dose (ADD) of the toxic elements discovered by consumption and skin contact in the population was calculated in this study to assess exposure.

8.1.3. Toxicity/Dose–Response Assessment

Dose–response analysis determines the toxicity of PTEs based on their exposure strengths. The two critical toxicity indices used are a carcinogen intensity factor called the cancer slope factor (SF) and a non-carcinogenic level termed the reference dose (RfD).

8.1.4. Characterization of Risk

Risk characterization aids in predicting the potential cancerous and non-cancerous health hazards that the vast majority of the population in the research area is vulnerable to; this is done by combining all the data obtained to create statistical evidence of exposure to cancer and the threat proportion [45]. The sudden rise in pollution of priority metals [46] can have disastrous health repercussions on communities and habitats, disrupt crop productivity, and make animal and human drinking water unsafe. Toxic metal, in particular, can build up inside important organs, causing short- and long-term damage. Metals are commonly encountered in corrosive effluents, where water forms electrostatic dipolar ion interactions with priority metals [47]. They can be produced in massive amounts and come from a variety of production processes.

Every day, almost 2 million tons of uncontrolled chemical, wastewater, and farm runoff are released into water bodies [48]. The amount of toxic metal ions released into the environment has increased dramatically over time [49]. Even treated industrial effluent contains potentially toxic metal ions, which have a negative impact on human health and the environment.

9. Advanced Treatment of Pharmaceutical Industrial Effluents

The principal technique used for enhanced treatment of industrial wastewater is based on physicochemical mechanics, which has become the focal point of scientific study and technical applications over time [50,51]. Examples of physical or chemical procedures used to treat industrial effluents are improved oxidation techniques, flotation, activated carbon adsorption, membrane separation, coagulation, and sedimentation, and the characteristics of each process are highlighted and described below.

9.1. Coagulation and Flocculation

Coagulation is a chemical water treatment process used to remove solids by adding coagulant agents and modifying the electrostatic charges of suspended particles in water to form large flocs, which then settle down as sludge. For improved treatment of industrial effluents, it is essential to squeeze and remove bound water from around hydrophilic colloids. As a result, determining the coagulation impact is heavily dependent on the flocculant nature. Polymers and metal salts are frequently used as flocculants. This procedure is effective for removing chromaticity as well as hazardous organic materials [52].

9.2. Flotation

Excluding sedimentation, flotation can eliminate dissolved particles from secondary effluents. The technology creates a large number of tiny bubbles by introducing air into effluent, creating floating floc with a lower density than the wastewater. It can also float to the top of the effluent to separate it. Ion flotation could be a suitable option for removing hazardous metal ions from industrial effluents. Ion flotation is based on the application of surfactant to make ionic metal species in effluent hydrophobic, followed by the removal of these hydrophobic species by gas bubbles [53].

9.3. Activated Carbon Adsorption

Activated carbon offers several benefits as an adsorbent: it features a large surface area, a multi-level pore volume, adsorption ability, and a constant chemical property. As a result, it is commonly utilized as an adsorbent to eliminate contaminants [54]. Activated carbon is used in the treatment of water and wastewater, including hazardous and difficult-to-meet discharge standards pharmaceutical wastewater. This is also a significantly advanced treatment method for industrial wastewater. There are two different types of activated carbon adsorption: chemical and physical. There is no adsorption selectivity, and physical adsorption is reversible. When saturated with adsorbates, activated carbon desorbs easily.

On the other hand, chemical adsorption is irreversible and challenging to desorb since it only adsorbs one or a few different adsorbates [55,56]. Saturation of activated carbon restores its adsorption property through a renewal process for cyclic use. Because of its ability to be reprocessed, greater treatment effect, and wide applicability, this technology is extensively employed for advanced water treatment. However, numerous drawbacks, including high relative prices, low regeneration efficiency, and complicated operation, limit its use [54,57].

9.4. Photocatalytic Oxidation

Ultraviolet photocatalytic oxidation, also known as photochemical oxidation, is a way of combining ultraviolet radiation as a catalyst with a UV-sensitive oxidant. When exposed to ultraviolet light, the breakdown of oxidants produces free radicals with a higher oxidative ability, allowing them to oxidize more and more difficult-to-decompose organic pollutants using only oxidants. Based on the kinds of oxidants used, photochemical oxidation can be classified as UV/O_3, UV/H_2O, $UV/H_2O_2/O_3$, etc. Photochemical reactions take place through diverse and complex reaction mechanisms depending on radical generation principles that are susceptible to experimental settings, and their effectiveness might be harmed by the presence of quenching chemicals in wastewater [58].

9.5. Electrochemical Oxidation

The electrochemical approach is well recognized for the removal of poisonous and hazardous pollutants in water and wastewater. In the field of water treatment, electrochemistry is a novel approach. The technique works on the following principle: during the electrochemical reaction, the reactant loses electrons and is oxidized at the anode surface. The chemical at the cathode surface, on the other hand, will lose electrons and get reduced. In particular, the oxidation of the anode causes the elimination of refractory organic materials. Electrolytic recovery, electrochemical oxidation, electrolytic air flotation, electrodialysis, and micro-electrolysis are examples of traditional electrochemical effluent treatment systems. When compared to other techniques, the electrochemical process, often known as the "environmentally friendly" process, offers a significant benefit. The electrochemical process, for example, is normally carried out at room temperature and pressure, has high efficiency, may be used alone or in conjunction with other procedures, covers a limited area, produces no secondary pollution, and has a reasonably high level of automation. Future electrochemical reaction research will focus on the anode and electrochemical reactor development [59].

9.6. Ultrafiltration (UF)

The ultrafiltration process is utilized in the tertiary treatment of water and wastewater for additional cleaning and treatment. The difference in pressure between the membranes on either side is what propels the movement; the membrane for ultrafiltration serves as the filtering medium. When a fluid travels through the surface of the membrane under a specific amount of pressure, water, inorganic ions, and tiny molecules pass through, while other components are prevented from crossing the membrane barrier.

9.7. Reverse Osmosis (RO)

Reverse osmosis efficiently removes metal ions from water, making it suitable for drinking [60,61]. RO membranes can be classified as either cellulose ester or aromatic polyamide. A variety of organic materials and dissolved inorganic salts can be removed using RO methods. It also has a high rate of water recycling and salt extraction efficiency.

9.8. Electrodialysis (ED)

Electrodialysis (ED) is among the new techniques for removing and recovering metals. Recently, metal-contaminated water has been treated using the emerging technique of electrodialysis [62]. The electrodialysis process (ED) combines electrolysis with dialysis. The dissolved salts in the wastewater are subsequently transported to the anode and the cathode by the direct current electric field. In this manner, separation and recovery are accomplished as the number of anions and cations in the intermediate phase gradually decreases. Numerous benefits include less environmental pollution, less energy and chemical usage, ease of use, and automation. However, because desalination effectiveness is lower than RO, it is limited to the removal of salt from the water. Overview of the benefits and drawbacks of some selected metal treatment technologies is provided by Table 3.

Table 3. Overview of the benefits and drawbacks of some selected metal treatment technologies.

s/n	Technology	Benefits	Drawbacks	Reference
1	Coagulation and flocculation	-Coagulants are relatively economical and easy to operate -It does not call for distinct processes	-Partial removal of metal	[47] [52]
2	Adsorption (activated carbon)	-Numerous adsorbents are accessible -Easy to use and affordable	-Waste products are created -Regenerating agents can be difficult	[63] [54]
3	Electrodialysis	-High effectiveness in separating ions with opposite charges -Can handle metal concentrations that are low -Treatment setup is easy	-High level of energy use -Operational characteristics have an impact on separation effectiveness -Creation of auxiliary streams	[62]
4	Reverse osmosis	-High metal separation effectiveness -Purifies contaminated water	Running cost is high due to high forces -It is important to regenerate	[60] [61]

10. Management of Potentially Toxic Elements in Pharmaceutical Industrial Effluents

Waste management and sewage treatment have recently become major issues for humans. Pharmaceutical metabolites have been found in the water cycle in the past 10 years, which has caused policy makers, water distributors, and communities at large to worry about the potential implications for human well-being. The World Health Organization (WHO) and the Environmental Protection Agency (EPA) both support a significant risk management strategy called the "water quality approach". It implies that, to avoid any associated danger, particularly health problems, pharmaceutical effluents should be managed to a level that meets precise water quality guidelines [64]. With this method, countries are encouraged to take into account their unique cultural, social, and environmental circumstances when developing and implementing effective risk management strategies [65]. However, it is well acknowledged that the degree of wastewater treatment should correspond to the intended application. For this reason, cost-effectiveness studies are crucial for assessing various risk management choices for concrete proof of decision making, treatment preference selection, and distribution of resources, particularly in economically developing nations [66].

If there is no landfill leachate collection, pharmaceuticals improperly disposed of in domestic waste wind up in landfills, where they may eventually be transmitted into water bodies [67]. Regarding the known and conceivable threats to humans and ecosystems, the presence of pharmaceuticals in surface water has sparked alarm among freshwater suppliers, policy makers, and the general public. It has been established that several pharmaceutical effluents have unfavorable impacts on ecosystems, including abnormalities, behavior, fertility, and death. Due difficulties faced in biodegrading toxic metals, which are naturally present in the earth's crust, these materials have contributed to a significant number of environmental pollutants [5,51]. To stop environmental deterioration, the regulatory bodies for the pharmaceutical companies must demand constant supervision and the application of regulations governing the proper discharge of pharmaceutical pollutants. The World Health Organization has listed the issue as among the top ten potential health risks [68]. The rapidly expanding pharmaceutical sector, which serves as a reservoir for antibiotics in the surroundings, plays a vital role; effective treatment of pharmaceuticals in surface water will demand:

- Evaluation of the occurrence, distribution, and environmental effects of additional effluent.
- Uniform data gathering, processing, and storage practices.
- Monitoring priorities for toxins, water bodies, and ecosystems.
- Improved knowledge of the various exposure paths.
- Determining what medicines are present in treated water (septic tanks) compared to untreated water.

11. Evaluation and Monitoring of Possibly Toxic Elements in the Environment

To assess and control pollution, monitoring and evaluation of environmental toxic metal concentrations are required. It is important to regularly evaluate the concentrations of possibly hazardous metals and metalloids in the resident biota as well as in the various environments, such as freshwater, sediments, and soils [69,70]. The prevalence, main sources, and eventual outcome of these toxic metals in the environment, as well as their bioaccumulation in food chains, will all be valuable information that may be gleaned from this environmental investigation. This analysis is additionally employed to determine how dangerous these substances are to both public health and animals [71].

12. Sustainable Developmental Goal on Water Management

To preserve human health and boost water availability, increasing water contamination must be addressed at its source and treated. All areas of life and sustainable development require access to fresh water in quantities and conditions that are both sufficient and of high quality. To set investment priorities, trustworthy surface water resource monitoring is necessary. Enhancing and accelerating efforts to address water quality issues requires a better awareness of the possible links between water contamination and Sustainable Development Goals [72].

Water contamination poses a health threat to living creatures and the ecosystem [72,73]. Sustainable Development Goal 3 aims to safeguard everyone's good health and wellness at every phase of their lives. Evidence suggests that improper wastewater treatment and disposal, which contaminates water supplies, reduces health benefits [74].

According to SDG 6, everyone must have access to water and sanitation services that are managed sustainably. This goal specifically obligates member states to establish widespread access to clean water and sanitation, enhance water quality by lowering water contamination, manage water resources holistically, and safeguard ecosystems reliant on water. Additionally, enhancing sanitation should consider every link in the "sanitation chain", including the user's experience, the management of sewage, and its reuse or disposal. These links include technical elements such as the toilet or wastewater treatment plants as well as human experience [75].

SDG 14 focuses on the sustainable management of water bodies to prevent detrimental effects; as a result, it aims to control fisheries, address ocean acidification, minimize its consequences, and prevent all forms of marine pollution from entering the ocean. SDG 14 calls for improving the protection and wise use of the oceans' natural resources by putting into practice international law, which establishes the necessary legal framework. The largest ecosystem on the globe is the ocean, and it is vital to our survival that the oceans and seas remain healthy. To achieve the Sustainable Development Goals, we must take an integrated approach that prioritizes protecting our oceans and marine resources [76].

13. Management of Pharmaceutical Effluent and Policy Recommendations

Pharmaceutical wastes should be released into the environment more frequently if proper steps are taken to mitigate the dangers as global pharmaceutical demand is stoked by aging populations, improvements in healthcare, and increased livestock production. To ensure successful implementation, in collaboration with relevant governmental organizations, regional governments, and other stakeholders, a nationwide pharmaceutical policy and action plan to reduce environmental dangers should be developed. To maintain freshwater habitats and water quality, regulations that cost-effectively manage pharmaceutical effluents rely on two methods [76].

13.1. Suggestions to Enhance Awareness, Comprehension, and Feedback

1. To lay the groundwork for future pollution reduction initiatives, information, understanding, and reporting on the prevalence, distribution, and hazards to human health and the environment posed by pharmaceutical effluents in groundwater must be improved.
2. Utilize strategic and focused assessment methods to identify any environmental concerns associated with current and future active pharmaceutical components.
3. Promote the use of novel modeling techniques, monitoring techniques, and different resolutions to better comprehend and predict risks. Place a higher priority on chemicals and water bodies.
4. When there is a lack of solid scientific data and serious potential repercussions from inaction, take precautions. Train and interact with communities to control perceptions and take action [76].

13.2. Utilization-Focused Strategies

Reduce the illegal and wasteful usage and disposal of medications by imposing, rewarding, or advocating reductions in these behaviors. They aim is to educate and influence the attitudes and actions of doctors, veterinary doctors, pharmacists, consumers, and farmers. Reduce the consumption and supply of medications that are not essential. Enhance tests and postpone prescribing medications if they are not crucial. Consider banning or limiting the use of synthetic compounds as food supplements in the animal and seafood sectors.

5. Minimize the self-prescription of high-risk medications (such as antibiotics and drugs that affect the hormonal balance) and unauthorized drug sales.
6. Encourage and improve the best methods for handling and storing wastewater and manure from animals given pharmacological treatment.

14. Conclusions

Authorities, water service providers, and communities at large have all expressed concern over the prevalence of pharmaceuticals in the environment. Pharmaceuticals used for human health, such as penicillin, painkillers, stimulants, and chemotherapy drugs, as well as those used in veterinary medicine, are of particular importance, but it has been demonstrated through research that these pharmaceuticals have unplanned, undesirable deleterious effects on living organisms. A significant contributor to global environmental degradation today is emerging pollutants. It is imperative that the organizations in charge of environmental control and reporting, in addition to actions taken to protect global health, respond to the treatment and removal of new pollutants from pharmaceutical industries. Potentially toxic metals (PTEs) are retained after absorption, where they build up in the body. Harmful metals that enter the body through the food chain have a wide range of negative effects on various human organs and tissues. Without effective management and treatment, pharmaceutical effluent discharge could constitute substantial health and environmental dangers to the people who rely on surface and groundwater for drinking and domestic purposes. Potentially toxic metals should be addressed because they can lead to acute or persistent symptoms. To preserve public health and the environment, a comprehensive study on the environmental evaluation of carcinogenic substances, particularly toxic elements and metalloids, should be supported and advocated for. To protect living organisms and the welfare of their consumers, efforts should be made to reduce the presence of PTEs on land and water.

Author Contributions: Conceptualization, H.K.O. and M.M.O.; methodology, H.K.O., M.M.O., F.A.O., T.M.M.-S. and C.O.O.; investigation, F.A.O. and C.Z.; resources, H.K.O., A.G.A. and J.C.N.; visualization, M.M.O.; supervision, H.K.O., M.M.O. and C.O.O.; project administration, H.K.O. and J.C.N.; funding acquisition, H.K.O., M.M.O. and C.O.O. All authors have read and agreed to the published version of the manuscript.

Funding: Partial support was received through University of Ilorin TETFUND IBR intervention.

Institutional Review Board Statement: Not applicable.

Informed Consent Statement: Not applicable.

Data Availability Statement: All the data and materials are available.

Conflicts of Interest: The authors declare no conflict of interest.

References

1. Gadipelly, C.; Pérez-González, A.; Yadav, G.D.; Ortiz, I.; Ibáñez, R.; Rathod, V.K.; Marathe, K.V. Pharmaceutical industry wastewater: Review of the technologies for water treatment and reuse. *Ind. Eng. Chem. Res.* **2014**, *5329*, 11571–11592. [CrossRef]
2. Pal, P. Treatment and Disposal of Pharmaceutical Wastewater: Toward the Sustainable Strategy. *Sep. Purif. Rev.* **2018**, *47*, 179–198. [CrossRef]
3. Emenike, E.C.; Adeleke, J.; Iwuozor, K.O.; Ogunniyi, S.; Adeyanju, C.A.; Amusa, V.T.; Okoro, H.K.; Adeniyi, A.G. Adsorption of crude oil from aqueous solution: A review. *J. Water Process. Eng.* **2022**, *50*, 103730. [CrossRef]
4. Orosun, M.M.; Tchokossa, P.; Nwankwo, L.I.; Lawal, T.O.; Bello, S.A.; Ige, S.O. Assessment of heavy metal pollution in drinking water due to mining and smelting activities in Ajaokuta. *Niger. J. Technol. Dev.* **2016**, *13*, 30–38. [CrossRef]
5. Okoro, H.K.; Ige, J.O.; Ngila, C.J. Characterization and Evaluation of Heavy Metals Pollution in River Sediments from South Western Nigeria. *J. Kenya Chem. Soc.* **2018**, *11*, 18–27.
6. Emenike, E.C.; Iwuozor, K.O.; Anidiobi, S.U. Heavy metal pollution in aquaculture: Sources, impacts and mitigation techniques. *Biol. Trace Elem. Res.* **2021**, *200*, 4476–4492. [CrossRef]
7. Ogunkunle, C.O.; Mustapha, K.; Oyedeji, S.; Fatoba, P.O. Assessment of metallic pollution status of surface water and aquatic macrophytes of earthen dams in Ilorin, north-central of Nigeria as indicators of environmental health. *J. King Saud. Univ. Sci.* **2016**, *28*, 324–331. [CrossRef]
8. Qiu, Y.W. Bioaccumulation of heavy metals both in wild and mariculture food chains in Daya Bay, South China. *Estuar. Coast. Shelf Sci.* **2015**, *163*, 7–14. [CrossRef]
9. Tóth, G.; Hermann, T.; Szatmári, G.; Pásztor, L. Maps of heavy metals in the soils of the European Union and proposed priority areas for detailed assessment. *Sci. Total Environ.* **2016**, *565*, 1054–1062. [CrossRef]
10. WHO. *World Health Organization Guidelines for Drinking-Water*, 4th ed.; WHO: Geneva, Switzerland, 2011. Available online: http://whqlibdoc.who.int/publications/2011/9789241548151_eng.pdf (accessed on 12 December 2022).
11. Lapworth, D.; Baran, N.; Stuart, M.E.; Ward, R.S. Emerging organic contaminants in groundwater: A review of sources, fate and occurrence. *Environ. Pollut.* **2012**, *163*, 287–303. [CrossRef]
12. Monteiro, S.; Boxall, A. Occurrence and Fate of Human Pharmaceuticals in the Environment. *Rev. Environ. Contam. Toxicol.* **2010**, *53*–154. [CrossRef]
13. Kümmerer, K. The presence of pharmaceuticals in the environment due to human use—Present knowledge and future challenges. *J. Environ. Manag.* **2009**, *90*, 2354–2366. [CrossRef] [PubMed]
14. Glassmeyer, S.T.; Hinchey, E.K.; Boehme, S.E.; Daughton, C.G.; Ruhoy, I.S.; Conerly, O.; Daniels, R.L.; Lauer, L.; McCarthy, M.; Nettesheim, T.G. Disposal practices for unwanted residential medications in the United States. *Environ. Int.* **2009**, *353*, 566–572. [CrossRef] [PubMed]
15. Tchounwou, P.B.; Yedjou, C.G.; Patlolla, A.K.; Sutton, D.J. Heavy Metals Toxicity and the Environment. *EXS* **2012**, *101*, 133–164. [CrossRef] [PubMed]
16. Singh, R.; Gautam, N.; Mishra, A.; Gupta, R. Heavy metals and living systems: An overview. *Indian J. Pharmacol.* **2011**, *43*, 246–253. [CrossRef] [PubMed]
17. Islam, N.; Huq, E.M.; Islam, N.; Islam, A.; Ali, R.; Khatun, R.; Haque, M.; Islam, R.; Akhter, S.N. Human Health Risk Assessment for Inhabitants of Four Towns of Rajshahi, Bangladesh due to Arsenic, Cadmium and Lead Exposure. *Environ. Asia* **2018**, *102*, 168–182.
18. Gbogbo, F.; Arthur-Yartel, A.; Bondzie, J.A.; Dorleku, W.P.; Dadzie, S.; Kwansa-Bentum, B. Risk of heavy metal ingestion from the consumption of two commercially valuable species of fish from the fresh and coastal waters of Ghana. *PLoS ONE* **2018**, *13*, e0194682. [CrossRef]
19. Amadid, O. Quality assessment of Aba River using heavy metal pollution index. *Am. J. Environ. Eng.* **2012**, *2*, 45–49.
20. Fatima, S.; Hussain, I.; Rasool, A.; Xiao, T.; Farooqi, A. Comparison of two alluvial aquifers shows the probable role of river sediments on the release of arsenic in the groundwater of district Vehari, Punjab, Pakistan. *Environ. Earth Sci.* **2018**, *77*, 382. [CrossRef]

21. Ojedokun, A.T.; Bello, O.S. Sequestering heavy metals from wastewater using cow dung. *Water Resour. Ind.* **2016**, *13*, 7–13. [CrossRef]
22. Li, J.-J.; Pang, L.-N.; Wu, S.; Zeng, M.-D. Advances in the effect of heavy metals in aquatic environment on the health risks for bone. *Earth Environ. Sci.* **2018**, *186*, 012057. [CrossRef]
23. Ali, H.; Khan, E.; Ilahi, I. Environmental chemistry and ecotoxicology of hazardous heavy metals: Environmental persistence, toxicity, and bioaccumulation. *J. Chem.* **2019**, *2019*, 6730305. [CrossRef]
24. Nizami, G.; Rehman, S. Assessment of heavy metals and their effects on quality of water of rivers of Uttar Pradesh, India: A review. *J. Environ. Chem. Toxicol.* **2018**, *2*, 22.
25. Okoro, H.K.; Ige, J.O.; Iyiola, O.A.; Ngila, J.C. Fractional profile, mobility patterns and correlations of heavy metals in estuary sediments from olonkoro river, in tedecatachment of western region, Nigeria. *Environ. Nanotechnol. Monit. Manag.* **2017**, *8*, 53–62.
26. Emenike, E.C.; Ogunniyi, S.; Ighalo, J.O.; Iwuozor, K.O.; Okoro, H.K.; Adeniyi, A.G. Delonix regia biochar potential in removing phenol from industrial wastewater. *Bioresour. Technol. Rep.* **2022**, *19*, 101195. [CrossRef]
27. Brodin, T.; Fick, J.; Jonsson, M.; Klaminder, J. Dilute concentrations of a psychiatric drug alter behavior of fish from natural populations. *Science* **2013**, *339*, 814–815. [CrossRef]
28. Horky, P. Methamphetamine pollution elicits addiction in wild fish. *J. Exp. Biol.* **2021**, *224*, jeb242145. [CrossRef]
29. Ghorani-Azam, A.; Riahi-Zanjani, B.; Balali-Mood, M. Effects of air pollution on human health and practical measures for prevention in Iran. *J. Res. Med. Sci.* **2016**, *21*, 65. [CrossRef]
30. Luo, L.; Wang, B.; Jiang, J.; Huang, Q.; Yu, Z.; Li, H. Heavy metal contaminations in herbal medicines: Determination. Comprehensive risk assessments. *Front. Pharm.* **2020**, *11*, 595335. [CrossRef]
31. Mortada, W.I.; Sobh, M.A.; El-Defrawy, M.M.; Farahat, S.E. Study of lead exposure from automobile exhaust as a risk for nephrotoxicity among traffic policemen. *Am. J. Nephrol.* **2001**, *21*, 274–279. [CrossRef]
32. Engwa, G.A.; Ferdinand, P.U.; Nwalo, F.N.; Unachukwu, M.N. Mechanism and Health Effects of Heavy Metal Toxicity in Humans. In *Poisoning in the Modern World—New Tricks for an Old Dog?* Karcioglu, O., Arslan, B., Eds.; IntechOpen: London, UK, 2019. [CrossRef]
33. Gupta, D.K.; Tiwari, S.; Razafindrabe, B.H.N.; Chatterjee, S. Arsenic Contamination from Historical Aspects to the Present. In *Arsenic Contamination in the Environment*; Gupta, D., Chatterjee, S., Eds.; Springer: Cham, Switzerland, 2017; pp. 1–12. [CrossRef]
34. Bundschuh, J.; Armienta, M.A.; Morales-Simfors, N.; Alam, M.A.; Lopez, D.L.; Delgado Quezada, V.; Dietrich, S.; Schneider, J.; Tapia, J.; Sracek, O.; et al. Arsenic in Latin America: New findings on source, mobilization and mobility in human environments in 20 countries based on decadal research 2010–2020. *Crit. Rev. Environ. Sci. Technol.* **2020**, *51*, 1727–1865. [CrossRef]
35. Orosun, M.M. Assessment of Arsenic and Its Associated Health Risks Due to Mining Activities in Parts of North-Central Nigeria: Probabilistic Approach Using Monte Carlo. *J. Hazard. Mater.* **2021**, *412*, 125262. [CrossRef] [PubMed]
36. Jaishankar, M.; Tseten, T.; Anbalagan, N.; Mathew, B.B.; Beeregowda, K.N. Toxicity, mechanism and health effects of some heavy metals. *Interdiscip. Toxicol.* **2014**, *7*, 60–72. [CrossRef] [PubMed]
37. Okoro, H.K.; Pandey, S.; Ogunkunle, C.O.; Ngila, C.J.; Zvinowanda, C.; Jimoh, I.; Lawal, I.A.; Orosun, M.M.; Adeniyi, A.G. Nanomaterial-based biosorbents: Adsorbent for efficient removal of selected organic pollutants from industrial wastewater. *Emerg. Contam.* **2022**, *8*, 46–58. [CrossRef]
38. Richter, P.; Faroon, O.; Pappas, R.S. Cadmium and cadmium zincratios and tobacco-related morbidities. *Int. J. Environ. Res. Public Health* **2017**, *14*, 1154. [CrossRef]
39. Jiang, J.H.; Ge, G.; Gao, K.; Pang, Y.; Chai, R.C.; Jia, X.H. Calcium Signaling Involvement in Cadmium-Induced Astrocyte Cytotoxicity and Cell Death Through Activation of MAPK and PI3K/Akt Signaling Pathways. *Neurochem. Res.* **2015**, *40*, 1929–1944. [CrossRef]
40. Cao, Z.R.; Cui, S.M.; Lu, X.X.; Chen, X.M.; Yang, X.; Cui, J.P. Effects of occupational cadmium exposure on workers. *Chin. J. Ind. Hyg. Occup. Dis.* **2018**, *36*, 474–477. [CrossRef]
41. Ogunlalu, O.; Oyekunle, I.P.; Iwuozor, K.O.; Aderibigbe, A.D.; Emenike, E.C. Trends in the mitigation of heavy metal ions from aqueous solutions using unmodified and chemically-modified agricultural waste adsorbents. *Curr. Res. Green Sustain. Chem.* **2021**, *4*, 100188. [CrossRef]
42. Cefalu, W.T.; Hu, F.B. Role of chromium in human health and in diabetes. *Diabetes Care* **2004**, *27*, 2741–2751. [CrossRef]
43. Achmad, R.T.; Budiawan, B.; Auerkari, I.E. Effects of chromium on human body. *Annu. Res* **2017**, *13*, 1–8. [CrossRef]
44. Vincent, J.B. Effect of chromium supplementation on body composition, human and animal health and insulin and glucose metabolism. *Curr. Opin. Clin. Nutr. Metab. Care* **2019**, *22*, 483–489. [CrossRef]
45. United State Environmental Protection. Agency Guidance for Characterizing Background Chemicals in Soil at Superfund Sites. Office of Emergency and Remedial Response, Washington, DC. *OSWER Dir.* **2021**, *9285*, 7–41.
46. Qin, G.; Niu, Z.; Yu, J.; Li, Z.; Ma, J.; Xiang, P. Soil heavy metal pollution and food safety in China: Effects, sources & removing technology. *Chemosphere* **2021**, *267*, 129205. [CrossRef] [PubMed]
47. Abdullah, N.; Yusof, N.; Lau, W.; Jaafar, J.; Ismail, A.F. Recent trends of heavy metal removal from water/wastewater by membrane technologies. *J. Ind. Eng. Chem.* **2019**, *76*, 17–38. [CrossRef]
48. Bolisetty, S.; Peydayesh, M.; Mezzenga, R. Sustainable technologies for water purification from heavy metals: Review and analysis. *Chem. Soc.* **2019**, *48*, 463–487. [CrossRef]

49. Zhou, Q.; Yang, N.; Li, Y.; Ren, B.; Ding, X.; Bian, H.; Yao, X. Total concentrations and sources of heavy metal pollution in global river and lake water bodies from 1972 to 2017. *Glob. Ecol. Conserv.* **2020**, *22*, e00925. [CrossRef]
50. Orosun, M.M.; Oniku, A.S.; Adie, P.; Orosun, O.R.; Salawu, N.B.; Louis, H. Magnetic susceptibility measurement and heavy metal pollution at an automobile station in Ilorin, North-Central Nigeria. *Environ. Res. Commun.* **2020**, *2*, 015001. [CrossRef]
51. Okoro, H.K.; Orosun, M.M.; Victor, A.; Zvinowanda, C. Health risk assessment, chemical monitoring and spatio-temporal variations in concentration levels of phenolic compounds in surface water collected from River Oyun, Republic of Nigeria. *Sustain. Water Resour. Manag.* **2022**, *8*, 189. [CrossRef]
52. Yiping, G.; Yu, B. *Advanced Treatment and Recycling Technology of Wastewater Treatment Plant*; China Architecture Press: Beijing, China, 2010; pp. 198–206.
53. Polat, M.; Polat, H.; Chander, S. Physical and chemical interactions in coal flotation. *Int. J. Miner. Process.* **2007**, *72*, 199–213. [CrossRef]
54. Orosun, M.M.; Adewuyi, A.D.; Salawu, N.B.; Isinkaye, M.O.; Orosun, O.R.; Oniku, A.S. Monte Carlo approach to risks assessment of heavy metals at automobile spare part and recycling market in Ilorin, Nigeria. *Sci. Rep.* **2020**, *10*, 22084. [CrossRef]
55. Orosun, M.M.; Usikalu, M.R.; Oyewumi, K.J.; Onumejor, C.A.; Ajibola, T.B.; Valipour, M.; Tibbett, M. Environmental Risks Assessment of Kaolin Mines and Their Brick Products Using Monte Carlo Simulations. *Earth Syst. Environ.* **2022**, *6*, 157–174. [CrossRef]
56. Iwuozor, K.O.; Emenike, E.C.; Aniagor, C.O.; Iwuchukwu, F.U.; Ibitogbe, E.M.; Temitayo, O.B.; Omuku, P.E.; Adeniyi, A.G. Removal of pollutants from aqueous media using cow dung-based adsorbents. *Curr. Res. Green Sustain. Chem.* **2022**, *5*, 100300. [CrossRef]
57. Louis, H.; Gber, T.E.; Charlie, D.E.; Egemonye, T.C.; Orosun, M.M. Detection of hydroxymethanesulfonate (HMS) by transition metal-anchored fullerene nanoclusters. *J. Iran. Chem. Soc.* **2023**, *20*, 713–729. [CrossRef]
58. Inglezakis, V.J.; Amzebek, A.; Kuspangaliyeva, B.; Sabrassov, Y.; Balbayeva, G.; Yerkinova, A. Treatment of municipal Solidwaste landfill leachate by use of combined biological, physical and photochemical processes. *Desalin. Water Treat.* **2018**, *112*, 218–231. [CrossRef]
59. Maxwell, O.; Oghenerukevwe, O.F.; Olusegun, O.A.; Joel, E.S.; Daniel, O.A.; Oluwasegun, A.; Jonathan, H.O.; Samson, T.O.; Adeleye, N.; Michael, O.M.; et al. Sustainable nano-sodium silicate and silver nitrate impregnated locally made ceramic filters for point-of-use water treatments in sub-Sahara African households. *Heliyon* **2021**, *7*, e08470. [CrossRef]
60. Cui, Y.; Ge, Q.; Liu, X.Y.; Chung, T.S. Novel forward osmosis process to effectively remove heavy metal ions. *J. Membr. Sci.* **2014**, *467*, 188–194. [CrossRef]
61. Maher, A.; Sadeghi, M.; Moheb, A. Heavy metal elimination from drinking water using nanofiltration membrane technology and process optimization using response surface methodology. *Desalination* **2014**, *352*, 166–173. [CrossRef]
62. Liu, Y.; Ke, X.; Zhu, H.; Chen, R.; Chen, X.; Zheng, X.I.; Jin, Y.; Van der Bruggen, B. Treatment of raffinate generated via copper ore hydrometallurgical processing using a bipolar membrane electrodialysis system. *Chem. Eng. J.* **2020**, *382*, 122956. [CrossRef]
63. Ahmaruzzaman, M. Industrial wastes as low-cost potential adsorbents for the treatment of wastewater laden with heavy metals. *Adv. Colloid Interface Sci.* **2011**, *166*, 36–59. [CrossRef]
64. Victor, R.; Kotter, R.; O'Brien, G.; Mitropoulos, M.; Panayi, G. *WHO Guidelines for the Safe Use of Wastewater, Excreta and Greywater*; World Health Organization: Geneva, Switzerland, 2008; Volumes 1–4.
65. Sato, T.; Qadir, M.; Yamamoto, S.; Endo, T.; Zahoor, A. Global, regional, and country level need for data on wastewater generation, treatment, and use. *Agric. Water Manag.* **2013**, *130*, 1–13. [CrossRef]
66. Wichelns, D.; Drechsel, P.; Qadir, M. *Wastewater: Economic Asset in an Urbanizing World*; Springer: Berlin/Heidelberg, Germany, 2015.
67. Saad, W. Drug product immobilization in recycled polyethylene/polypropylene reclaimed from municipal solid waste: Experimental and numerical assessment. *Environ. Technol.* **2017**, *38*, 3064–3073. [CrossRef] [PubMed]
68. WHO. *WHO Guidance for Climate-Resilient and Environmentally Sustainable Health Care Facilities*; World Health Organization: Geneva, Switzerland, 2020; p. 6.
69. Adeniyi, A.G.; Emenike, E.C.; Iwuozor, K.O.; Okoro, H.K.; Ige, O.O. Acid mine drainage: The footprint of the Nigeria mining industry. *Chem. Afr.* **2022**, *56*, 1907–1920. [CrossRef]
70. Omeje, M.; Orosun, M.M.; Adewoyin, O.O.; Joel, E.S.; Usikalu, M.R.; Olagoke, O.; Ehinlafa, O.E.; Omeje, U.A. Radiotoxicity Risk Assessments of ceramic tiles used in Nigeria: The Monte Carlo Approach. *Environ. Nanotechnol. Monit. Manag.* **2021**, *17*, 100618. [CrossRef]
71. Elzwayie, A.; Afan, H.A.; Allawi, M.F.; El-Shafie, A. Heavy metal monitoring, analysis and prediction in lakes and rivers: State of the art. *Environ. Sci. Pollut. Res.* **2017**, *24*, 12104–12117. [CrossRef] [PubMed]
72. United Nation-Water. Water quality and sanitation. In *Word Water Development Report*; United Nations Educational, Scientific and Cultural Organization: Paris France, 2018.
73. Corcoran, E.; Nellemann, C.E.; Baker, R.; Bos, D.; Osborn, H.S. *Sick Water? The Central Role of Waste-Water Management in Sustainable Development: A Rapid Response Assessment*; United Nations Environment Programme, UN-HABITAT, GRID-Arendal. Teaterplassen 3; UNEP/Earthprint: Arendal, Norway, 2010; p. 4836.
74. Lam, S.; Nguyen-Viet, H.; Tuyet-Hanh, T.; Nguyen-Mai, H.; Harper, S. Evidence for public health risks of wastewater and excreta management practices in Southeast Asia: A Scoping Review. *Int. J. Environ. Res. Public Health* **2015**, *12*, 12863–12885. [CrossRef]

75. Tilley, E.; Ulrich, L.; Lüthi, C.; Reymond, P.; Zurbrügg, C. *Compendium of Sanitation Systems and Technologies*, 2nd ed.; Swiss Federal Institute of Aquatic Science and Technology: Duebendorf, Switzerland, 2014; p. 180.
76. Comber, S. Active pharmaceutical ingredients entering the aquatic environment from wastewater treatment works: A cause for concern? *Sci. Total Environ.* **2018**, *613–614*, 538–547. [CrossRef]

Disclaimer/Publisher's Note: The statements, opinions and data contained in all publications are solely those of the individual author(s) and contributor(s) and not of MDPI and/or the editor(s). MDPI and/or the editor(s) disclaim responsibility for any injury to people or property resulting from any ideas, methods, instructions or products referred to in the content.

Article

Investigation of Efficient Adsorption of Toxic Heavy Metals (Chromium, Lead, Cadmium) from Aquatic Environment Using Orange Peel Cellulose as Adsorbent

Aminur Rahman [1,*], Kazuhiro Yoshida [2], Mohammed Monirul Islam [1] and Genta Kobayashi [3]

1. Department of Biomedical Sciences, College of Clinical Pharmacy, King Faisal University, Al-Ahsa 31982, Saudi Arabia
2. Laboratory of Phycology and Benthology, Faculty of Agriculture, Saga University, 1 Honjo-Cho, Saga 840-8502, Japan
3. Laboratory of Applied Microbiology, Faculty of Agriculture, Saga University, 1 Honjo-Cho, Saga 840-8502, Japan
* Correspondence: marahman@kfu.edu.sa; Tel.: +966-(0)-547757460

Abstract: Heavy metals in the environment cause adverse effects on living organisms. Agro-wastes have the potential to remove heavy metals from aqueous solutions. In this study, the orange peel cellulose (OPC) beads were utilized as adsorbents to remove metals from wastewater. The surface of the adsorbent was studied by Fourier transform infrared spectroscopy (FT-IR) and scanning electron microscopy coupled with energy dispersive x-ray spectroscopy (SEM-EDS). The concentrations of the metals before and after adsorption were measured using inductively coupled plasma mass spectrometry. The removal of the metal ions (i.e., Cr^{6+}, Cd^{2+}, and Pb^{2+}) using the OPC was investigated by varying the pH, contact time, and adsorbent dosages parameters. The maximum removal efficiency obtained for the metal ions occurred at pHs 4–8. The use of the Langmuir isotherm and Freundlich isotherm models demonstrated the statistical significance of the heavy metal adsorption processes ($R^2 > 0.96$). At a neutral pH, the OPC adsorption order was $Pb^{2+} > Cd^{2+} > Cr^{6+}$ with % removal values of 98.33, 93.91, and 33.50, respectively. The adsorption equilibrium for Cr^{6+} was reached after 36 h. For Cd^{2+} and Pb^{2+}, equilibrium was reached after 8 and 12 h, respectively. The FT-IR and SEM-EDS confirmed the presence of many functional groups and elements on the adsorbent. The adsorption of heavy metals using the OPC is a low-cost, eco-friendly, and innovative method for the removal of metals in aquatic environments. The findings of this study will be highly significant for the public in the affected areas worldwide that have credible health concerns due to water contamination with heavy metals.

Keywords: orange peel cellulose; adsorption; agro-wastes; bioremediation; chemical modification; heavy metals; polluted water; adsorption isotherms

1. Introduction

Water is the most vital and valuable resource in human civilization. However, water sources such as lakes and rivers have been contaminated by agricultural, industrial, and domestic wastes [1]. Water pollution in developed and developing countries has increased due to anthropogenic activities, such as fertilizer preparation, electroplating, leather making, sugar milling, textile making, mining, metallurgical processing, and municipal waste processing [2–5]. Toxic metals are known to cause severe impairment to marine organisms, terrestrial plants and animals, and human beings; thus, their release into water systems due to such activities is of significant concern [6,7]. In industrial wastewater, lead, cadmium, arsenic, mercury, and chromium ions are the most commonly found toxic heavy metal ions [8–10]. The quality of food and vegetables grown on soil contaminated with metals is significantly impacted, and their consumption may have negative health effects on human

nutrition levels [11]. Heavy metals are able to block specific cellular functions of certain biomolecules, such as proteins and enzymes [12]. They can also gather in living tissues and have been associated with many diseases (e.g., cancer) [13].

In particular, lead (Pb) causes several health issues, including mephitis, hypertension, stomach pain, constipation, vomiting, nausea, speaking difficulties, etc. [14]. Long-term cadmium exposure has been related to various types of cancer, including lung, kidney, breast, prostate, pancreas, etc. [15]. Additionally, the increased concentration of hexavalent chromium is poisonous, mutagenic, and carcinogenic to living things; as a result, it is categorized as a priority pollutant [16], whereas trivalent chromium is much less toxic. Therefore, it is essential to remove heavy metals from water bodies for safe water drinking and human activities [2,17,18].

Water bodies are particularly vulnerable to heavy metal contamination. The slow increase in metals in marine environments has become a significant health concern worldwide [19]. The monitoring and removal of metal ions using effective technologies are essential tasks for the safety assessment of the overall environment [20]. Various physico-chemical approaches have been established to remove toxic substances from contaminated water bodies. These methods include electrochemical treatment, ion exchange, reverse osmosis, evaporation, precipitation, adsorption on activated coal, entrapment, encapsulation, microbial biomass, and so on [21–23]. However, most methods are expensive and inefficient, especially for metals at low concentrations in large solution volumes [18]. In addition, they often have poor filtration and adsorption capabilities.

Biological techniques that are less expensive and more environmentally friendly should be taken into consideration as alternative methods for the remediation of heavy metals [24]. Adsorption is one of the most promising approaches for the elimination of heavy metal ions in wastewater sources which has attracted the interest of chemists and ecologists [25]. However, the creation of more efficient adsorbents and their preparation to be safer and more ecologically friendly is a challenge [26]. Plant biomass-based activated carbons can be used for the sequestration of metal ions [27]. Chemically modified chitosan is a great adsorbent for heavy metals [28,29] as well as bacteria-based biomass which could be used for the removal of heavy metals [17,23]. Agricultural wastes and non-edible plant parts are rich in natural polymers (e.g., hemicelluloses, cellulose, pectin, and lignin), which are known to have an unusual strength and attractive mechanical properties [30–32]. Lignocellulosic materials are biobased and biodegradable; thus, their use and disposal as bioadsorbents contribute to the enrichment and isolation of environmental pollution. Lignocellulosic waste has been proposed as a bioadsorbent for the removal of heavy metal ions in wastewater [33–35]. Thus, it is hypothesized that beads formulated with agricultural lignocellulosic waste can bind sufficiently and eliminate heavy metal pollutants in environmental wastewater. Among the various bioadsorbents, fruit peels have received global attention as they are readily available. Meanwhile, orange peels are also adsorbents, which are resource saving and environmentally friendly. They are also available abundantly and inexpensively. They contain pectin, cellulose, hemicelluloses, chlorophyll pigments, lignin, and numerous low-molecular-weight hydrocarbons, which are appropriate adsorptive materials [36]. Thus, the adsorption of heavy metal ions by orange peel cellulose (OPC) is considered as an effective, low-cost, and innovative method to remove the heavy metal ions from aquatic environments.

Although adsorbents are environmentally friendly and low-cost, most raw bioadsorbents show a low metal ion sorption capacity because of the absence of suitable functional groups for effective adsorption. A lignocellulosic adsorbent has been modified by chemical treatment [37–39]. In the first step, the cellulose fiber from orange fruit peels is obtained using an acid (e.g., hydrochloric acid (HCl) or sulphuric acid (H_2SO_4)) or an alkaline solution (e.g., sodium hydroxide (NaOH)). The use of a NaOH solution is known as the soda process. A bleaching process (e.g., with sodium hypochlorite (NaClO) or sodium chlorite ($NaClO_2$)) is then performed to eradicate the residual lignin and to obtain a purified cellulose fiber. Extracted cellulose fibers contain multiple functional groups, such as hydroxyl (-OH), carbonyl (=CO), carboxyl (-COOH) [40], amidoxime [41], and amide [42]. They are used in various applications (e.g., as adsorbents) [43,44].

This work presents an eco-friendly, low-cost, and straightforward heavy metal cleanup technique. The adsorption capacity of the OPC was thoroughly examined to determine the best approach. This study aimed to use the agricultural waste material (i.e., OPC) as an excellent source of bioadsorbent for the removal of heavy metal ions and to apply the bioadsorbent for the elimination of contaminants from wastewater, thus contributing to the reduction in environmental pollution.

2. Materials and Methods

2.1. Reagents

The chemicals used in this study were obtained from Fujifilm Wako Pure Chemical Corporation (Osaka, Japan) unless otherwise mentioned. The chemicals were American Chemical Society grade and used as received without any purification procedure. Gum guar was obtained from Sigma (Steinheim, Germany). Whatman filter paper pore size 45 μm from Advantec Toyo (Tokyo, Japan) was used to filter the cellulose. Standard stock solutions of 1 M potassium chromate (K_2CrO_4), cadmium chloride ($CdCl_2$), and lead acetate $Pb(C_2H_3O_2)_2$ were prepared with Milli Q water and stored at 4 °C. The 100 and 1.0 mM working solutions were made by diluting the standard stock solution with purified water from a Milli Q system from Millipore (Merck, Darmstadt, Germany), with Ʊ set at 18.2. The pHs of the metal solutions were adjusted, adding with 0.1 M HCl or 0.1 M NaOH. All the adsorption experiments were performed at 28 °C.

2.2. Extraction of Cellulose from Orange Fruit Peels

Oranges were bought from local grocery stores in Saga, Japan. First, the fruits were thoroughly washed with tap water to remove any dust particles on the surface of the fruits. Then, the peels were removed and cut into ~0.25 cm^2 pieces. The cut peels were then washed with deionized water to remove extra particles. The washed peels were dried in an air oven at 70 °C for 24 h. The dried peels were then crushed to a fine powder by using a mechanical blending machine. The particles with sizes of <240 μm were separated using a mesh strainer. OPC fibers were extracted from the powdered sample by following a previous procedure [45]. Briefly, 30 g of powder was soaked in a mixture of 150 mL of ethyl alcohol (99.5%), 75 mL of 0.8 M NaOH, and 75 mL of 0.8 M calcium chloride ($CaCl_2$). The resulting mixture was incubated for 20 h at room temperature on the lab bench. The fibers were filtered and rinsed using deionized water until a neutral pH was obtained. The fibers were then dried in an air oven at 70 °C for 24 h. Finally, the obtained OPC fibers were used to prepare beads with gum guar, which is an exo-polysaccharide composed of galactose and mannose. Gum guar has stabilizing and thickening properties, which are advantageous in food, feed, and industrial applications [46]. A total of 5 g of OPC powder and 2.5 g gum guar were mixed into 5 mL of 2% acetic acid (CH_3COOH) to make a slurry. The slurry was heated and stirred in a water bath at 60 °C for 30 min. The round-shaped beads were prepared by hand, and they were then dried overnight at 50 °C. The beads prepared were approximately 3 mm in diameter (Figure 1). The obtained OPC beads were stored in a ziplocked bag at room temperature until further use.

2.3. Fourier Transform Infrared (FT-IR) Spectroscopy

The FT-IR spectroscopy thoroughly describes the detailed information on the functional groups existing in a sample. Therefore, the structural information of the bioadsorbents was assessed by FT-IR in order to discover the chemical functional groups and binding processes involved in the metal adsorption process using a model VERTEX 70v from Bruker (Osaka, Japan). FT-IR absorbance spectra of untreated and treated OPC samples with three metal ions (i.e., Cr^{6+}, Cd^{2+}, and Pb^{2+}) were in the wavelength range of 400–4000 cm^{-1}.

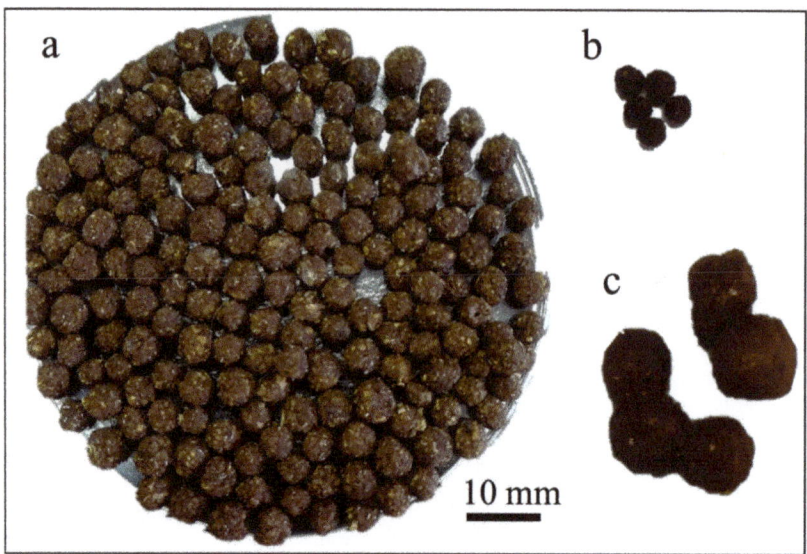

Figure 1. The orange peel cellulose beads (OPC). Beads before exposure to a mixture of three metal solutions (**a**,**b**). Beads after exposure to a combination of three metal solutions (Cr^{6+}, Cd^{2+}, and Pb^{2+}) (**c**).

2.4. Scanning Electron Microscopy (SEM) and Energy Dispersive Spectroscopy (EDS)

The surface of the prepared OPC was investigated by SEM that was equipped with an EDS using a model 3400N from Hitachi (Tokyo, Japan). The OPC beads were exposed to a 1.0 mM cocktail of the three metal ions for 12 h at room temperature with continuous shaking at 160 rpm. The OPC beads were collected and washed carefully using deionized water. Before imaging, the samples were coated with platinum. An OPC bead sample for SEM was mounted on an aluminum stub with double-sided carbon tape and covered with 10 nm of gold-palladium with a vacuum magnetron sputtering equipment model MSP-1S Sputtering Targets Manufacturer™ (Ibaraki, Japan). The surface morphology was analyzed using the EDS at 15.0–25.0 kV. EDS investigations were accomplished to determine the existence of the ions on the adsorbent beads. Samples for EDS were coated with 15 nm of carbon using a model 500× carbon coater attachment on the sputter coater from Emitech (Mahwah, NJ, USA). High-resolution spectra were fitted and calculated using the AVANTAGE software provided by Thermo Fisher Scientific (XPS, Wilmington, DE, USA), where wt% is expressed as the amount of the element in terms of mass fraction of the element in the sample. The EDS analyses were performed by removing the other elements, except chromium, cadmium, and lead. However, oxygen and carbon were preserved as the signal elements. Our system provided information on the quantitative element percentages, including the atomic percentage and weight percentage of each element studied.

2.5. Inductively Coupled Plasma Mass Spectroscopy (ICP-MS) Analyses

After the adsorption of the metal ions, using the prepared OPC beads, the concentration of the metal ions was measured by using an ICP-MS instrument model 7900 from Agilent Technologies (Santa Clara, CA, USA). The OPC beads were exposed to a mixture of the three metal ions for 12 h with continuous shaking at 160 rpm and a temperature of 28 °C. Control samples were shaken similarly but without the OPC beads. A cellulose-free supernatant was collected by centrifugation for 10 min at 12,298× *g*. The cellulose-free liquid was filtered using a 0.45 μm filter and syringe to remove all small particles. All the solutions were acidified with 2M HNO_3. Appropriate dilutions were performed before analysis. The samples were made to a constant volume before determining metal ion contents. Standard solutions of mixed metals of 0, 0.01, 0.1, 1, 10, 20, and 50 ppm were prepared. A standard curve was prepared before measuring the metal concentrations of the control and samples. All measurements were performed in triplicate and the results were expressed as the mean ± the standard error. The adsorption percentage was quantified using Equations (1) and (2).

$$C_{ad} = C_0 - C_t \quad (1)$$

$$\text{Adsorption}\% = \frac{C_{ad}}{C0} \times 100 \quad (2)$$

where C_0, C_t, and C_{ad} (ppm) are the initial concentration, the concentration at a time (t), and the concentration adsorbed, respectively.

Equation (3) was used to calculate the adsorption capacity q_e (mg g^{-1}) after the equilibrium.

$$q_e = \frac{C_{ad}}{W(g)} \cdot \frac{V(mL)}{1000} \quad (3)$$

where W, V, and q_e are the amount (g) of OPC used, volume (mL) of the mixed metal solution, and adsorption when equilibrium was attained, respectively [29].

An adsorption isotherm describes useful information on the adsorption capacity, binding affinity, and surface characteristics of the adsorbent, which may help to know the binding mechanism of the adsorbate with the adsorbent [47]. In this study, the equilibrium adsorption properties of OPC beads for metals uptake were clarified using Langmuir adsorption isotherm. Equation (4) represents Langmuir's isotherm.

$$q_e = \frac{q_{max} K_L C_e}{1 + K_L C_e} \quad (4)$$

where q_{max} represents the highest adsorption capacity (mg g^{-1}), and K_L is Langmuir's isotherm constant that illustrates the binding affinity between metals and test beads. The isotherm constants can be calculated from the intercepts and slopes of linear plots. Equation (5) represents the linear form of Langmuir's isotherm.

$$\frac{1}{q_e} = \frac{1}{K_L q_{max}} \cdot \frac{1}{C_e} + \frac{1}{q_{max}} \quad (5)$$

The separation factor (R_L) was calculated using Equation (6).

$$R_L = \frac{1}{1 + C_i K_L} \quad (6)$$

where R_L indicates the adsorption opportunity is either favorable (0 < R_L > 1), unfavorable (R_L > 1), linear (R_L = 1), or irreversible (R_L = 0) [48].

Additionally, adsorption data from solutions are most frequently represented using the Freundlich isotherm. The linear logarithmic Equation (7) can be used to express the Freundlich model [29].

$$\ln q_e = \frac{1}{n} \ln c_e + \ln K_F \quad (7)$$

where K_F (mg g^{-1}) stands for the adsorption intensity-related Freundlich characteristic constants, and n stands for the adsorption intensity. Particularly, it denotes the favorable metal adsorption when 1/n is in the range of 0.1–1.0 [49].

2.6. Effect of pH on Adsorption

The adsorption by OPC was assessed at different pHs of a metal ion solution. In a 3 mL tube, 10 mg of OPC and 1.0 mL of a 1.0 mM metal ion solution were mixed. The pH was adjusted to 3.0, 4.0, 5.0, 6.0, 7.0, or 8.0 ± 0.1. Adsorption was accomplished at 28 °C for 24 h with nonstop shaking at 160 rpm. All experiments were performed in triplicate.

2.7. Effect of Time on Adsorption

To determine how contact time affects bioadsorption, five beads of OPC (~100 mg) were added to a 10 mL cocktail solution of the three ions, each at a concentration of 1.0 mM. The pH was maintained to 7.0 ± 0.2. Bioadsorption was performed at 28 °C for 2, 4, 8, 12, 24, 36, and 48 h with continuous shaking at 160 rpm.

2.8. Bioadsorbent Dosages

The quantity of adsorbent used plays a crucial role in bioadsorption. Therefore, a 10 mL cocktail solution of the three metal ions, each at a concentration of 1.0 mM, was incubated with an increasing amount of OPC beads 1, 2, 3, and 5 (~20–100 mg). A metal solution was adjusted to pH 7.0 ± 0.2 and incubated at 28 °C with shaking for 24 h at 160 rpm.

2.9. Statistical Analysis

The OriginPro 2021 software from OriginLab corporation (Northampton, MA, USA) was used to complete the statistical analysis. The results of three duplicate tests ($n = 3$) were reported as the mean value with error bars denoting the standard deviation. The coefficient of determination (R^2) was used to evaluate how well the bioadsorption model fit the data. A linear fit correlation was performed to assess the relationship between metal concentration and adsorbent amount. The amount of adsorbent and metal concentration significantly correlated positively, $R^2 = 0.999$. The slope deviates significantly from zero at the 0.05 threshold. Mendeley software (Elsevier, Mendeley Ltd., London, UK) was used for reference management.

3. Results and Discussion

Orange fruit peel is a low-cost adsorbent widely available as waste material. Therefore, it can be an effective adsorbent for the removal and recovery of metal ions from wastewater. This study investigated the adsorption of selected metal ions onto the OPC in batch experiments. The adsorption was found to depend on the adsorbent dosage, exposure of time, and pH of the sample. After a comparison with similar published studies, the developed OPC showed a better adsorption capacity for the metal ions studied.

3.1. FT-IR Analysis

The FT-IR investigation was accomplished to determine the main functional groups present on the OPC surface that may be playing a significant role in the adsorption process on the surfaces of adsorbents [50]. The FT-IR spectra of the prepared adsorbents were obtained in the scanning range of 400–4000 cm^{-1} and are presented in Figure 2. The broad and intense peak was observed at around 3371 cm^{-1}, which suggested the presence of carboxylic, ketone, and phenolic functional groups in the cellulose, hemicellulose, pectin, and lignin. These functional groups have been shown to assist in metal ion adsorption in aqueous solutions [51]. The most intense peak was observed at 1629 cm^{-1}, which indicated the strong presence of carboxylate anions (Figure 2). Another major peak at 2917 cm^{-1} was attributed to the CH stretching vibrations of the methyl and methoxy substituent groups [52]. A simple peak was found at 2355 cm^{-1} in both the control and metal-treated OPC due to the N-H or C=O functional group, whereas the vibrational peak at 2850 cm^{-1} may be due to the C–H asymmetric stretching vibration [53]. The peaks at 1633 and 1423 cm^{-1} indicated asymmetric and symmetric vibrations of ionic carboxylic groups (-COOA), respectively [51].

The peaks at 1100 and 1383 cm^{-1} were referred to as the C–O–C and C-H groups, respectively (Figure 2). Deprotonated carboxyl and hydroxyl groups have been shown to coordinate with metal ions [54]. The peak of the OPC slightly shifted from 1612 to 1629 cm^{-1} (C=O stretching vibration of carboxyl groups), thus revealing the possible involvement of metal ions adsorption. These changes have been related to carboxylate and hydroxylate ions, which contribute to the metal uptake [51]. The FT-IR results recommended that the increasing number of carboxylate ligands enhanced the metal-binding ability to the adsorbent [55]. Therefore, the FT-IR study revealed the possible involvement of the major functional groups in the adsorbent, such as hydroxyl and carboxyl, which contributed to the ion exchange with metal ions during the bioadsorption process. The oxygen-containing functional groups, including hydroxyl, carboxyl, phenolic, and carbonyl, played essential roles in the Cd(II) ion adsorption [56]. According to the FT-IR analysis, the carboxyl and hydroxyl functional groups were mostly responsible for the Pb(II) ion adsorption [57].

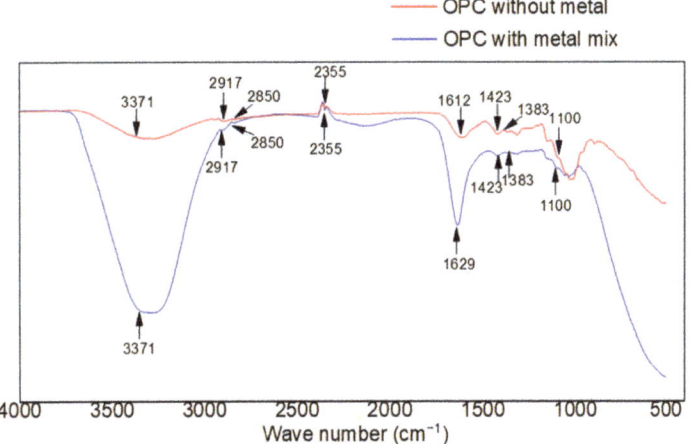

Figure 2. The FT−IR spectra of orange fruit peel cellulose before and after being treated with a combination of three metals (Cr^{6+}, Cd^{2+}, and Pb^{2+}). Conditions for adsorption parameters: concentration of metal ions = 1 mM, pH = 7, at room temperature for 2 h.

3.2. SEM-EDS Analysis

The morphology of the OPC surface was observed by using SEM-EDS. An analysis was completed before and after exposure to the cocktail solution of the three metals considered. This was performed to determine the metal ions distribution on the surface of the adsorbents (OPC) and the physical morphology of the adsorbents. A representative SEM of the rough surface of an OPC sample after bioadsorption is shown in Figure 3. At higher magnifications (i.e., 500×, 1000×, and 2000×), the OPC sample displayed an uneven surface, which is made of valleys and hills that formed dimples of different sizes (Figure 3a–c). The surface of the OPC sample was heterogeneous; therefore, the bioadsorption performance of this cellulose was likely better than smooth cellulose. After incubation for 12 h, the OPC appeared swollen and larger at the 500× magnification (Figure 1c).

The EDS results suggested that the surface of each cellulose sample consisted mainly of carbon and oxygen. The ions of interest were not found in the EDS spectrum of the OPC before the bioadsorption (control samples). On the other hand, after the bioadsorption, the ions were found throughout the surface of the sample (Figure 3e).

The mapping and overlapping of the images were accomplished to validate the distribution of the metal ions on the surface of the OPC. The image mapping was performed using 64 images of the corresponding metals. This approach detected the distribution of the metal ions on the surface of the prepared OPC. The results are presented in Figure 4, where the metal ions are depicted in different colors. The results suggested the uptake and accumulation of all three metal ions on the surface of the OPC, which were due to the exposure of the OPC to the metal solution.

The EDS intensity levels for all the metal ions continued persistently during the mapping. The image mapping further confirmed the results found in the ion imaging of the surface of the orange peel cellulose. The bioadsorption performances were compared based on the quantification by ICP-MS. Nevertheless, they were able to confirm the presence of the metal ions on the surface of the adsorbents. Similarly, the OPC was exposed to a solution without the metal ions. This was used as the control. Metal ions in the control samples were not detected.

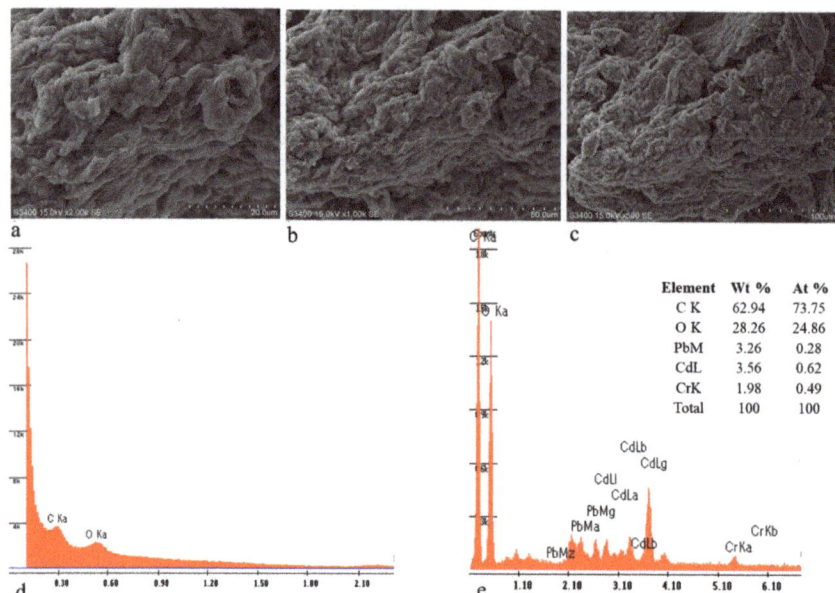

Figure 3. Images were taken using a scanning electron microscope (SEM) of the surface of OPC after bioadsorption. Micrographs at 2000×, 1000×, and 500× are shown in (**a**–**c**), respectively. (**d**) EDS spectrograms of OPC at 500× in absence of metals (control). (**e**) EDS spectrograms of OPC at 500×. One bead was incubated in a cocktail solution containing three metals (Cr^{6+}, Cd^{2+}, and Pb^{2+}) at 1 mM, pH 7.0, 28 °C, with continuous shaking for 12 h at 160 rpm.

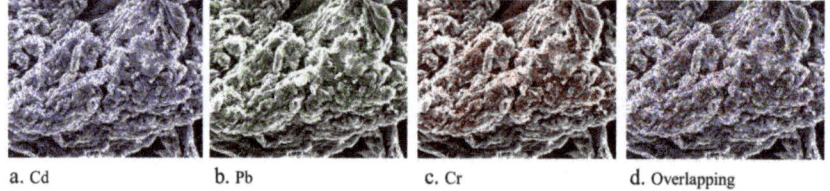

Figure 4. SEM-EDS images of mapping of OPC after bioadsorption of metal ions. (**a**) SEM with Cd, (**b**) SEM with Pb, (**c**) SEM with Cr, and (**d**) SEM with all the metal ions overlapped. Magnification at 500×.

3.3. ICP-MS Analyses

To further confirm the ability of the OPC to accumulate the selected metal ions, the ICP-MS analyses were performed. After the bioadsorption, the metal ions in the control and exposed samples were determined by the ICP-MS varying conditions, such as the pH, adsorbent dosages, and contact time.

3.4. Effect of pH

For the bioadsorption of metal ions using biological materials, pH is an important parameter that influences the protonation of functional groups and controls the metal chemistry of the material [50,58]. The bioadsorption was studied individually at varying pHs in the range of 3–8, and the removal of metals was investigated. The results are summarized in Figure 5.

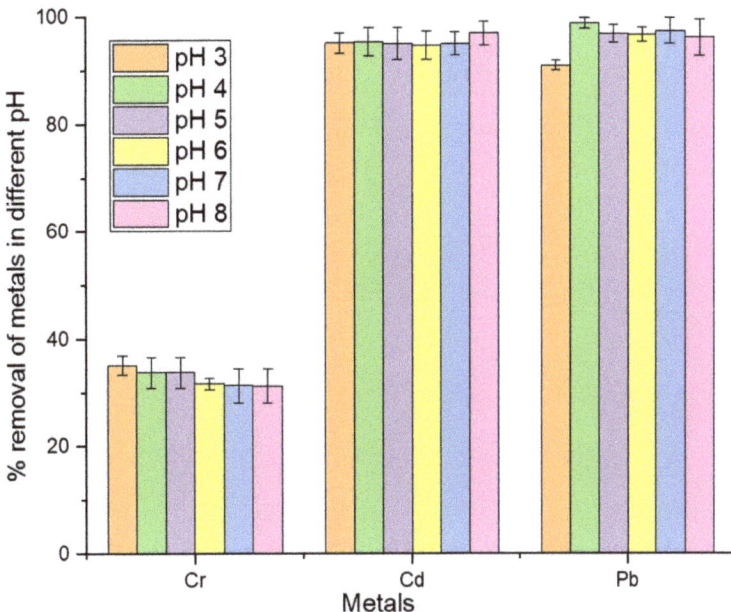

Figure 5. Effect of initial pH on removal percentage of metals by the OPC of CrO_4^{2-}, Cd^{2+}, and Pb^{2+} ions. One bead (~20 mg) was incubated in 1 mL individual metal solution at initial pH ranges of 3–8. The incubation conditions: metal concentration in 1 mM, shaking at 160 rpm for 12 h, and temperature 28 °C. The error bars show the mean ± SE for N = 3.

The removal of the studied ions increased markedly in the pH range of 4–8. The maximum bioadsorption was observed at pH 3–5, 3–8, and 4 for CrO_4^{2-}, Cd^{2+}, and Pb^{2+}, respectively. The bioadsorption of CrO_4^{2-} decreased with an increasing pH. At an acidic pH, the electronegative functional groups on the surface were protonated; thus, they were most suitable for binding with anions [35]. The bioadsorption difference was insignificant for the other metal ions at pH values between 4 and 8. A nearly identical pattern of metal ions accumulation was seen at various pHs. Similar observations have been reported in the removal of Cu^{2+}, Pb^{2+}, Cd^{2+}, Ni^{2+}, and Zn^{2+} using a modified orange peel in which the bioadsorption reached an equilibrium at pH values between 5.0 and 5.5 [59]. In a similar study by Nathan et al., bioadsorption using Kiwi beads showed that Cd^{2+}, Cr^{6+}, Cu^{2+}, and Ni^{2+} were stable in the pH range studied [35]. Several studies have shown that the maximum bioadsorption of Cr^{6+} was achieved at various pH values of 2 [60,61], 3 [62], and 5 [63].

At an acidic pH, chromium ions occur in two forms, namely as chromic acid (H_2CrO_4) and hydrogen chromate ions ($HCrO_4^-$) at pH ranges of 1–2 and 3–7, respectively [64]. The high concentration of H^+ and H_3O^+ protonates the carbonyl and hydroxyl groups, when the pH is low. There is little to no adsorption because of the competition between these ions and the aqueous heavy metal ions for the available binding sites in adsorbents. [65]. At a basic pH, some metals are precipitated [66]. According to the World Health Organization guideline, drinking water is neutral in pH. The OPC beads could be applied to purify drinking water and wastewater. Therefore, all the experiments were conducted at pH 7 in a cocktail solution of the metal ions.

3.5. Effect of Adsorbent Dosage

The dosage of the adsorbent significantly affected the adsorption process, removal efficiency, adsorption capacity, and other studied parameters. The adsorbent dosages of the OPC beads studied were 1, 2, 3, and 5 (~20–100 mg), containing a 1.0 mM concentration of each of the three metal ions. The results are shown in Figure 6. The adsorption of the metal ions increased with the increase in the adsorbent concentration (i.e., from 1 to 5 beads for all the metal ions).

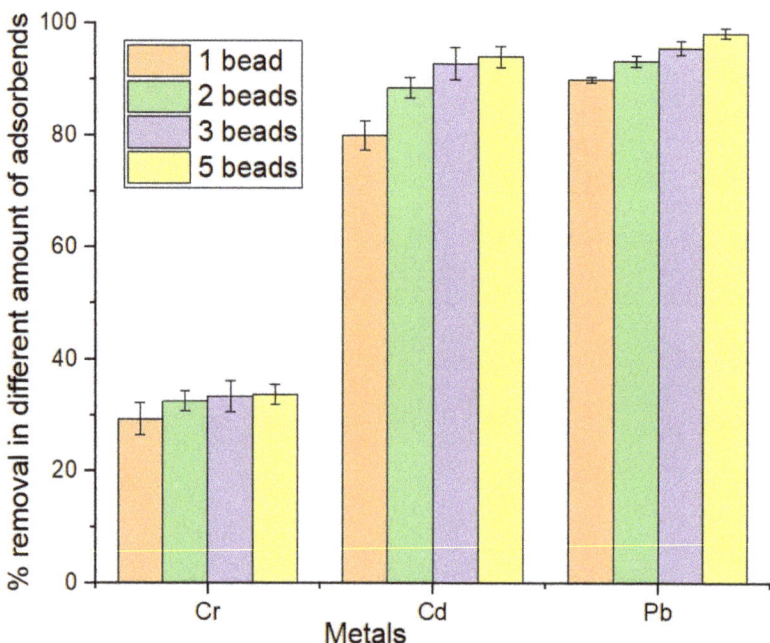

Figure 6. Effect of bioadsorbent amount of OPC on percentages removal of metal ions. One to five beads (~20–100 mg) were incubated in a cocktail solution containing three ions (Cr^{6+}, Cd^{2+}, and Pb^{2+}) at 1 mM at pH 7.0, 28 °C, with continuous shaking for 24 h at 160 rpm. The error bars show the mean ± SE for N = 3.

As shown in Figure 6, there is a rising trend in the metal adsorption capacity of the OPC beads. However, due to their surface and strong affinity for metals, the OPC beads demonstrated an outstanding adsorption capacity. Meanwhile, several studies have shown that an increase in the adsorbent dosage improved the removal efficiency [35,62,63,67,68]. Additionally, more surfaces and functional groups were available on the adsorbent at higher adsorbent doses, which improved the overall adsorption of metals [69,70].

The unit adsorption of the metal ions was calculated based on the amount of adsorbent used. The Langmuir adsorption isotherm is the most used linear model for monolayer adsorption and is frequently used to calculate the adsorption parameters [71]. On heterogeneous surfaces, multilayer adsorption is modeled using the Freundlich isotherm [71]. The equilibrium values were well-fitted by the Langmuir and Freundlich isotherm models. Figure 7 displays the estimated model parameters together with the linear regression coefficient (R^2) for the Langmuir and Freundlich isotherm models. The R^2 values are calculated for the experimental linear relationship to be statistically significant. Table 1 contains a list of the corresponding adsorption parameters.

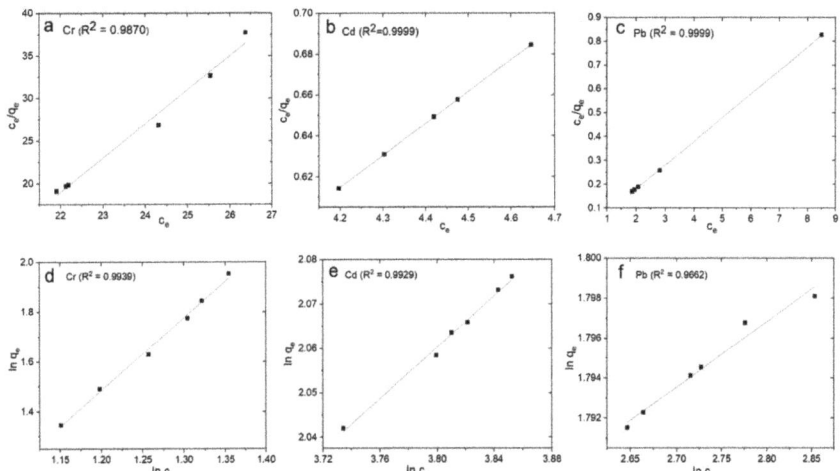

Figure 7. Images of Langmuir isotherm plots for the adsorption of metal ions using OPC beads, (**a**) Cr, (**b**) Cd, (**c**) Pb, and Freundlich isotherm plots for the metal ions, (**d**) Cr, (**e**) Cd, (**f**) Pb.

Table 1. Langmuir and Freundlich isotherm model parameters for the removal of metal ions by OPC.

Models	Parameters	Cr^{6+}	Pb^{2+}	Cd^{2+}
Langmuir	q_{max} (mg g^{-1})	4.90	50.10	29.00
	K_L (L/mg)	0.0509	0.2015	0.2204
	R_L	0.6574	0.0013	0.0037
	R^2	0.9879	0.9999	0.9999
Freundlich	K_F (mg g^{-1})	4.18	47.46	25.79
	$1/n$	0.2892	0.3291	0.2896
	R^2	0.9939	0.9662	0.9929

The R^2 values of all the metal ions were close to 1, confirming the Langmuir model's outstanding applicability to the adsorption processes [72]. The expressions of the straight lines were initiated by means of a mathematical transformation of the isotherm equation. The maximum adsorption capacities of the OPC beads for Pb, Cd, and Cr were observed to be 50.10, 29, and 4.9 mg g^{-1}, respectively. The detailed results are shown in Table 1, which illustrates the linear regression coefficient values, Freundlich and Langmuir's constant and adsorption possibilities. The values of R_L were found to be between 0 and 1, which confirmed the favorable uptake of the heavy metal ions [71].

However, the unit adsorption decreased with the increase in the adsorbent dosages. For example, the unit adsorption of Pb was reduced from 50.10 to 10.94 mg g^{-1} as the adsorbent dosage was increased from 1 to 5 beads (~20–100 mg/10 mL). Similarly, the unit adsorption values of Cd and Cr were reduced from 29 to 6.82 and 4.9 to 1.12 mg g^{-1}, respectively (Figure 8). The possible explanation for this may be due to the overlapping or aggregation of the adsorbent surface area, which was available to the metal ions in the solution. Therefore, when the amount of adsorbent is more, some surface areas of the adsorbent may be occupied with each other, and metals cannot be adsorbed on those sites of the adsorbent. A similar study was performed by Yang and Cui where the alkali-treated tea residue (ATTR) was used as an adsorbent. The adsorption of Pb was decreased from 2.09 mg g^{-1} to 0.63 mg g^{-1} with the increase in the ATTR dosage from 1 to 5 g L^{-1} [57].

In addition, for a more concentrated solution or effluent, a given mass of the adsorbent is able to purify a smaller volume of the effluent [69]. Therefore, in this study, a 100 mg (5 beads) adsorption dosage was selected as suitable for conducting the adsorption procedure. It was noticed that the ratio of the adsorption rates was not equal to the adsorbent dosages. An increase in the adsorbent dosage reduces the available metal ions for the adsorption. However, at the optimal amount of adsorbent, the number of sites available were sufficient for an interaction with the metal ions in the solution. Therefore, an excess adsorbent is not suitable for bioadsorption. Several studies have selected an optimal dosage for different bioadsorbent materials in the removal of metal ions from contaminated waters [35,62,63,67,68]. However, the dosages differed from the results obtained in this work. This was because of the different conditions, such as the bioadsorbent source, metal ions studied, and concentration of metal ions.

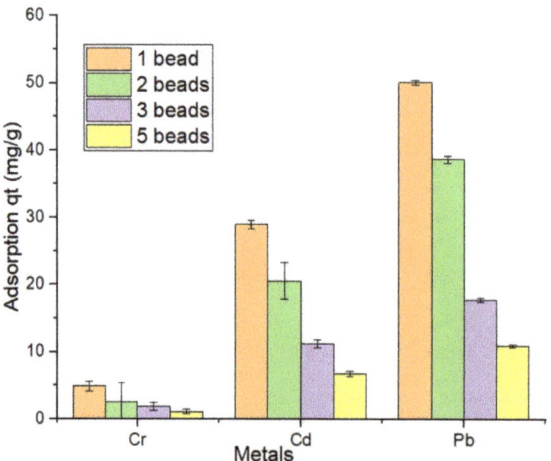

Figure 8. The unit adsorption of metal ions at different dosages of adsorbents. One to five beads (20–100 mg) were incubated in a cocktail solution containing three ions (Cr^{6+}, Cd^{2+}, and Pb^{2+}) at 1 mM at pH 7.0, 28 °C, with continuous shaking for 24 h at 160 rpm. The error bars show the mean ± SE for N = 3.

3.6. Influence of Contact Time

The exposure time is the most crucial parameter in the development of surface charges on the bioadsorbent, which is used for the bioadsorption of metal ions. To study the effect of the exposure time on the bioadsorption, seven time points between 2 and 48 h were set. The results are presented in Figure 9. The bioadsorption of Cr^{6+} increased linearly in the first 24 h. It then slightly increased for Cr^{6+} at 36 h. The adsorption of Cr^{6+} was not significantly high. A similar study has shown that Cr^{6+} was not adsorbed significantly by apple beads [67]. For Cd^{2+}, better bioadsorption was observed after 4 h. Then, the bioadsorption rate was slow until 36 h. After that, the rate was almost constant. Therefore, the maximum bioadsorption was 36 h for the Cd^{2+} ions. For Pb^{2+}, the bioadsorption increased in the first 36 h and then the adsorption rate was constant.

In this study, the bioadsorption rate was faster during the initial stages. The faster initial removal rate followed by a slower rate was likely due to the availability of the binding sites on the OPC surface during the initial phases [31,73]. For Cr^{6+}, the bioadsorption equilibrium was reached after 36 h (Figure 9). For Cd^{2+} and Pb^{2+}, the equilibrium was reached after 8 and 12 h, respectively.

As anticipated, the maximum adsorption capacities were different for the individual metal ion solutions under the optimized circumstances. After 48 h of exposure, the percentage concentration of the Pb, Cd, and Cr ions decreased by 98.33, 93.91, and 33.50, respectively (Figure 10). However, the OPC adsorbents which contain cellulose, hemicellulose, and lignin would exhibit a distinct mechanism of adsorption with the removal of various metal ions in a different process. This process might be involved in the complex formation between phenolic, hydroxyl, and carboxylic groups with heavy metal ions as well as an electrostatic attraction. A similar study using orange peel activated carbon has shown the removal efficiency order trend of $Pb^{2+} > Cr^{3+} > Cd^{2+}$, which disagrees with the current research, except for Pb^{2+} [69]. This was due to many circumstances, such as the bioadsorbent supply, the examined metal ions, and the concentration of metal ions. Another study using a kiwi peel bead has shown that the decreasing order of the bioadsorption was Cd > Pb > Cr, with approximately 92%, 67%, and 34%, and the simultaneous removal of ions, respectively [35]. The bioadsorption performance of Cd and Cr agreed with the current study.

The maximal adsorption capacities when using orange peel which are reported for the absorption of several metals are listed in Table 2 along with the appropriate references.

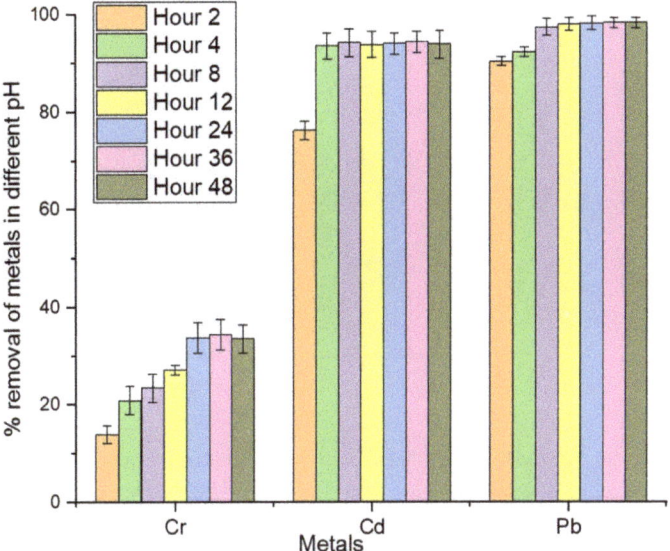

Figure 9. Effect of contact time on heavy metal removal percentage by OPC. Five beads (100 ± 5 mg) were incubated in a cocktail solution containing three ions (Cr^{6+}, Cd^{2+}, and Pb^{2+}) for 2–48 h with continuous shaking at 160 rpm (temp: 28 °C; volume: 10 mL; pH: 7 and concentration: 1 mM; time: 2–48 h). The error bars show the mean ± SE for N = 3.

Table 2. Summary of the types of adsorbents, types of metal, and maximum metal removal percentage capacity.

Biosorbents	Cr %	Cd %	Pb %	Reference
Orange peel	66.8			[62]
Orange peel			95.1	[74]
Orange peel	89.6	77.2	80	[69]
Orange peel	80			[60]
Modified orange peel		90	99	[59]
Orange peel		44.42		[68]
Orange peel			85	[75]
Orange peel			64.3	[45]
Orange peel		91	98	[36]
Orange peel		48.4		[68]
Modified orange peel		91		[76]
Dried orange peel		97.75		[77]
Orange peel cellulose	33.50	93.91	98.33	This study

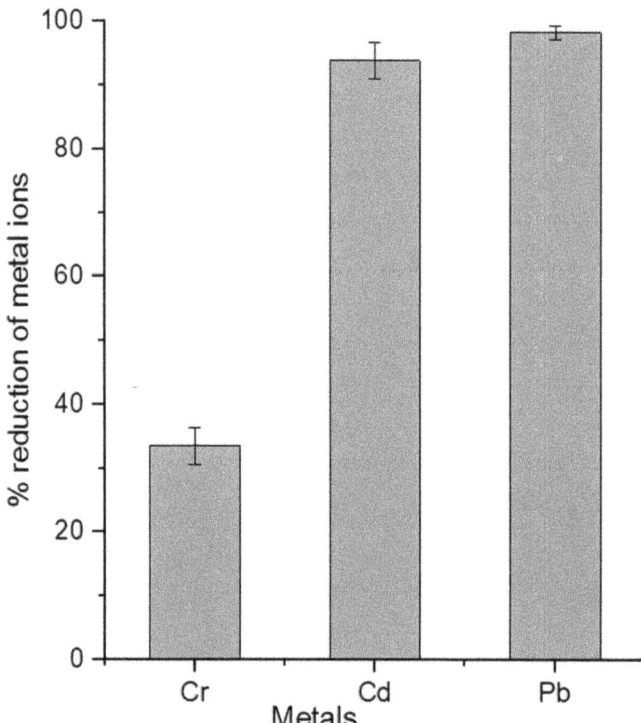

Figure 10. Percentage removal of metal ions after 48 h of OPC exposure. Five beads (100 ± 5 mg) were incubated in a cocktail solution containing three ions (Cr^{6+}, Cd^{2+}, and Pb^{2+}) with continuous shaking at 160 rpm (temp: 28 °C, volume: 10 mL, pH: 7, and concentration: 1 mM). The error bars show the mean ± SE for N = 3.

3.7. Future Research, Practice, and Policy

The adsorption of the chromate from the aqueous phase was observed by 33.50%, which is relatively low. Thus, these beads at the present formulation are likely less useful for the Cr removal from polluted water. However, the high removal efficiencies of these metals could be possible by changing the formulation and controlling the parameters, such as the mass of the adsorbent, pH of the reaction system, and contact time. It would also be possible to learn more about the type of ion attaching to the surface of the beads, the energy changes, and the viability of the reaction via a further investigation into bioadsorption isotherms and thermodynamics. More research is required to determine to what extent the metal ions may be reduced to 100% using the same adsorbent materials.

4. Conclusions

The present study developed low-cost, environmentally friendly, and greater adsorption-featured beads from orange peel cellulose (OPC). Considering the findings of this study, the OPC was confirmed as a great substance to adsorb heavy metals such as Pb, Cd, and Cr under experimental conditions. The SEM-EDS confirmed the presence and distribution of metal ions on the surface of the OPC exposed in a metal solution. Moreover, the FT-IR spectra of the treated OPC have revealed numerous functional groups in its surfaces that can efficiently adsorb metal ions. Therefore, compared to previously reported adsorbents, our orange peel-based adsorbents have shown effectiveness in a multi-ion solution, which is more like the water we drink. Overall, the findings from this study indicate that the developed OPC can be used in many broad-scale and alternative applications for cleaning wastewater. Nevertheless, we think this work is a significant advancement in the field and will be useful to a wide range of scientists working in the analytical and environmental sciences. Thus, the use of the material as a biosorbent may play an important role in reducing the pollution caused by direct anthropogenic activities and bring considerable economic benefits.

Author Contributions: A.R.: conceptualization, experiments, manuscript writing, data analysis, fund acquisition. K.Y.: ICP-MS assistance and proofreading. M.M.I.: data analysis, proofreading. G.K.: resources, data analysis, reviewing. All authors contributed to the revisions of the manuscript and approved it for publication. All authors have read and agreed to the published version of the manuscript.

Funding: The authors extend their appreciation to the Deputyship for Research and Innovation, the Ministry of Education in Saudi Arabia, for funding this research work (grant number GRANT 2898) at King Faisal University, Al-Ahsa 31982, Saudi Arabia.

Institutional Review Board Statement: This article does not contain any studies with human participants or animals performed by any of the authors.

Informed Consent Statement: No personal information was used in the study.

Data Availability Statement: The data used in this study are available from the corresponding author upon reasonable request.

Acknowledgments: We thank Masatoshi Goto and Kei Kimura for their kind suggestions in this study. In addition, we are thankful to Noriko Ryuda for assisting with the SEM-EDS imaging and Keisuke Tsuge for helping with the ICP-MS analysis.

Conflicts of Interest: The authors declare that the research was conducted without any potential conflict of interest.

References

1. Babu, D.J.; Prasanna, P.K.Y. Optimization of Cu (II) Biosorption onto Sea Urchin Test Using Response Surface Methodology and Artificial Neural Networks. *Int. J. Environ. Sci. Technol.* **2019**, *16*, 1885–1896. [CrossRef]
2. Rahman, A.; Nahar, N.; Nawani, N.N.; Jass, J.; Desale, P.; Kapadnis, B.P.; Hossain, K.; Saha, A.K.; Ghosh, S.; Olsson, B.; et al. Isolation and Characterization of a Lysinibacillus Strain B1-CDA Showing Potential for Bioremediation of Arsenics from Contaminated Water. *J. Environ. Sci. Health Part A Toxic/Hazard. Subst. Environ. Eng.* **2014**, *49*, 1349–1360. [CrossRef]
3. Asokan, N.M.R.S.; Sundari, N.S. Bioremediation of Chromium (VI) by Stenotrophomonas Maltophilia Isolated from Tannery Effluent. *Int. J. Environ. Sci. Technol.* **2018**, *15*, 207–216. [CrossRef]
4. Akhtar, N.; Syakir Ishak, M.I.; Bhawani, S.A.; Umar, K. Various Natural and Anthropogenic Factors Responsible for Water Quality Degradation: A Review. *Water* **2021**, *13*, 2660. [CrossRef]
5. Rahman, A. *Bioremediation of Toxic Metals for Protecting Human Health and the Ecosystem*; Örebro University: Örebro, Sweden, 2016; ISBN 9789175291468.
6. Briffa, J.; Sinagra, E.; Blundell, R. Heavy Metal Pollution in the Environment and Their Toxicological Effects on Humans. *Heliyon* **2020**, *6*, e04691. [CrossRef]
7. Moraes, R.R.L.V.J.S. Removal of Organic Pollutants from Wastewater Using Chitosan: A Literature Review. *Int. J. Environ. Sci. Technol.* **2019**, *16*, 1741–1754. [CrossRef]
8. Balali-Mood, M.; Naseri, K.; Tahergorabi, Z.; Khazdair, M.R.; Sadeghi, M. Toxic Mechanisms of Five Heavy Metals: Mercury, Lead, Chromium, Cadmium, and Arsenic. *Front. Pharmacol.* **2021**, *12*, 1–19. [CrossRef]
9. Ghori, N.H.G.T.; Imadi, M.Q.H.S.R.; Altay, A.G.V. Heavy Metal Stress and Responses in Plants. *Int. J. Environ. Sci. Technol.* **2019**, *16*, 1807–1828. [CrossRef]
10. Lupa, L.; Maranescu, B.; Visa, A. Equilibrium and Kinetic Studies of Chromium Ions Adsorption on Co (II)-Based Phosphonate Metal Organic Frameworks. *Sep. Sci. Technol.* **2018**, *53*, 1017–1026. [CrossRef]
11. Khan, A.; Khan, S.K.M.A.; Ullah, M.A.H.; Rehman, J.N.I.U. Heavy Metals Effects on Plant Growth and Dietary Intake of Trace Metals in Vegetables Cultivated in Contaminated Soil. *Int. J. Environ. Sci. Technol.* **2019**, *16*, 2295–2304. [CrossRef]
12. Krstić, V.; Urošević, T.; Pešovski, B. A Review on Adsorbents for Treatment of Water and Wastewaters Containing Copper Ions. *Chem. Eng. Sci.* **2018**, *192*, 273–287. [CrossRef]
13. Lim, J.T.; Tan, Y.Q.; Valeri, L.; Lee, J.; Geok, P.P.; Chia, S.E.; Ong, C.N.; Seow, W.J. Association between Serum Heavy Metals and Prostate Cancer Risk–A Multiple Metal Analysis. *Environ. Int.* **2019**, *132*, 105109. [CrossRef]
14. Gupta, V.K.; Carrott, P.J.M.; Ribeiro Carrott, M.M.L.; Suhas. Low-Cost Adsorbents: Growing Approach to Wastewater Treatmenta Review. *Crit. Rev. Environ. Sci. Technol.* **2009**, *39*, 783–842. [CrossRef]

15. Genchi, G.; Graziantono, L.; Carocci, A.; Catalano, A. The Effects of Toxicity. *Int. J. Environ. Res. Public Health* **2020**, *17*, 3782. [CrossRef]
16. Tripathi, M.; Upadhyay, S.K.; Kaur, M.; Kaur, K. Toxicity Concerns of Hexavalent Chromium from Tannery Waste. *J. Biotechnol. Bioeng.* **2018**, *2*, 40–44.
17. Prithviraj, D.; Deboleena, K.; Neelu, N.; Noor, N.; Aminur, R.; Balasaheb, K.; Abul, M. Biosorption of Nickel by Lysinibacillus Sp. BA2 Native to Bauxite Mine. *Ecotoxicol. Environ. Saf.* **2014**, *107*, 260–268. [CrossRef]
18. Rahman, A.; Nahar, N.; Nawani, N.N.; Jass, J.; Hossain, K.; Saud, Z.A.; Saha, A.K.; Ghosh, S.; Olsson, B.; Mandal, A. Bioremediation of Hexavalent Chromium (VI) by a Soil-Borne Bacterium, Enterobacter Cloacae B2-DHA. *J. Environ. Sci. Heal. Part A Toxic/Hazard. Subst. Environ. Eng.* **2015**, *50*, 1136–1147. [CrossRef]
19. Nawani, N.; Rahman, A.; Nahar, N.; Saha, A.; Kapadnis, B.; Mandal, A. Status of Metal Pollution in Rivers Flowing through Urban Settlements at Pune and Its Effect on Resident Microflora. *Biologia* **2016**, *71*, 494–507. [CrossRef]
20. Yewale, P.P.; Rahman, A.; Nahar, N.; Saha, A.; Jass, J.; Mandal, A.; Nawani, N.N. Sources of Metal Pollution, Global Status, and Conventional Bioremediation Practices. *Handb. Met. Interact. Bioremediation* **2017**, 25–40. [CrossRef]
21. Acharya, J.; Kumar, U.; Rafi, P.M. International Journal of Current Engineering and Technology Removal of Heavy Metal Ions from Wastewater by Chemically Modified Agricultural Waste Material as Potential Adsorbent-A Review. *Int. J. Curr. Eng. Technol.* **2018**, *8*, 526–530.
22. Yelebe, E.O.Z.R.; Nelson, B.O.E.S. Clean-up of Crude Oil-Contaminated Soils: Bioremediation Option. *Int. J. Environ. Sci. Technol.* **2020**, *17*, 1185–1198. [CrossRef]
23. Nawani, N.; Rahman, A.; Mandal, A. Microbial Biomass for Sustainable Remediation of Wastewater. *Biomass Biofuels Biochem.* **2022**, *12*, 271–292.
24. Obey, G.; Adelaide, M.; Ramaraj, R. Biochar Derived from Non-Customized Matamba Fruit Shell as an Adsorbent for Wastewater Treatment. *J. Bioresour. Bioprod.* **2022**, *7*, 109–115. [CrossRef]
25. Jjagwe, J.; Olupot, P.W.; Menya, E.; Kalibbala, H.M. Synthesis and Application of Granular Activated Carbon from Biomass Waste Materials for Water Treatment: A Review. *J. Bioresour. Bioprod.* **2021**, *6*, 292–322. [CrossRef]
26. Nabipour, H.; Rohani, S.; Batool, S.; Yusuff, A.S. An Overview of the Use of Water-Stable Metal-Organic Frameworks in the Removal of Cadmium Ion. *J. Environ. Chem. Eng.* **2023**, *11*, 109131. [CrossRef]
27. Ugwu, E.I.; Agunwamba, J.C. A Review on the Applicability of Activated Carbon Derived from Plant Biomass in Adsorption of Chromium, Copper, and Zinc from Industrial Wastewater. *Environ. Monit. Assess.* **2020**, *192*, 240. [CrossRef]
28. Popa, A.; Visa, A.; Maranescu, B.; Hulka, I.; Lupa, L. Chemical Modification of Chitosan for Removal of Pb(II) Ions from Aqueous Solutions. *Materials* **2021**, *14*, 7894. [CrossRef] [PubMed]
29. Rahman, A.; Haque, A.; Ghosh, S.; Shinu, P.; Attimarad, M. Modified Shrimp-Based Chitosan as an Emerging Adsorbent Removing Heavy Metals (Chromium, Nickel, Arsenic, and Cobalt) from Polluted Water. *Sustainability* **2023**, *15*, 2431. [CrossRef]
30. Kanamarlapudi, S.L.R.K.; Chintalpudi, V.K.; Muddada, S. Application of Biosorption for Removal of Heavy Metals from Wastewater. *Biosorption* **2018**. [CrossRef]
31. Basu, M.; Guha, A.K.; Ray, L. Adsorption of Cadmium on Cucumber Peel: Kinetics, Isotherm and Co-Ion Effect. *Indian Chem. Eng.* **2018**, *60*, 179–195. [CrossRef]
32. Abdullah-Al-Mamun, M.; Hossain, M.S.; Debnath, G.C.; Sultana, S.; Rahman, A.; Hasan, Z.; Das, S.R.; Ashik, M.A.; Prodhan, M.Y.; Aktar, S.; et al. *Unveiling Lignocellulolytic Trait of a Goat Omasum Inhabitant Klebsiella Variicola Strain HSTU-AAM51 in Light of Biochemical and Genome Analyses*; Springer International Publishing: Berlin/Heidelberg, Germany, 2022; ISBN 0123456789.
33. Afroze, S.; Sen, T.K. A Review on Heavy Metal Ions and Dye Adsorption.Pdf. *Water Air Soil Pollut.* **2018**, *229*, 1–50. [CrossRef]
34. Thakur, V.; Sharma, E.; Guleria, A.; Sangar, S.; Singh, K. Modification and Management of Lignocellulosic Waste as an Ecofriendly Biosorbent for the Application of Heavy Metal Ions Sorption. *Mater. Today Proc.* **2020**, *32*, 608–619. [CrossRef]
35. Nathan, R.J.; Barr, D.; Rosengren, R.J. Six Fruit and Vegetable Peel Beads for the Simultaneous Removal of Heavy Metals by Biosorption. *Environ. Technol.* **2022**, *43*, 1935–1952. [CrossRef] [PubMed]
36. Liang, S.; Guo, X.; Feng, N.; Tian, Q. Application of Orange Peel Xanthate for the Adsorption of Pb^{2+} from Aqueous Solutions. *J. Hazard. Mater.* **2009**, *170*, 425–429. [CrossRef] [PubMed]
37. Johar, N.; Ahmad, I.; Dufresne, A. Extraction, Preparation and Characterization of Cellulose Fibres and Nanocrystals from Rice Husk. *Ind. Crops Prod.* **2012**, *37*, 93–99. [CrossRef]
38. Ravindran, L.; Sreekala, M.S.; Thomas, S. Novel Processing Parameters for the Extraction of Cellulose Nanofibres (CNF) from Environmentally Benign Pineapple Leaf Fibres (PALF): Structure-Property Relationships. *Int. J. Biol. Macromol.* **2019**, *131*, 858–870. [CrossRef]
39. Cruz-Lopes, L.; Macena, M.; Esteves, B.; Santos-Vieira, I. Lignocellulosic Materials Used as Biosorbents for the Capture of Nickel (II) in Aqueous Solution. *Appl. Sci.* **2022**, *12*, 933. [CrossRef]
40. Wang, N.; Jin, R.N.; Omer, A.M.; Ouyang, X. kun Adsorption of Pb(II) from Fish Sauce Using Carboxylated Cellulose Nanocrystal: Isotherm, Kinetics, and Thermodynamic Studies. *Int. J. Biol. Macromol.* **2017**, *102*, 232–240. [CrossRef]
41. Hokkanen, S.; Bhatnagar, A.; Sillanpää, M. A Review on Modification Methods to Cellulose-Based Adsorbents to Improve Adsorption Capacity. *Water Res.* **2016**, *91*, 156–173. [CrossRef]
42. Liu, J.; Chen, T.W.; Yang, Y.L.; Bai, Z.C.; Xia, L.R.; Wang, M.; Lv, X.L.; Li, L. Removal of Heavy Metal Ions and Anionic Dyes from Aqueous Solutions Using Amide-Functionalized Cellulose-Based Adsorbents. *Carbohydr. Polym.* **2020**, *230*, 115619. [CrossRef]

43. Mo, J.; Yang, Q.; Zhang, N.; Zhang, W.; Zheng, Y.; Zhang, Z. A Review on Agro-Industrial Waste (AIW) Derived Adsorbents for Water and Wastewater Treatment. *J. Environ. Manage.* **2018**, *227*, 395–405. [CrossRef] [PubMed]
44. Sharma, A.; Thakur, M.; Bhattacharya, M.; Mandal, T.; Goswami, S. Commercial Application of Cellulose Nano-Composites–A Review. *Biotechnol. Rep.* **2019**, *21*, e00316. [CrossRef]
45. Feng, N.C.; Guo, X.Y. Characterization of Adsorptive Capacity and Mechanisms on Adsorption of Copper, Lead and Zinc by Modified Orange Peel. *Trans. Nonferrous Met. Soc. China* **2012**, *22*, 1224–1231. [CrossRef]
46. Mudgil, D.; Barak, S.; Khatkar, B.S. Guar Gum: Processing, Properties and Food Applications-A Review. *J. Food Sci. Technol.* **2014**, *51*, 409–418. [CrossRef]
47. Ayub, A.; Irfan, M.; Rizwan, M.; Irfan, A. International Journal of Biological Macromolecules Development of Sustainable Magnetic Chitosan Biosorbent Beads for Kinetic Remediation of Arsenic Contaminated Water. *Int. J. Biol. Macromol.* **2020**, *163*, 603–617. [CrossRef] [PubMed]
48. Ashfaq, A.; Nadeem, R.; Bibi, S.; Rashid, U.; Hanif, A.; Jahan, N.; Ashfaq, Z.; Ahmed, Z.; Adil, M.; Naz, M. Efficient Adsorption of Lead Ions from Synthetic Wastewater Using Agrowaste-Based Mixed Biomass (Potato Peels and Banana Peels). *Water* **2021**, *13*, 3344. [CrossRef]
49. He, C.; Lin, H.; Dai, L.; Qiu, R.; Tang, Y.; Wang, Y.; Duan, P.G.; Ok, Y.S. Waste Shrimp Shell-Derived Hydrochar as an Emergent Material for Methyl Orange Removal in Aqueous Solutions. *Environ. Int.* **2020**, *134*, 105340. [CrossRef]
50. Sohail, A.; Javed, S.; Khan, M.U.; Umar, A. Biosorption of Heavy Metals onto the Bark of Prosopis Spicigira: A Kinetic Study for the Removal of Water Toxicity. *Am. Eurasian J. Toxicol. Sci.* **2015**, *7*, 300–315. [CrossRef]
51. Reddy, N.A.; Lakshmipathy, R.; Sarada, N.C. Application of Citrullus Lanatus Rind as Biosorbent for Removal of Trivalent Chromium from Aqueous Solution. *Alex. Eng. J.* **2014**, *53*, 969–975. [CrossRef]
52. Iqbal, M.; Saeed, A.; Zafar, S.I. FTIR Spectrophotometry, Kinetics and Adsorption Isotherms Modeling, Ion Exchange, and EDX Analysis for Understanding the Mechanism of Cd^{2+} and Pb^{2+} Removal by Mango Peel Waste. *J. Hazard. Mater.* **2009**, *164*, 161–171. [CrossRef]
53. Zhang, Y.; Chen, B.; Zhang, L.; Huang, J.; Chen, F.; Yang, Z.; Yao, J.; Zhang, Z. Controlled Assembly of Fe3O4 Magnetic Nanoparticles on Graphene Oxide. *Nanoscale* **2011**, *3*, 1446–1450. [CrossRef] [PubMed]
54. D'Halluin, M.; Rull-Barrull, J.; Bretel, G.; Labrugère, C.; Le Grognec, E.; Felpin, F.X. Chemically Modified Cellulose Filter Paper for Heavy Metal Remediation in Water. *ACS Sustain. Chem. Eng.* **2017**, *5*, 1965–1973. [CrossRef]
55. Abd-Talib, N.; Chuong, C.S.; Mohd-Setapar, S.H.; Asli, U.A.; Pa'ee, K.F.; Len, K.Y.T. Trends in Adsorption Mechanisms of Fruit Peel Adsorbents to Remove Wastewater Pollutants (Cu (II), Cd (II) and Pb (II)). *J. Water Environ. Technol.* **2020**, *18*, 290–313. [CrossRef]
56. Chen, S.; Xue, C.; Wang, J.; Feng, H.; Wang, Y.; Ma, Q.; Wang, D. Adsorption of Pb (II) and Cd (II) by Squid Ommastrephes Bartrami Melanin. *Bioinorg. Chem. Appl.* **2009**, *2009*, 901563. [CrossRef]
57. Yang, X.; Cui, X. Adsorption Characteristics of Pb (II) on Alkali Treated Tea Residue. *Water Resour. Ind.* **2013**, *3*, 1–10. [CrossRef]
58. Lo, S.F.; Wang, S.Y.; Tsai, M.J.; Lin, L.D. Adsorption Capacity and Removal Efficiency of Heavy Metal Ions by Moso and Ma Bamboo Activated Carbons. *Chem. Eng. Res. Des.* **2012**, *90*, 1397–1406. [CrossRef]
59. Guo, X.Y.; Liang, S.; Tian, Q.H. Removal of Heavy Metal Ions from Aqueous Solutions by Adsorption Using Modified Orange Peel as Adsorbent. *Adv. Mater. Res.* **2011**, *236*, 237–240. [CrossRef]
60. Ugbe, F.A.; Pam, A.A.; Ikudayisi, A.V. Thermodynamic Properties of Chromium (III) Ion Adsorption by Sweet Orange (*Citrus sinensis*) Peels. *Am. J. Anal. Chem.* **2014**, *05*, 666–673. [CrossRef]
61. Jisha, T.J.; Lubna, C.H.; Habeeba, V. Removal of Cr (VI) Using Orange Peel as an Adsorbent. *Int. J. Adv. Res. Innov. Ideas Educ.* **2017**, *3*, 276–283.
62. Tejada-Tovar, C.; Gonzalez-Delgado, A.D.; Villabona-Ortiz, A. Removal of Cr (VI) from Aqueous Solution Using Orange Peel-Based Biosorbents. *Indian J. Sci. Technol.* **2018**, *11*, 1–13. [CrossRef]
63. Gönen, F.; Serin, D.S. Adsorption Study on Orange Peel: Removal of Ni(II) Ions from Aqueous Solution. *African J. Biotechnol.* **2012**, *11*, 1250–1258. [CrossRef]
64. Sakulthaew, C.; Chokejaroenrat, C.; Poapolathep, A.; Satapanajaru, T.; Poapolathep, S. Hexavalent Chromium Adsorption from Aqueous Solution Using Carbon Nano-Onions (CNOs). *Chemosphere* **2017**, *184*, 1168–1174. [CrossRef] [PubMed]
65. Peng, W.; Li, H.; Liu, Y.; Song, S. A Review on Heavy Metal Ions Adsorption from Water by Graphene Oxide and Its Composites. *J. Mol. Liq.* **2017**, *230*, 496–504. [CrossRef]
66. Król, A.; Mizerna, K.; Bożym, M. An Assessment of PH-Dependent Release and Mobility of Heavy Metals from Metallurgical Slag. *J. Hazard. Mater.* **2020**, *384*, 121502. [CrossRef]
67. Singh, R.J.; Martin, C.E.; Barr, D.; Rosengren, R.J. Immobilised Apple Peel Bead Biosorbent for the Simultaneous Removal of Heavy Metals from Cocktail Solution. *Cogent Environ. Sci.* **2019**, *5*, 1673116. [CrossRef]
68. Akinhanmi, T.F.; Ofudje, E.A.; Adeogun, A.I.; Aina, P.; Joseph, I.M. Orange Peel as Low-Cost Adsorbent in the Elimination of Cd(II) Ion: Kinetics, Isotherm, Thermodynamic and Optimization Evaluations. *Bioresour. Bioprocess.* **2020**, *7*, 34. [CrossRef]
69. Ali, M.H.H.; Abdel-Satar, A.M. Removal of Some Heavy Metals from Aqueous Solutions Using Natural Wastes Orange Peel Activated Carbon. *IJRDO J. Appl. Sci.* **2017**, *3*, 13–30.
70. Memić, Š.A.M.; Sulejmanović, E.Š.J. Adsorptive Removal of Eight Heavy Metals from Aqueous Solution by Unmodified and Modified Agricultural Waste: Tangerine Peel. *Int. J. Environ. Sci. Technol.* **2018**, *15*, 2511–2518. [CrossRef]

71. Al-Qahtani, K.M. Water Purification Using Different Waste Fruit Cortexes for the Removal of Heavy Metals. *J. Taibah Univ. Sci.* **2016**, *10*, 700–708. [CrossRef]
72. Mallampati, R.; Xuanjun, L.; Adin, A.; Valiyaveettil, S. Fruit Peels as Efficient Renewable Adsorbents for Removal of Dissolved Heavy Metals and Dyes from Water. *ACS Sustain. Chem. Eng.* **2015**, *3*, 1117–1124. [CrossRef]
73. Xia, Z.; Zhang, S.; Cao, Y.; Zhong, Q.; Wang, G.; Li, T.; Xu, X. Remediation of Cadmium, Lead and Zinc in Contaminated Soil with CETSA and MA/AA. *J. Hazard. Mater.* **2019**, *366*, 177–183. [CrossRef] [PubMed]
74. Lima, J.M.S.; de Souza, H.D.P.; Cunha, J.R.M.S. Use of Orange Peel (*Citrus sinensis*) in the Bioabsorption of Potentially Toxic Metals from Water Resources through ICP-OES. *Ciência e Nat.* **2020**, *42*, e16. [CrossRef]
75. Amin, M.T.; Alazba, A.A.; Amin, M.N. Absorption Behaviours of Copper, Lead, and Arsenic in Aqueous Solution Using Date Palm Fibres and Orange Peel: Kinetics and Thermodynamics. *Polish J. Environ. Stud.* **2017**, *26*, 543–557. [CrossRef]
76. Lahieb Faisal, M.; Al-Najjar, S.Z.; Al-Sharify, Z.T. Modified Orange Peel as Sorbent in Removing of Heavy Metals from Aqueous Solution. *J. Green Eng.* **2020**, *10*, 10600–10615.
77. Yirga, A.; Werede, Y.; Nigussie, G.; Ibrahim, F. Dried Orange Peel: A Potential Bio Sorbent for Removal of Cu (II) and Cd (II) Ions from Aqueous Solution. *Chem. J.* **2020**, *7*, 2581–7507.

Disclaimer/Publisher's Note: The statements, opinions and data contained in all publications are solely those of the individual author(s) and contributor(s) and not of MDPI and/or the editor(s). MDPI and/or the editor(s) disclaim responsibility for any injury to people or property resulting from any ideas, methods, instructions or products referred to in the content.

MDPI
St. Alban-Anlage 66
4052 Basel
Switzerland
www.mdpi.com

Sustainability Editorial Office
E-mail: sustainability@mdpi.com
www.mdpi.com/journal/sustainability

Disclaimer/Publisher's Note: The statements, opinions and data contained in all publications are solely those of the individual author(s) and contributor(s) and not of MDPI and/or the editor(s). MDPI and/or the editor(s) disclaim responsibility for any injury to people or property resulting from any ideas, methods, instructions or products referred to in the content.

www.ingramcontent.com/pod-product-compliance
Lightning Source LLC
LaVergne TN
LVHW070433100526
838202LV00014B/1586